FUNDAMENTALS OF POWER ELECTRONICS

AUTHOR

Prof. Soumitra Kumar Mandal

Professor of Electrical Engineering

National Institute of Technical Teachers' Training & Research, Kolkata

REVIEWER

Dr. Amod C. Umarikar

Associate Professor, Electrical Engineering (EE), Indian Institute of Technology Indore, Madhya Pradesh

All India Council for Technical Education

Nelson Mandela Marg, Vasant Kunj, New Delhi, India

Dear Readers,

To prevent the piracy, this book is secured with HIGH SECURITY HOLOGRAM on the front title cover. In case you don't find the hologram on the front cover title, please write us to at contact@khannabooks.com or whatsapp us at +91-99109 09320 and avail special gift voucher for yourself.

Specimen of Hologram on front Cover title:

Moreover, there is a SPECIAL DISCOUNT COUPON for you with EVERY HOLOGRAM.

How to avail this SPECIAL DISCOUNT:

Step 1: Scratch the hologram

Step 2: Under the scratch area, your "coupon code" is available

Step 3: Logon to www.khannabooks.com

Step 4: Use your "coupon code" in the shopping cart and get your copy at a special discount

Step 5: Enjoy your reading!

ISBN: 978-93-5538-724-0

Book Code: DIP474EN

Fundamentals Of Power Electronics

by Prof. Soumitra Kumar Mandal

First Edition: 2026

Originally Published by:

All India Council for Technical Education, Ministry of Education, GOI

Exclusively Published & Distributed by:

Khanna Book Publishing Co. (P) Ltd.
4C/4344, Ansari Road, Daryaganj,
New Delhi (INDIA)

Visit us at: **www.khannabooks.com**
Write us at: **contact@khannabooks.com**

To view complete list of books,
Please scan the QR Code:

Printed in Bharat.

सत्यमेव जयते

प्रो. टी. जी. सीताराम
अध्यक्ष
Prof. T. G. Sitharam
Chairman

आज़ादी का अमृत महोत्सव

अखिल भारतीय तकनीकी शिक्षा परिषद्
(भारत सरकार का एक सांविधिक निकाय)
(शिक्षा मंत्रालय, भारत सरकार)
नेल्सन मंडेला मार्ग, वसंत कुंज, नई दिल्ली–110070
दूरभाष : 011–26131498
ई–मेल : chairman@aicte-india.org

ALL INDIA COUNCIL FOR TECHNICAL EDUCATION
(A STATUTORY BODY OF THE GOVT. OF INDIA)
(Ministry of Education, Govt. of India)
Nelson Mandela Marg, Vasant Kunj, New Delhi-110070
Phone : 011-26131498
E-mail : chairman@aicte-india.org

FOREWORD

Engineers are the backbone of any modern society. They are the ones responsible for the marvels as well as the improved quality of life across the world. Engineers have driven humanity towards greater heights in a more evolved and unprecedented manner.

The All India Council for Technical Education (AICTE), have spared no efforts towards the strengthening of the technical education in the country. AICTE is always committed towards promoting quality Technical Education to make India a modern developed nation emphasizing on the overall welfare of mankind.

An array of initiatives has been taken by AICTE in last decade which have been accelerated now by the National Education Policy (NEP) 2020. The implementation of NEP under the visionary leadership of Hon'ble Prime Minister of India envisages the provision for education in regional languages to all, thereby ensuring that every graduate becomes competent enough and is in a position to contribute towards the national growth and development through innovation & entrepreneurship.

One of the spheres where AICTE had been relentlessly working since past couple of years is providing high quality original technical contents at Under Graduate & Diploma level prepared and translated by eminent educators in various Indian languages to its aspirants. For students pursuing 2nd year of their Engineering education, AICTE has identified 87 books, which shall be translated into 12 Indian languages - Hindi, Tamil, Gujarati, Odia, Bengali, Kannada, Urdu, Punjabi, Telugu, Marathi, Assamese & Malayalam. In addition to the English medium, books in different Indian Languages are going to support the students to understand the concepts in their respective mother tongue.

On behalf of AICTE, I express sincere gratitude to all distinguished authors, reviewers and translators from the renowned institutions of high repute for their admirable contribution in a record span of time.

AICTE is confident that these outcomes based original contents shall help aspirants to master the subject with comprehension and greater ease.

(Prof. T. G. Sitharam)

ACKNOWLEDGEMENT

The authors are grateful to the authorities of AICTE, particularly Prof. (Dr.) T G Sitharam, Chairman; Dr. Abhay Jere, Vice-Chairman, Prof. Rajive Kumar, Member-Secretary, Dr. Ramesh Unnikrishnan, Advisor-II and Dr. Sunil Luthra, Director, Training and Learning Bureau for their planning to publish the books on Fundamental of Power Electronics. We sincerely acknowledge the valuable contributions of the reviewer of the book Prof. Amod C. Umarikar, IIT Indore. I have received immense co-operation and inspiration for writing this book from Prof. Debi Prasad Mishra, Director NITTTR, Kolkata, Prof. Gurnam Singh, Pumjab Engineering College, Chandigarh and my wife Malvika and my children Om and Puja.

This book is an outcome of various suggestions of AICTE members, experts and authors who shared their opinion and thought to further develop the engineering education in our country. Acknowledgements are due to the contributors and different workers in this field whose published books, review articles, papers, photographs, footnotes, references and other valuable information enriched us at the time of writing the book.

Prof. Soumitra Kumar Mandal

PREFACE

The book titled "Fundamental of Power Electronics" is an outcome of the rich experience of teaching of Power Electronics courses in Under Graduate and Post Graduate students. The initiation of writing this book is to expose Electrical Power and Basic Electronics to the engineering students, the Fundamentals of Power Electronics as well as enable them to get an insight of the subject. Keeping in mind the purpose of wide coverage as well as to provide essential supplementary information, we have included the topics recommended by AICTE, in a very systematic and orderly manner throughout the book. Efforts have been made to explain the fundamental concepts of the subject in the simplest possible way.

During the process of preparation of the manuscript, the various standard text books have been referred and accordingly. Different sections like critical questions, solved and supplementary problems, short answer type questions, long answer type questions and multiple choice questions etc. are incorporated in this book. The book covers all types of medium and advanced level problems and these have been presented in a very logical and systematic manner. The gradations of those problems have been tested over many years of teaching to a wide variety of students.

Apart from illustrations and examples as required, we have enriched the book with numerous solved problems in every unit for proper understanding of the related topics.

This book consists of five units. The unit-I on power electronics devices can be able to provide detail information regarding construction, working principle, V-I characteristics of power semiconductor devices which are in wide use in modern industry for the control and conversion of electrical power. This unit can helps students to get a primary idea about the construction, operating principle and applications of Power bipolar junction transistor (Power BJT), Power metal oxide field effect transistor (Power MOSFET), and Insulated gate bipolar transistor(IGBT). The comparison between Power BJT, Power MOSFET and IGBT is incorporated in tabular manner so that students can able to differentiate the BJT, MOSFT and IGBT very easily. Switching characteristics, safe operating area (SOA) and series and parallel operation of Power BJT, Power MOSFET and IGBT are also discussed in detail. The basic concept of single electron transistor (SET) and some aspects of Nano-technology are also described in detail to help students to know about the recent development of power semiconductor devices.

The unit-II on thyristor family devices can be able to provide detail information regarding construction, two transistor analogy, types, working, V-I characteristics, mounting and cooling of silicon-controlled rectifier (SCR) devices which are widely use in modern industry for the control and conversion of electrical power. This unit can help students to get knowledge about types of thyristors such as SCR, LASCR, SCS, GTO, UJT, PUT, DIAC and TRIAC, thyristor family devices: symbol, construction, operating principle and V-I characteristics. Turn-on and turn-off characteristics of thyristor are also discussed. The comparison between SCR, GTO, DIAC and TRIAC is incorporated in tabular manner so that students can be able to differentiate the SCR, GTO, DIAC and TRIAC very easily. Protection circuits such as over-voltage, over-current, snubber and crowbar for thyristor are explained in detail.

The unit-III "turn-on and turn-off methods of thyristors" can be able to provide detail information regarding different turn-on methods of SCR such as high voltage triggering, thermal triggering, illumination triggering, dv/dt triggering and gate triggering. Gate triggering is most commonly used in modern industry for the control and conversion of electrical power. This unit can help students to get a primary idea about gate trigger circuits, R and RC circuits, SCR triggering using UJT, PUT as Relaxation oscillator and synchronized UJT circuit and Pulse transformer and opto-coupler based triggering. The different SCR turn–off methods such as class A- resonant commutation circuit, class B-shunt resonant commutation circuit, class C-complimentary symmetry commutation circuit, class D-auxiliary commutation, class E-external pulse commutation and class F-line or natural commutation are described in detail to help students to know about the recent development of turn-off methods of thyristors.

The unit-IV on "phase controlled rectifiers" can be able to provide detail information regarding phase control: firing angle and conduction angle, single phase half controlled, full controlled and midpoint–controlled rectifier with R and RL load, circuit diagram, working principle, and input-output waveforms of phase controlled rectifiers. The equations of dc output voltage, performance parameters of converters, and the effect of freewheeling diode in phase controlled rectifiers are also incorporated in this unit. After study this unit, student can able to get knowledge about different configurations of bridge controlled rectifiers: full bridge, half bridge with common anode, common cathode and SCRs in one arm and diodes in another arm.

The unit-V on "Industrial Control Circuits" can be able to provide detail information regarding different industrial applications of power electronics devices. This unit can helps students to get a primary idea about burglar's alarm system, battery charger using SCR, emergency light system, and temperature controller using SCR. The illumination control/fan speed control using TRIAC, switch mode power supply (SMPS), UPS: off-line and online and SCR based AC and DC circuit breakers are also described elaborately to help students to know about the applications of power semiconductor devices.

It is important to note that in this book, all relevant laboratory practical are included. In addition, some essential information for the users are incorporated under the heading "Know More" section.

The subject matters are presented in a constructive manner so that an Engineering degree prepares students to work in different sectors or in national laboratories at the very forefront of technology.

We sincerely hope that the book will inspire the students to learn and discuss the ideas behind basic principles of fundamental of power electronics and will surely contribute to the development of a solid foundation of the subject. We would be thankful to all beneficial comments and suggestions which will contribute to the improvement of the future editions of the book. It gives us immense pleasure to place this book in the hands of the teachers and students. It was indeed a big pleasure to work on different aspects covering in the book.

Prof. Soumitra Kumar Mandal

OUTCOME BASED EDUCATION

For the implementation of an outcome based education the first requirement is to develop an outcome based curriculum and incorporate an outcome based assessment in the education system. By going through outcome based assessments, evaluators will be able to evaluate whether the students have achieved the outlined standard, specific and measurable outcomes. With the proper incorporation of outcome based education there will be a definite commitment to achieve a minimum standard for all learners without giving up at any level. At the end of the programme running with the aid of outcome based education, a student will be able to arrive at the following outcomes:

Programme Outcomes (POs) are statements that describe what students are expected to know and be able to do upon graduating from the program. These relate to the skills, knowledge, analytical ability attitude and behaviour that students acquire through the program. The POs essentially indicate what the students can do from subject-wise knowledge acquired by them during the program. As such, POs define the professional profile of an engineering diploma graduate.

National Board of Accreditation (NBA) has defined the following seven POs for an Engineering diploma graduate:

PO1. Basic and Discipline specific knowledge: Apply knowledge of basic mathematics, science and engineering fundamentals and engineering specialization to solve the engineering problems.

PO2. Problem analysis: Identify and analyses well-defined engineering problems using codified standard methods.

PO3. Design/ development of solutions: Design solutions for well-defined technical problems and assist with the design of systems components or processes to meet specified needs.

PO4. Engineering Tools, Experimentation and Testing: Apply modern engineering tools and appropriate technique to conduct standard tests and measurements.

PO5. Engineering practices for society, sustainability and environment: Apply appropriate technology in context of society, sustainability, environment and ethical practices.

PO6. Project Management: Use engineering management principles individually, as a team member or a leader to manage projects and effectively communicate about well-defined engineering activities.

PO7. Life-long learning: Ability to analyse individual needs and engage in updating in the context of technological changes.

COURSE OUTCOMES

By the end of the course the students are expected to learn:

CO-1: How to select power electronic devices for specific applications.

CO-2: To maintain the performance of Thyristors.

CO-3: To troubleshoot turn-on and turn-off circuits of Thyristors.

CO-4: To maintain phase controlled rectifiers.

CO-5: To maintain industrial control circuits.

Mapping of Course Outcomes with Programme Outcomes to be done according to the matrix given below:

Course Outcomes	Expected Mapping with Programme Outcomes (1- Weak Correlation; 2- Medium correlation; 3- Strong Correlation)						
	PO-1	PO-2	PO-3	PO-4	PO-5	PO-6	PO-7
CO-1	3	3	3	3	1	1	3
CO-2	3	2	2	2	1	1	3
CO-3	3	2	2	2	1	1	3
CO-4	3	2	2	3	1	1	3
CO-5	3	3	3	3	1	1	3

GUIDELINES FOR TEACHERS

To implement Outcome Based Education (OBE) knowledge level and skill set of the students should be enhanced. Teachers should take a major responsibility for the proper implementation of OBE. Some of the responsibilities (not limited to) for the teachers in OBE system may be as follows:

- Within reasonable constraint, they should manoeuvre time to the best advantage of all students.
- They should assess the students only upon certain defined criterion without considering any other potential ineligibility to discriminate them.
- They should try to grow the learning abilities of the students to a certain level before they leave the institute.
- They should try to ensure that all the students are equipped with the quality knowledge as well as competence after they finish their education.
- They should always encourage the students to develop their ultimate performance capabilities.
- They should facilitate and encourage group work and team work to consolidate newer approach.
- They should follow Bloom's taxonomy in every part of the assessment.

Bloom's Taxonomy

Level	Teacher should Check	Student should be able to	Possible Mode of Assessment
Create	Students ability to create	Design or Create	Mini project
Evaluate	Students ability to justify	Argue or Defend	Assignment
Analyse	Students ability to distinguish	Differentiate or Distinguish	Project/Lab Methodology
Apply	Students ability to use information	Operate or Demonstrate	Technical Presentation/ Demonstration
Understand	Students ability to explain the ideas	Explain or Classify	Presentation/Seminar
Remember	Students ability to recall (or remember)	Define or Recall	Quiz

GUIDELINES FOR STUDENTS

Students should take equal responsibility for implementing the OBE. Some of the responsibilities (not limited to) for the students in OBE system are as follows:

- Students should be well aware of each UO before the start of a unit in each and every course.
- Students should be well aware of each CO before the start of the course.
- Students should be well aware of each PO before the start of the programme.
- Students should think critically and reasonably with proper reflection and action.
- Learning of the students should be connected and integrated with practical and real life consequences.
- Students should be well aware of their competency at every level of OBE.

ABBREVIATIONS AND SYMBOLS

List of Abbreviations

General Terms

Abbreviations	Full form
ASCR	Asymmetrical silicon-controlled rectifier
BJT	Bipolar Junction Transistor
CE	Common emitter
CNT	Carbon nanotubes
COMFET	Conductivity modulated field effect transistor
DIAC	Diode for Alternating current
FBSOA	Forward bias safe operating area
FET	Field effect transistor
FB	Forward blocking
FC	Forward conduction
GTO	Gate Turn-off thyristor
HVDC	High voltage direct current
IGBT	Insulated gate bipolar transistor
IC	Integrated circuit
LASCR	Light activated silicon controlled rectifier
LED	Light emitting diode
MOSFET	Metal Oxide Semiconductor Field Effect Transistor
MCT	MOS controlled thyristors
MOSIGT	Metal oxide insulate
NED	Nanotube emissive displays
NED	Nanotube Emissive Displays
ODF	Overdrive factor
PUT	Programmable unijunction transistor
RB	Reverse blocking
RBSOA	Reverse bias safe operating area
RTD	Temperature detector resistor
SCR	Silicon controlled rectifier
SCS	Silicon controlled switch
SET	Single electron transistor
SMPS	Switch mode power supply

Abbreviations	Full form
SIT	Static induction transistor
SITH	Static induction thyristor
SOA	Safe Operating Area
TB	Tera-byte
TRIAC	Triode for alternating current
TUF	Transformer utilization factor
UJT	Unijunction transistor
UPS	Uninterruptible power supplies
VLSI	Very large scale integrated circuit

List of Symbols

Symbols	Description
A	Anode
B	Base
C	Collector
C, K	Cathode
C_{gs}	Gate to source capacitance
D	Drain
dV_{DS}/dt	Rate of change of drain to source voltage
E	Emitter
G	Gate
I_A	Anode current
I_B	Base current
I_C	Collector current
I_D	Drain current
I_E	Emitter current
I_T	Total current
I_g	Gate current
I_L	Latching current
I_H	Holding current

Symbols	Description
J_1, J_2, J_3	Junctions
N_A, N_D	Doping intensity
P	Power dissipation
$P_{on-state}$	Power dissipation in on-state
P_{on}	Power dissipation in on-state
$P_{Total-loss}$	Total power loss
$P_{turn-off}$	Power dissipation during turn-off state
P_{gav}	Average gate power dissipation
Q	Charge
$R_{Channel}$	Channel resistance
R_d	Resistance of drift region
R_{E1}, R_{E2}	Emitter resistance
R_D	Drain resistance
R_{drift}	Resistance of drift region
R_{DS}	Drain to source resistance
R_s	Resistance of source voltage
R_s, C_s	Snubber resistance and capacitance
S	Source
t_d	Delay time
t_{on}	Switching on time
t_{off}	Switching off time
t_r	Rise time
$t_{turn-on}$	Turn-on time
t_s	Storage time
t_f	fall time
t_p	spread time
t_q	Turn-off time of thyristor
t_{rr}	reverse time

Symbols	Description
t_{gr}	Gate recovery time
$t_{turn-off}$	Turn-off time
T_j	Internal junction temperature
V_{AK}	Voltage across anode to cathode
V_{BD}	Reverse breakdown voltage
V_{BE}	Base emitter voltage
V_{BR}	Reverse breakdown voltage
V_{BO}	Forward breakover voltage
V_{CB}	Collector base voltage
V_{CE}	Collector emitter voltage
V_D	Anode-gate junction voltage
V_{DWM}	Peak forward blocking voltage
V_{DS}	Drain to source voltage
V_{drift}	Voltage drop across drift region
V_{DRM}	Peak repetitive forward blocking voltage
V_G	Gate voltage
V_g	Input voltage at gate
V_{GS}	Gate to source voltage
V_{gmin}	Minimum gate voltage
V_{GE}	Gate to emitter voltage
V_P	Peak point of voltage
V_{RWM}	Peak reverse voltage
V_{RRM}	Repetitive peak reverse voltage
V_S	Gate to source voltage
V_{SF}	Voltage safety factor
V_{Th}	Threshold voltage
V_T	On-state voltage drop
α	Common base dc current gain

Symbols	Description
β	Common emitter dc current gain
β_A	current gain of auxiliary transistor
β_M	current gain of main transistor
β_f	Forced current gain
η	intrinsic stand-off ratio

LIST OF FIGURES

LIST OF TABLES

TABLE OF CONTENT

1 Power Electronic Devices

UNIT SPECIFIES

Through this unit we have discussed the following aspects:

- Different Power electronic devices
- Power transistor: construction, working principle, V-I characteristics and uses.
- IGBT: Construction, working principle, V-I characteristics and uses.
- Concept of single electron transistor (SET)
- Aspects of Nano- technology.

The practical applications of the topics are discussed for generating further curiosity and creativity as well as improving problem solving capacity of learners.

Not only a large number of multiple choice questions are incorporated in this chapter but also questions of short and long answer types marked in two categories following lower and higher order of Bloom's taxonomy, assignments through a number of numerical problems. A list of references and suggested readings are illustrated in this unit so that learners can go through them for practice and get detailed knowledge. It is also be noted that some QR codes have been incorporated in different sections of this unit for getting more information on various topics of interest.

After discussion the content related to theory, there is a "Know More" section at end of this unit which is related to laboratory experiments. This section has been carefully designed so that the supplementary information provided in this section becomes beneficial for the earners. Usually, this section highlights the initial activity, examples of some interesting facts, analogy, and history of the development of the power electrics devices focusing the salient observations and finding, timelines starting from the development of the concerned topics up to the recent time, applications of the subject matter for our day-to-day life, and industrial applications of power electronics devices.

RATIONALE

This unit on power electronics devices can be able to provide detail information regarding construction, working principle, V-I characteristics of power semiconductor devices which are in wide use in modern industry for the control and conversion of electrical power. This unit can help students to get a primary idea about the construction, operating principle and applications of Power bipolar junction transistor (Power BJT), Power metal oxide field effect transistor (Power MOSFET), and Insulated gate bipolar transistor(IGBT). The comparison between Power BJT, Power MOSFET and IGBT is incorporated in tabular manner so that students can able to differentiate the BJT, MOSFT and IGBT very easily. Switching characteristics, safe operating area (SOA) and series and parallel operation of

Power BJT, Power MOSFET and IGBT are also discussed in detail. The basic concept of single electron transistor (SET) and some aspects of Nano-technology are also described in detail to help students to know about the recent development of power semiconductor devices.

PRE-REQUISITES

Semiconductor physics

Analog and digital electronics

Electrical circuit theory

Electrical Machines and power system

UNIT OUTCOMES

List of outcomes of this unit is as follows:

U1-O1: To define power electronics

U1-O2: To list all power electronics devices

U1-O3: To learn about construction, working principle, V-I characteristics of power Transistors

U1-O4: To give a list of applications of power transistors

U1-O5: To discuss about construction, working principle, V-I characteristics of IGBT

U1-O6: To provide a list of applications of IGBT

U1-O7: To understand the concept of single electron transistor (SET)

U1-O8: To explain aspects of Nano- technology

Unit-1 Outcomes	**EXPECTED MAPPING WITH COURSE OUTCOMES (*1-Weak Correlation; 2- Medium Correlation; 3-Strong Correlation*)**				
	CO-1	**CO-2**	**CO-3**	**CO-4**	**CO-5**
U1-O1	3	-	-	-	-
U1-O2	3	-	-	-	-
U1-O3	3	3	2	3	-
U1-O4	3	2	2	-	3
U1-O5	3	3	2	-	-
U1-O6	3	2	2	3	3
U1-O7	-	-	-	-	2
U1-O8	-	-	-	-	2

1.1 INTRODUCTION TO POWER ELECTRONICS

Power Electronics is the technology which can be applied for conversion and control of electrical power. In fact, the prime work of power electronics is to process and control the flow of electrical energy by providing voltage and current in such a way that these parameters are optimally matched with the consumer demand. This said conversion of electric power is called as *power electronics converter* or *power converter or switching converter* or *power modulator.*

The schematic block diagram of basic structure of a power electronics converter based system consists of power electronics converter and controller including the analogous measurement and interface circuits (Fig.1.1). Usually power converters are used to convert electric power one form (AC or DC) to another form (variable AC or DC) using power semiconductor devices and circuit elements like inductance (L) and capacitance (C). Controllers are needed to generate control signals to turn on and turn off of power switching devices. Consequently, the optimally required output voltage and current at specified frequency will be available at output terminals.

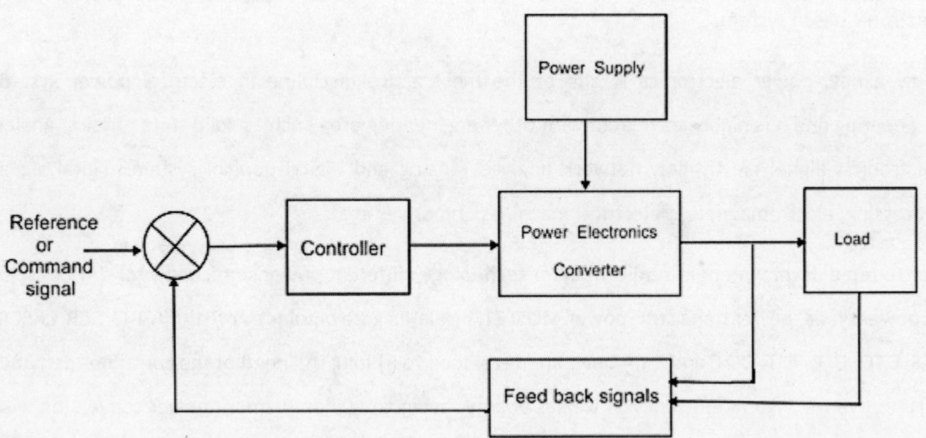

Fig.1.1 Schematic block diagram of basic structure of a power electronic based system

Power Source: The *source of electrical power* is dc power (for example, dc generator, battery, and solar photovoltaic cell) and ac power (for example, alternator and induction generator). The controlled ac/dc power flows from source to load through a *power converter.*

Power Electronics Converter: The power electronics converters are used to convert electric power in one form to another form as per requirement of load. The output of a *power converter* can be a variable dc or a variable ac with variable voltage, current and frequency. Generally, the output of a power converter depends on the nature of *load.* For example, if the load is induction motor, the converter output must be variable ac voltage as well as frequency. When the load is dc motor, the converter output will be variable dc voltage.

Controller: The controller input signals are *feedback signals* which are the measured parameters of the load or source such as voltage, current, speed and position etc. Initially the command (reference) signal is applied to the controller. Subsequently the feedback signals are compared with the command signal and the control signals are generated by the controller to turn-on or turn-off the power semiconductor devices of power converter. As a result, the required output will be available at the particular load.

Usually, *the controller* generates triggering pulses for switching ON/OFF power semiconductor devices of converter. A synchronizing circuit is required for dc to ac converter and ac to ac converter circuits.

Load: Electrical motors such as dc motors (separately excited, shunt, series and compound), induction motors (squirrel-cage, wound rotor and linear), synchronous motors, brushless dc motors, stepper motor and switched reluctance motors etc. are most commonly used as load in any a power electronic based system.

Now a day, power electronics is one of the most active discipline in electrical power system engineering and it is highly correlated with other engineering area such as solid state physics, analog electronics, digital electronics, network analysis, analog and digital control system, digital signal processing, electromagnetics, electrical machines and drives etc.

Due to rapid development in semiconductor technology, different power semiconductor devices such as power diode, power transistor, power MOSFET, insulated gate bipolar transistor (IGBT), SCR, LASCR, SCS, GTO, UJT, PUT, DIAC and TRIAC etc. are developed to (i) fulfil the need of the consumer demand, (ii) improve the system efficiency as well as energy saving and (iii) implement power conversion with minimum volume, minimum weight, and less cost.

For more on Introduction to Power Electronics

1.2 DEVELOPMENT OF POWER ELECTRONICS DEVICES

The terminology "*Power electronics*" originated at the beginning of nineteenth century with development of *mercury-arc rectifiers* which is used to convert high-voltage AC into DC voltage. The mercury-arc rectifiers were generally used to provide dc supply in industrial motors and drives, electric locomotives and high-voltage direct current (HVDC) power transmission.

In 1900, initially the mercury-arc rectifiers with glass envelope were developed. Subsequently, grid controlled mercury-arc rectifiers and mercury-arc rectifiers with metal envelope were developed in 1903 and 1908 respectively.

In 1914, *Thyratrons* was developed by Langmuir. Thyratron is a type of gas filled tube which is used as a high voltage electrical switch and a controlled rectifier. Generally, thyratron are manufactured as triode, tetrode and pentode.

Selenium rectifier was developed without a glass tube in 1925 and it is also known as *metal rectifier*. Copper oxide or selenium is used as semiconductor in selenium rectifier. The selenium rectifier device can be used in controlled converters, inverters, cyclo-converters and battery chargers.

In 1930, Rissik developed *Cyclo-converters* i.e. a variable frequency output voltage from fixed frequency input voltage. In 1933, Lenz developed the ac voltage regulator or controller using solid state power devices. The point-contact transistor was developed by W. H. Brattain, J. Bardeen and W. Shockley in 1947.

In 1948, bipolar junction transistor (BJT) using germanium was invented and this development was the beginning of new age of semiconductor electronics due to reduction in size, cost, and power consumption of solid state power devices.

In 1953, high current rating about 100A germanium *power diode* was developed. A new revolution in semiconductor electronics is in progress with the innovation of a four-layer silicon PNPN device i.e. thyristor in 1956. The SCR was introduced by General Electric in late 1957 and the power electronics really began a new era.

In 1964, the *power FET* was developed by Zuleeg and Teszner. The development of power FET has significant effect on the power semiconductor industry.

The *insulated-gate bipolar transistor (IGBT),* a three-terminal power-semiconductor device is developed in early 1980 to combine high efficiency and fast switching. It consists of four alternating layers PNPN that are controlled by a metal oxide semiconductor gate structure. It can be used as electrical power switching applications such as automobiles, trains, air conditioners, and refrigerators etc.

1.3 POWER ELECTRONICS DEVICES

In power electronics converter circuits, power semiconductor devices are used as *switches* in power converter to convert the power one form to another and also control the electric power. Generally, the following power semiconductor devices are manufactured and most commercially available in the market for industrial use:

- Power diodes
- Power Transistors i.e. Power BJT, Power MOSFET
- Silicon-controlled rectifier (SCR)
- Asymmetrical SCR (ASCR)
- Bidirectional diode thyrisitor (DIAC)
- Bidirectional triode thyrisitor (TRIAC)
- Gate turn-off thyristor (GTO)
- Insulated –gate bipolar transistors (IGBT)
- Light-activated SCR (LASCR)
- Programmable unijunction transistor (PUT)
- Light-activated programmable unijunction transistor (LAPUT)
- Reverse-blocking diode thyristor (RBDT)
- Reverse-blocking tetrode thyristor (SCS silicon-controlled switch)
- Reverse-conducting diode thyristor (RCDT)
- Reverse-conducting triode thyristor (RCTT)
- Silicon bilateral switch (SBS)
- Silicon unilateral switch (SUS)
- Static Induction Transistor (SIT)

- Static Induction Thyristors (SITH)
- Reverse Conducting thyristor (RCT)
- MOS controlled thyristors (MCT)

From the above solid state power devices, the most commonly used power semiconductor devices are power diodes, power BJT, power MOSFET, SCR, DIAC, TRIAC and IGBT etc. When these devices are in ON state, switches are closed and these devices behave just like ordinary switches and current flows from source to load through power semiconductor devices as well as load. In the same way, when these devices operate in OFF state, switch is opened and no current flows from source to load. Power semiconductor devices are available in wide range of voltage rating from few volts to several kV and current rating from few amperes to several kA.

Power semiconductor devices should have the following essential properties: (i) breakdown voltage should be very high, (ii) on state voltage drop must be low, (iii) on state resistance is low, (iv) turn-on and turn-off process will be very fast and (v) power dissipation capacity is high. Table 1.1 shows the symbol, voltage/current rating and operating frequency of power semiconductor devices. Switching on time, switching off time and *on-state resistance* of power semiconductor devices are illustrated in Table 1.2.

Table 1.1 Symbols voltage/current rating and operating frequency of power semiconductor devices

Device Name	**Symbol**	**Voltage/current rating (maximum)**	**Operating frequency**
Power Diode	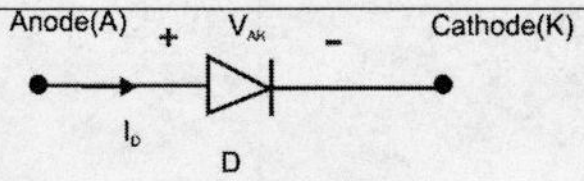 Anode(A) + V_{AK} - Cathode(K) I_D D	5kV/5kA	100-300kHz
Power BJT npn	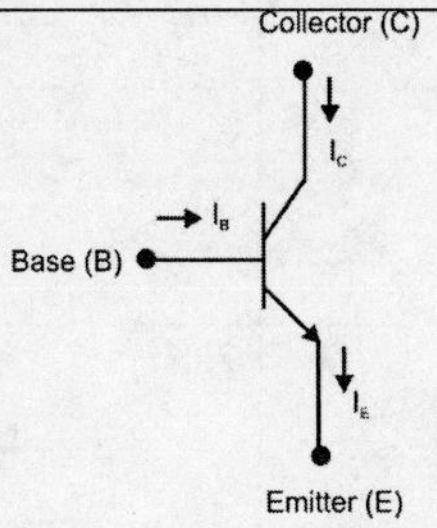 Collector (C) I_C I_B Base (B) I_E Emitter (E)	1400V/400A	10kHz

Device Name	Symbol	Voltage/current rating (maximum)	Operating frequency
Power MOSFET	Drain (D) I_D Gate (G) Source (S)	1kV/50A	100kHz- 1MHz
Thyristors (SCR)	Anode(A) + V_{AK} - Cathode(K) I_A SCR Gate(G)	10kV/5kA	1kHz
Light Activated SCR	Anode(A) + V_{AK} - Cathode(K) I_A Gate(G) LASCR (Light activated SCR)	6kV/3kA	1kHz
IGBT	Collector (C) I_C Gate (G) I_E Emitter (E)	3.3kV/1200A	50kHz
GTO	Anode(A) + V_{AK} - Cathode(K) I_A Gate(G) GTO or Anode(A) + V_{AK} - Cathode(K) I_A Gate(G) GTO	5kV/3kA	2kHz

Device Name	Symbol	Voltage/current rating (maximum)	Operating frequency
TRIAC	MT2 IA MT1 Gate(G) TRIAC	1200V/300A	0.5kHz

Table 1.2 Switching on time, switching off time and *on-state resistance* of power semiconductor devices.

Device Name	Switching on time (t_{on})	Switching off time (t_{off})	On-state resistance
Power Diode	50 to 100 μs	50 to 100 μs	0.3 to 0.6 milli-ohm
Power BJT	2 μs	9 to 30 μs	4 to 10 milli-ohm
Power MOSFET	0.1 μs	1 to 2 μs	1 to 2 milli-ohm
Thyristors	2 to 5 μs	20 to 100 μs	0.25 to .75 milli-ohm
IGBT	0.2 μs	2-5 μs	2 to 40 milli-ohm
GTO	3 to 5 μs	10-25 μs	2.5 milli-ohm
TRIAC	2 to 5 μs	200 to 400 μs	3.5 milli-ohm

For more on Power Electronics Devices

1.4 POWER TRANSISTOR

Power transistors are fully controlled device as the turn-on and turn-off switching of power transistors are controllable. Whenever a current signal is applied to the base of bipolar junction transistor (BJT), this device will be turned on and operate in ON state as long as the control signal is present. If the control signal is removed from the base of BJT, the power transistor will be turned off and operate in off-state. In the same way, if a voltage signal is applied to gate of MOSFET, this device will be ON and when a voltage signal is removed from gate, MOSFET operates in OFF state. In general, power transistors are four types such as

- Bipolar junction transistor (BJT)
- Metal oxide semiconductor field-effect transistor (MOSFET)
- Insulated gate bipolar transistor (IGBT) and
- Static induction transistor (SIT)

In this section, the construction, working principle, V-I characteristics, safe operating area, series and parallel operation, switching characteristics and uses of power BJT are explained. The structure of power MOSFET, working principle, V-I characteristics and uses are also discussed in this section. The structure, operating principle, and V-I characteristics of IGBT are discussed elaborately. The concept of single electron transistor (SET) and aspects of Nano- technology are also explained in this unit.

1.4.1 BIPOLAR JUNCTION TRANSISTOR (BJT)

A bipolar junction transistor (BJT) has a silicon or germanium crystal in which a layer of *n*-type semiconductor is sandwich between two layers of *p*-type semiconductor and it is called as *p-n-p* transistor. In the same way, BJT consists of a layer of *p*-type semiconductor between two layers of *n*-type semiconductor materials and this one is called as an *n-p-n* transistor. Actually the sandwich of semiconductor is very small and sealed inside either a metal or plastic case to protect from moisture.

The construction of a *p-n-p* transistor and an *n-p-n* transistor are depicted in Fig.1.2 and Fig.1.3 respectively. It can be seen from Fig.1.1 that each BJT has three regions such as emitter (E), base (B) and collector (C).

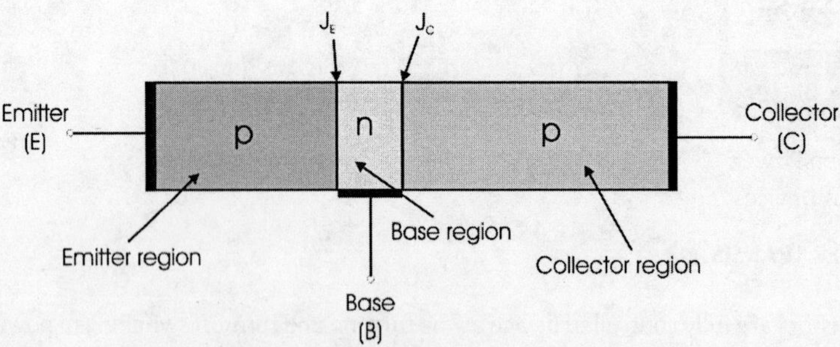

Fig.1.2 Construction of p-n-p transistors

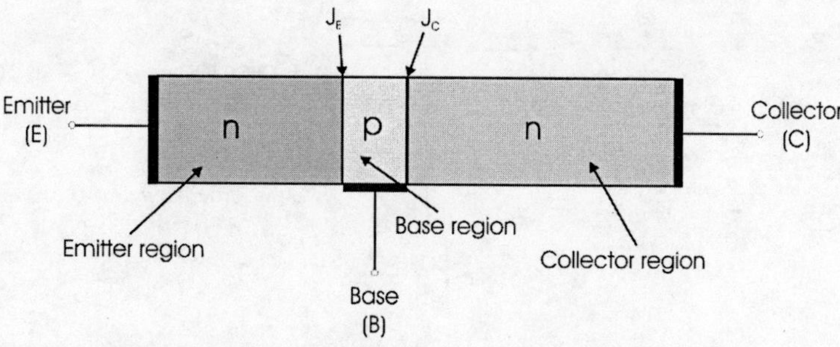

Fig.1.3 Construction of n-p-n transistors

Emitter (E): The emitter region provides charge carriers either electrons or holes to the base and collector regions. Always emitter region is heavily doped about 10^{19} per cm^3 donor or acceptor atoms.

Base (B): The base region is placed in the middle of two *"n"* or *"p"* regions and hence forms two *p-n* junctions in the transistor. Usually, the base of transistor is thin compared to emitter and collector region and this region is lightly doped about 10^{16} per cm^3.

Collector (C): The collector region is placed in the other end of the transistor which is opposite side to the emitter. This region collects charge carriers either electrons or holes. The collector region is always larger than the emitter region and base region of transistor. The doping level of collector is less compared to the doping level of emitter, but more than the doping level of base region about 10^{17} per cm^3.

BJT has two p-*n* junction namely *emitter-base junction* (J_E) and *collector-base junction* (J_C). The junction between emitter and base regions is known as *emitter-base junction* (J_E). Correspondingly, the junction between collector and base regions is called as *collector-base junction* (J_C).

The symbols of *p-n-p* and *n-p-n* transistors are depicted in Fig.1.4 1nd 1.5 respectively. The arrow on the emitter terminal represents the direction of current flow while the base emitter junction is forward biased.

The emitter, base and collector currents are represented by I_E, I_B and I_C respectively (Fig.1.4 and Fig.1.5). The V_{BE} represents the voltage drop across base to emitter of transistor. Consistently, V_{CB} and V_{CE} are the voltage across the collector base and collector emitter respectively.

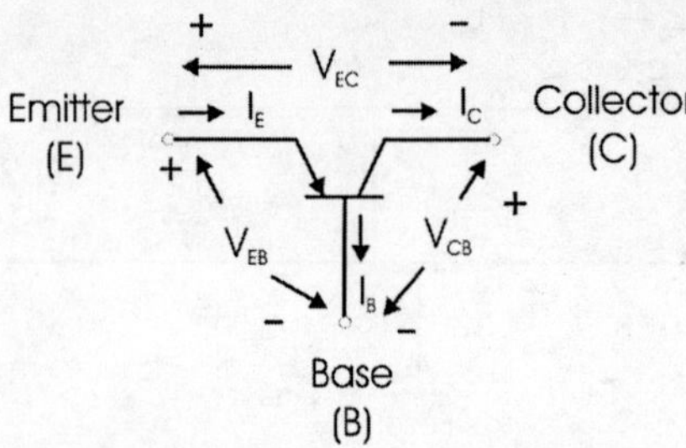

Fig.1.4 Symbol of p-n-p transistors

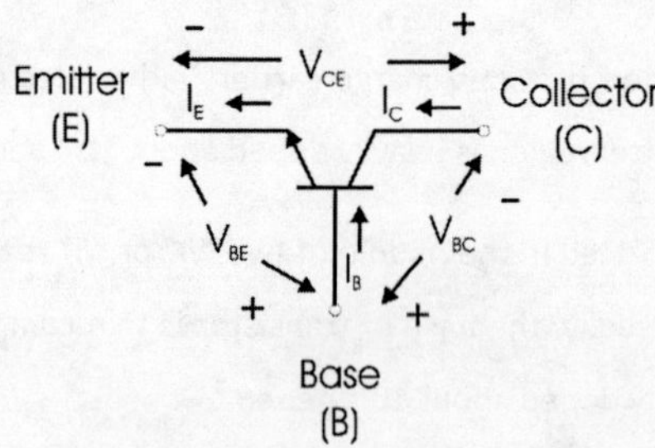

Fig.1.5 Symbol of n-p-n transistors

1.4.2 Relation Between α and β

The *common base dc current gain* (α) is defined as the ratio of the collector current (I_C) to the emitter current (I_E) and it is represented by α or h_{FB} or α_{dc}. The dc current gain α can be expressed as

$$\alpha = \frac{I_C}{I_E} \tag{1.1}$$

Since the collector current is always less than the emitter current, the dc current gain α is always less than unity. Usually the value of α is about 0.98. The dc current can be closer to unity by controlling the width and doping level of base region as small as possible. Practically, α varies in between 0.95 to 0.998.

The *common emitter dc current gain* (β) is defined as the ratio of the collector current (I_C) to the base current (I_B). It is represented by β or h_{FE} or β_{dc} and it can be expressed as

$$\beta = \frac{I_C}{I_B} \tag{1.2}$$

The common emitter dc current gain β of a transistor is called as *large signal common emitter current gain*. Since the collector current of bipolar junction transistor is always greater than the base current, the value of dc current gain (β) must be greater than unity. For example, if the collector current I_C is equal to 5mA and the base current I_B is 0.05-mA, the common emitter dc current gain β is equal to 100. Consequently, the collector current is 100 times that of base current. Generally, β varies in the range from 20 to 250.

The emitter current (I_E) of a transistor is sum of the base current (I_B) and collector current (I_C) and it can be expressed as

$$I_E = I_B + I_C \tag{1.3}$$

After dividing the both sides of the equation (1.3) by I_C, we get

$$\frac{I_E}{I_C} = \frac{I_B}{I_C} + 1 \tag{1.4}$$

As the current gain $\alpha = \frac{I_C}{I_E}$ and the current gain $\beta = \frac{I_C}{I_B}$, the equation (1.4) can be expressed as

$$\frac{1}{\alpha} = \frac{1}{\beta} + 1 = \frac{1+\beta}{\beta} \tag{1.5}$$

The equation (1.5) can be expressed as $\alpha = \frac{\beta}{\beta+1}$ (1.6)

The equation (1.6) can also be written as

$$\beta = \alpha(\beta+1) = \alpha\beta + \alpha$$

Thus, $\beta = \frac{\alpha}{1-\alpha}$ (1.7)

Example 1.1 A bipolar junction transistor (BJT) has an α of 0.97. Determine the value of β

Solution

The value of $\beta = \frac{\alpha}{1-\alpha} = \frac{0.97}{1-0.97}$ = 32.33 (as α =0.97)

Example 1.2 A BJT has a β of 99. Find the value of α.

Solution

The value of $\alpha = \dfrac{\beta}{\beta+1} = \dfrac{99}{99+1}$ = 0.99 (as β=99)

Example 1.3 The collector current of a bipolar junction transistor is about 90 mA and it's β is 99. Calculate the value of base current and emitter current.

Solution

The value of $\alpha = \dfrac{\beta}{\beta+1} = \dfrac{99}{99+1}$ = 0.99 (As β=99)

Since the current gain $\alpha = \dfrac{I_C}{I_E}$, the emitter current is equal to

$$I_E = \alpha I_C = 0.99 \times 90 \text{ mA=89.1 mA}$$

As the current gain $\beta = \dfrac{I_C}{I_B}$, the base current is equal to

$$I_B = \frac{I_C}{\beta} = \frac{90}{99} \text{ mA=0.909 mA}$$

Example 1.4 A BJT has β=95 and the emitter current is 80 mA. Compute the value of base current and collector current.

Solution

The value of $\alpha = \dfrac{\beta}{\beta+1} = \dfrac{95}{95+1}$ = 0.989 (As β=95)

Since the current gain $\alpha = \dfrac{I_C}{I_E}$, the collector current is

$$I_C = \alpha I_E = 0.989 \times 80 \text{ mA=79.12 mA (As } I_E = 80 \text{ mA)}$$

As the current gain $\beta = \dfrac{I_C}{I_B}$, the base current is equal to

$$I_B = \frac{I_C}{\beta} = \frac{79.12}{90} \text{ mA} = 0.879 \text{ mA}$$

Example 1.5 The bipolar junction transistor has a base current I_B=125μA and the collector current (I_C) is equal to 15mA. (a) Calculate the value of β, α and emitter current. (b) if the base current changes by 25μA and the corresponding collector current change is 0.5mA, determine the new value of β.

Solution

(a) The current gain $\beta = \frac{I_C}{I_B} = \frac{15\times10^{-3}}{125\times10^{-6}}$ = 120 (As I_B=125μA and I_C=**15**mA)

The value of $\alpha = \frac{\beta}{\beta+1} = \frac{120}{120+1}$ = 0.99

The emitter current is

$$I_E = I_B + I_C = 125 \text{ μA} + 15\text{mA} = 5.125 \text{ mA}$$

(b) When the change in base current is ΔI_B =25μA, the change in collector current is equal to ΔI_C =0.5mA

Therefore, the new base current is I_B=125μA+25μA=150 μA

and **t**he new collector current is equal to I_C =**1**5mA+0.5mA= 15.5mA

The new value of β is

$$\beta = \frac{I_C}{I_B} = \frac{15.5\times10^{-3}}{150\times10^{-6}} = 103.33$$

Example 1.6 The common base dc current gain of a transistor is 0.98. If the emitter current is 100mA, determine the base current and collector current.

Solution

The common-base dc current gain is equal to $\alpha = \frac{I_C}{I_E}$

Therefore, the collector current is $I_C = \alpha I_E = 0.98\times100$mA = 98mA (As α =0.98, and I_E = 100 mA)

The value of $\beta = \frac{\alpha}{1-\alpha} = \frac{0.98}{1-0.98}$ = 49

Hence, the base current is equal to $I_B = \frac{I_C}{\beta} = \frac{98}{49} = 2\text{mA}$

1.4.3 CONSTRUCTION AND WORKING PRINCIPLE OF POWER BIPOLAR JUNCTION TRANSISTOR

NPN power transistor consists of four-layer $n^+ p\ n^- n^+$ structure as depicted in Fig.1.6. This transistor has three terminals such as base (B), emitter (E) and collector (C). Base is used as an input terminal and collector is used as the output terminal. In common emitter (CE) configuration, emitter is common between input and output terminals.

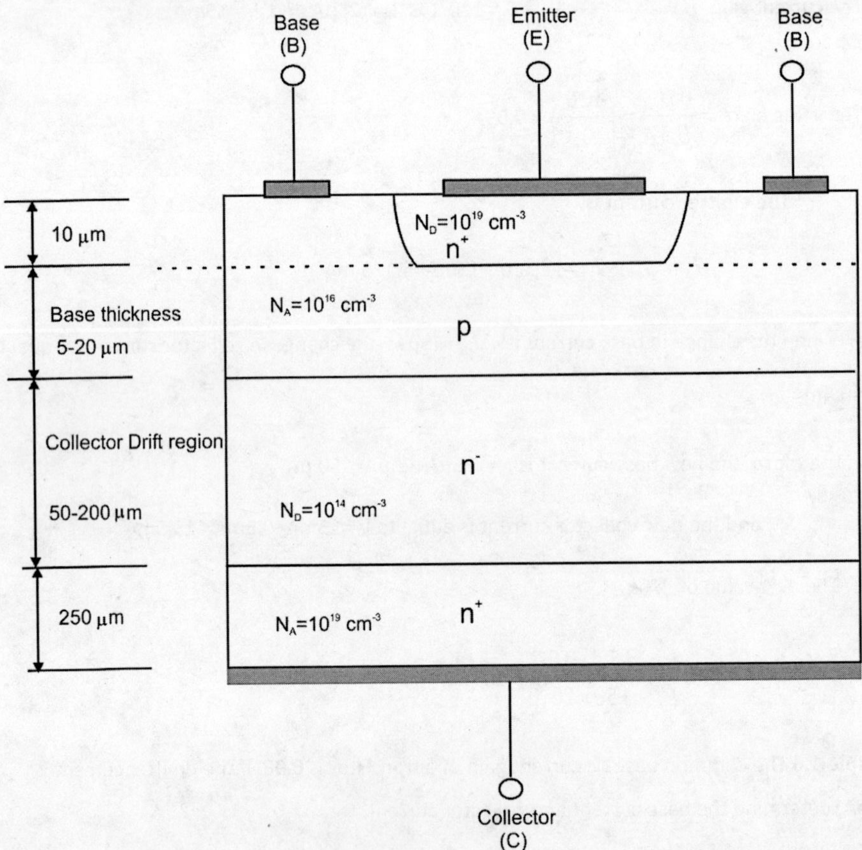

Fig.1.6 Structure of n-p-n type power BJT

It is clear from Fig.1.6 that the width of emitter, n^+ *layer* is about 10 μm and its doping intensity is about $N_A = 10^{19} cm^{-3}$. Generally, the base thickness of *p layer* is about 5 to 20 μm and the doping density of p type semiconductor materials is moderate and its value is about $N_D = 10^{16} cm^{-3}$. The width of collector drift region, n^- *layer* varies from 50 μm to 200 μm and the doping density is minimum in drift region and its value is about $N_A = 10^{14} cm^{-3}$. The width of *n⁺ collector region* is maximum and it is 250 μm. The doping density of n⁺ type semiconductor layer is about $N_A = 10^{19} cm^{-3}$. The doping density of semiconductor materials and width of each layer of power BJT are shown in Fig.1.6.

The base thickness of BJT is small to provide better amplification capability. Due to very small base thickness, the breakdown voltage capability of BJT has been reduced. Consequently, the base thickness of power transistors should be more compared to the logic level transistors to increase breakdown voltage capability. In case of power BJTs, the base thickness of power transistors should vary with in few tens of micrometres. Actually, the breakdown voltage of the power transistors depends on the thickness of the collector drift region. Consequently, the thickness of collector drift region varies from 50 μm to 200 μm depending upon the break down voltage.

In general, the vertical structure of power transistors provides the maximum cross-sectional area through which the current flows with in the device and the current density per unit area becomes minimum. As a result, the on-state resistance and the power loss of power transistors can be minimized. Due to large cross sectional area, the thermal resistance of power transistors is low and the power dissipation problems can be reduced significantly.

In case for a *p-n-p* transistor, the four-layer structure of power transistors consists of *p+ n p⁻ p+* and the doping level will be opposite of *n-p-n* transistor. In power electronics converter circuits, n-p-n transistors are used as power semiconductor switches.

The current gain of a power transistor is represented by $\beta = \frac{I_C}{I_B}$ and the value of β is small and it varies in the range 5 to 10. Darlington pair of power BJTs is used to increase the current gain. The monolithic design of a Darlington pair of power BJTs is depicted in Fig.1.7.

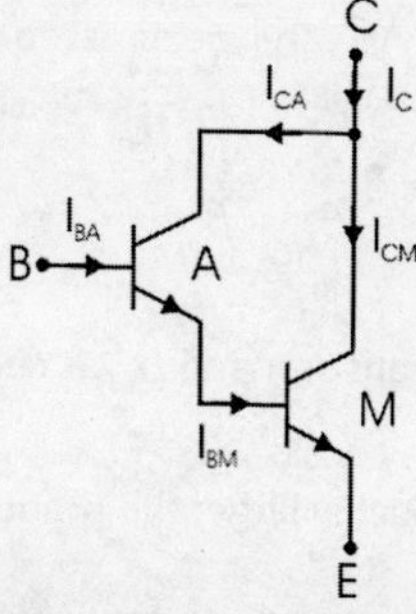

Fig.1.7 Darlington pair of power BJTs

In Fig.1.7, "A" represents the auxiliary transistor, "M" stands for the main transistor, I_{BA} is base current of auxiliary transistor, I_{CA} is collector current of auxiliary transistor, I_{EA} is emitter current of auxiliary transistor, I_{BM} is base current of main transistor and I_{CM} is collector current of main transistor. The relationship between collector current I_{CA} and base current I_{BA} of the auxiliary transistor is equal to

$$I_{CA} = \beta_A I_{BA} \tag{1.8}$$

where, β_A is current gain of auxiliary transistor.

It is clear from Fig.1.8 that the emitter current of auxiliary transistor I_{EA} is equal to the base current I_{BM} of the main transistor and I_{BM} is equal to

$$I_{BM} = (1+\beta_A) I_{BA} \tag{1.9}$$

The collector current of main transistor is

$$I_{CM} = \beta_M I_{BM} = \beta_M (1+\beta_A) I_{BA} \tag{1.10}$$

where, β_M is current gain of main transistor and $I_{BM} = (1+\beta_A) I_{BA}$

The total load current I_C is sum of collector current of auxiliary transistor I_{CA} and collector current of main transistor I_{CM}.

Subsequently, $I_C = I_{CA} + I_{CM} = \beta_A I_{BA} + \beta_M (1+\beta_A) I_{BA}$ (1.11)

$$= (\beta_A + \beta_M + \beta_M \beta_A) I_{BA} \tag{1.12}$$

As a result, the new current gain $\beta = \dfrac{I_C}{I_{BA}}$ can be expressed as

$$\beta = \beta_M \beta_A + \beta_M + \beta_A \tag{1.13}$$

where, β_A is current gain of auxiliary transistor and β_M is current gain of main transistor.

As the values of β_M and β_A are reasonably large, the product $\beta_M \beta_A$ can be very large compared to β_M and β_A. Consequently the estimated value of β is about $\beta_M \beta_A$ i.e. the product of the current gains of auxiliary and main transistors. As a result, the darlington pair of power BJTs needs very much smaller current control for its operation compared to single power transistor but the switching time of the darlington pair of power BJTs is more than the switching time of a single power BJT. The darlington pair of power BJTs with diodes D_1 and D_2 is shown in Fig.1.8. The diode D_1 is used to turn-off the main transistor firstly and the diode D_2 is used for half-bridge and full-bridge inverter circuits.

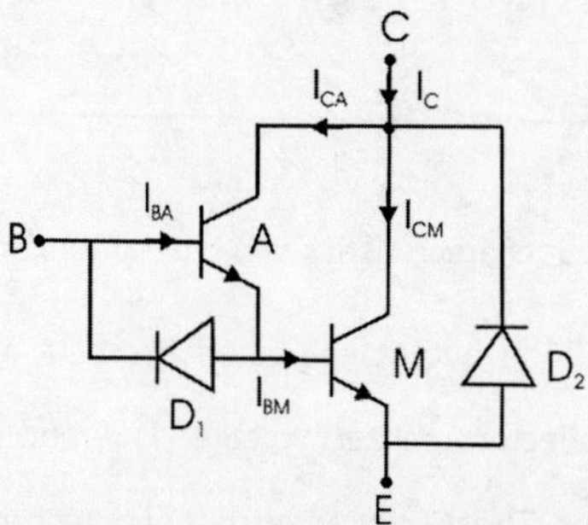

Fig.1.8 Darlington pair of power BJTs with diodes D_1 and D_2

1.4.4 V-I Characteristics of Power BJT

There are three possible configurations of power BJT such as common collector, common base and common emitter. The common emitter configuration is most commonly used in switching applications. Fig.1.9 shows the circuit diagram of common emitter configuration where BJT is used as a switch. The power BJT has three characteristics namely input characteristics, output characteristics and transfer characteristics.

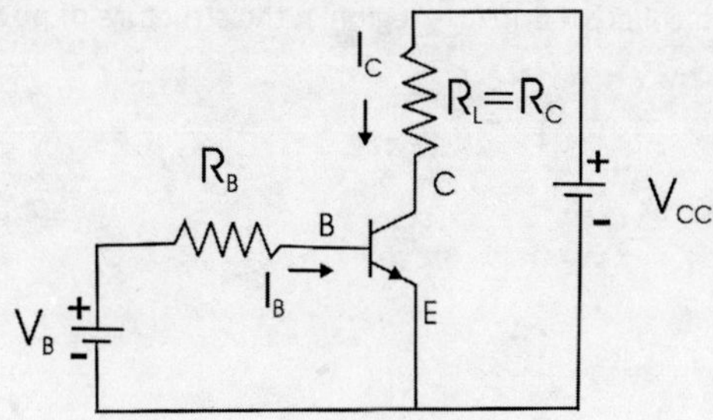

Fig.1.9 BJT is used as a switch

1.4.5.1 Input characteristics: The input characteristics of base current I_B with respect to base emitter voltage V_{BE} are depicted in Fig.1.10.

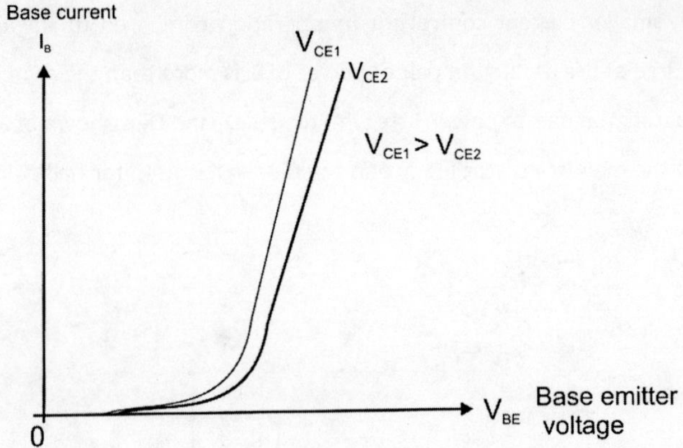

Fig.1.10 Input characteristics of a bipolar junction transistor

1.4.4.2 Output characteristics: The output characteristics of a n-p-n power BJT are the plot of collector current i_C versus collector emitter voltage V_{CE} and Fig.1.11 shows the V-I output characteristics of power transistor. The V-I characteristics of power transistor consists of three regions such as cut-off region, active region and saturation region. When the base current is not enough to turn it on, both the junctions are reverse biased and the transistor is in the off state and it operates in cut-off region. Transistor acts as an amplifier in the active region when the base current is amplified by a gain and collector-emitter voltage decreases with base current. The collector base junction is reverse biased and the base emitter junction is forward biased. When the base current is significantly high so that the collector-emitter voltage is low in saturation region and the power transistor acts a switch. In this case, both collector base junction and base emitter junction are forward biased.

However, the major difference between V-I characteristics of a power transistor and logic level transistor is the *quasi-saturation* which is present in power transistors. The quasi saturation is occurred due to the lightly doped collector drift n^- region in the structure of power transistors.

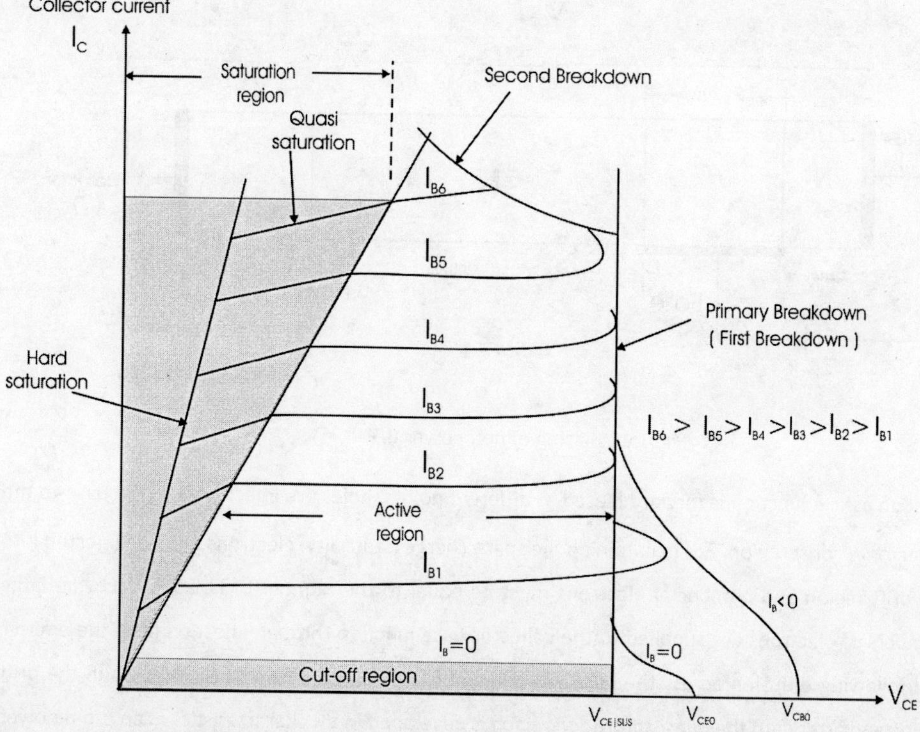

Fig.1.11 V-I characteristics of an npn power BJT

1.4.4.3 Quasi Saturation

To explain the quasi-saturation, a collector drift region (N^-) has been incorporated in the structure of power BJT as shown in Fig.1.12. At first, the transistor operates in the active region and the base current starts to increase. Due to increase in the base current, the collector current increases and the collector emitter voltage (V_{CE}) reduces as voltage drop across the collector load resistance, R_L increases. Accordingly, the voltage drops in the drift region increases. Subsequently the ohmic resistance drop also increases as the collector current (I_C) increases. Thus, the reverse bias voltage across collector base (*n⁻p*) junction becomes smaller and the junction gets forward biased at several points.

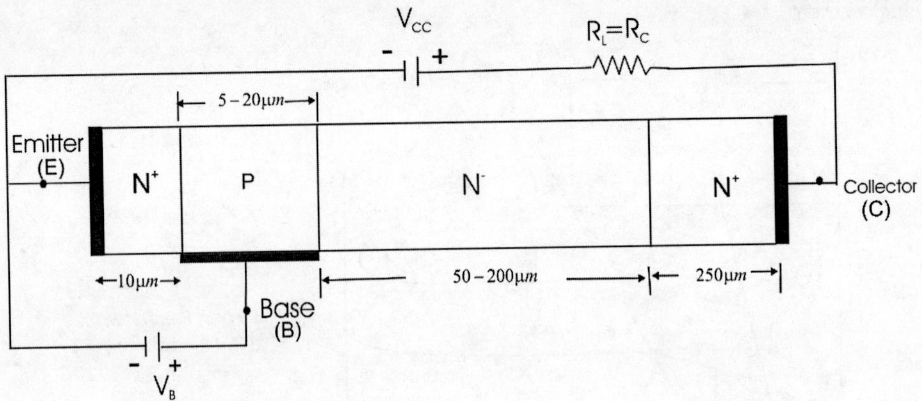

Fig. 1.12 Biasing of npn power BJT

As soon as $n^{-}p$ junction is forward biased at different points, holes are injected from the base to the collector n^{-} drift region. For maintaining the space charge neutrality, electrons must be injected into the drift region and number of electrons must be equal to the number of holes. In fact, the large numbers of electrons are supplied to the collector-base junction through injection from the emitter and following diffusion across the base. Consequently, the excess carriers can build up in the drift region and afterward the *quasi-saturation region* is developed in V-I characteristics of an n-p-n power BJT.

If the *ohmic resistance* of the n^{-} drift region is R_d , the collector current is

$$I_C = \frac{V_{CE}}{R_d} \tag{1.14}$$

Therefore, the *double injection* can be take place in the drift region in quasi saturation. The charge is accumulated in the n^{-} drift region from one side of drift region that is the collector-base *pn*$^{-}$ junction. The electron injection across the $n^{-}n^{+}$ junction is less measurable as there is much more abundant electrons at the *pn*$^{-}$ junction. When the injected carriers increase, the n^{-} drift region is shorted progressively and the voltage across the drift region is reduced even if the collector current (I_C) is large.

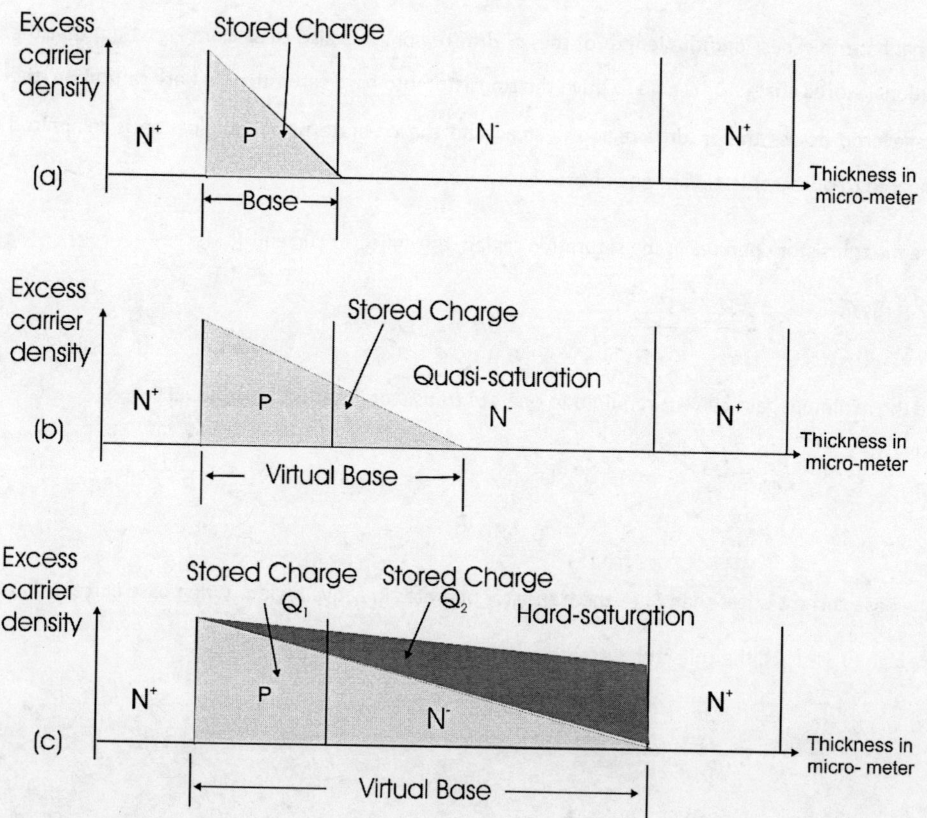

Fig.1.13 (a) Storage charge in the base region during operate in active mode

(b) Storage charge in the base region during operate in quasi-saturation

(c) Storage charge in the base region during operate in hard saturation

It is depicted in Fig.1.13(b) that the hole injection from the base to the collector-base pn^{-} junction has been introduced, the virtual base thickness is increased. Consequently, the effective value of β decreases and the magnitude of collector current for a specified base current should be decreased. The n^{-} drift region cannot be completely shorted out through high-level injection in the quasi-saturation. Therefore, the power dissipation of BJT in the quasi-saturation is greater larger than the power dissipation during hard saturation.

1.4.4.4 Hard saturation

The *hard saturation* is occurred whenever the excess carrier density reaches the end of n+ drift region. In hard saturation, it needs a minimum amount of stored charge Q_1 in the virtual base region as depicted in Fig.1.13(c). At that time the effective base thickness is increased and it is the sum of the normal base thickness and the length of the n^- drift region. It is also clear from Fig1.13(c) that the additional stored charge Q_2 is able to drive the transistor into hard saturation. In hard saturation, the voltage drop across the n^- drift region is small and the on-state power dissipation is minimized compared to quasi saturation of power transistor.

Once the transistor operates in the saturation region, the collector current I_{CS} is

$$I_{CS} = \frac{V_{CC} - V_{CES}}{R_C} \tag{1.15}$$

and the minimum base current required to operate transistor in saturation is equal to

$$I_{BS} = \frac{I_{CS}}{\beta} \tag{1.16}$$

If the base current is less than I_{BS}, the transistor operates in active region. When base current I_B is very high ($I_B > I_{BS}$), the collector current is equal to

$$I_{CS} = \frac{V_{CC}}{R_C} \quad \text{(as } V_{CES} \approx 0\text{)} \tag{1.17}$$

When the base current is high, transistor can be operating in *hard saturation*. The ratio of I_B and I_{BS} is called as overdrive factor (ODF) and it is equal to

$$ODF = \frac{I_B}{I_{BS}} \tag{1.18}$$

The value of ODF varies from 4 to 5. The forced current gain β_f is the ratio of I_{CS} to I_B and the value of β_f is always less than β.

The total power loss in two junctions of power BJT is equal to

$$P_{Total-loss} = V_{BE} I_B + V_{CE} I_C \tag{1.19}$$

The important features of V-I characteristics of power transistors are $V_{CE|sus}$, V_{CE0}, V_{CB0}, primary breakdown and second breakdown.

$V_{CE|sus}$: Whenever a sustainable collector current flows through power BJT, the said device will be able to withstand or sustained up to a maximum collector-emitter voltage. This maximum voltage across collector-emitter of power BJT is indicated by $V_{CE|sus}$.

V_{CE0}: When the base of power BJT become open circuit, the base current is zero ($I_B = 0$). The device operate in cut-off region and the power BJT can able to sustain up to a maximum collector emitter voltage V_{CE0}. The voltage V_{CE0} is known as collector emitter breakdown voltage at $I_B = 0$.

V_{CB0}: It is collector-base break down voltage when the emitter is open circuit. The value of V_{CB0} is greater than V_{CE0}. Fig.1.10 shows the $V_{CE|sus}$, V_{CE0} and V_{CB0}.

In the common emitter configuration, the breakdown voltage V_{CEO} is smaller than the breakdown voltage V_{CBO}. The V_{CEO} is function of V_{CBO} and β and it is equal to

$$V_{CEO} = \frac{V_{CBO}}{\beta^{\frac{1}{n}}} \tag{1.20}$$

where, n=4 for *n-p-n* power BJT and n=6 for *p-n-p* transistors.

The transistor with high breakdown voltage has small value of beta. The value of beta for high voltage *n-p-n* transistors varies between 10 to 20.

If $\beta = 15$ and n=4 for a typical *n-p-n* transistor,

$$V_{CEO} = \frac{V_{CBO}}{15^{\frac{1}{4}}} = \frac{V_{CBO}}{1.967} \tag{1.21}$$

and the value of V_{CEO} is approximately one-half of V_{CBO}.

Primary Breakdown

The primary breakdown of a power BJT is equivalent to conventional avalanche breakdown of collector-base junction due to large current flow. The primary breakdown can be avoided due to large power dissipation with in the device. The primary breakdown is well-known as *first breakdown*. In fact, the primary breakdown is the limit of collector emitter voltage at which power BJT can withstand or sustain with high collector current. If the junction temperature within the safe limit, transistor cannot be damaged and the device can return back to its original position when both collector current and collector-emitter voltage are reduced.

Secondary Breakdown

If the power BJT operates in an active region with a large collector current, there is a voltage drop across collector-emitter junction. When all of a sudden collector current increases considerably, there is large power dissipation within the power BJT. As the power dissipation is non-uniformly spread over the entire volume of the power BJT, local hot-spots are developed within the device and the transistor's breakdown occurs and the device can be completely damaged. This breakdown is known as *secondary breakdown*. By using the *snubber circuit*, maintaining the instantaneous voltage and current within forward biased safe operating area, and controlling the overall power dissipation within the device, the second breakdown can be avoided.

1.4.5 ON STATE LOSS of Power BJT

It is assumed that the power bipolar junction transistor operates at low and medium operating frequency and the power dissipation within the device is equal to the power loss during ON state. To compute on-state loss, usually we consider that the device operates in hard saturation region. Then power dissipation of BJT is equal to

$$P = V_{CE} I_C \tag{1.22}$$

Where, V_{CE} is collector-emitter saturation voltage during hard saturation and I_C is collector current

The different internal voltage drops in a power transistor is shown in Fig.1.14. The voltage drops across forward biased base emitter junction V_{BE}, voltage drops across forward biased collector-base junction V_{BC}, voltage drop across collector drift region V_D can be used to determine the on state voltage drop across collector emitter of power BJT and the V_{CE} is equal to

$$V_{CE} = V_{BE} - V_{BC} + V_D + I_C (R_E + R_C) \tag{1.23}$$

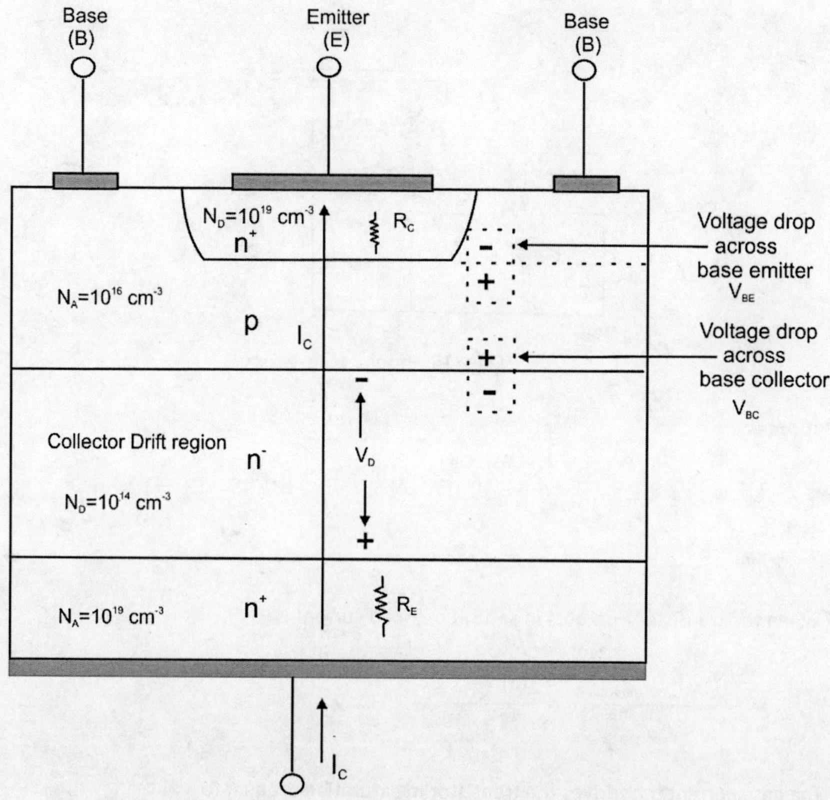

Fig.1.14 ON- state collector-emitter voltage of a power BJT

The difference between V_{BE} and V_{BC} voltages is about 0.1V to 0.2V. The collector base junction has very large area compared to the base-emitter junction. The doping levels across the collector-base junction are very low compared to the doping levels across the base-emitter junction. The voltage difference $V_{BE} - V_{BC}$ does not dependent on collector current.

R_E is the ohmic resistance of the heavily doped emitter and R_C is the ohmic resistance of the heavily doped collector region. When collector current is reasonably low, the voltage drops across R_E and R_C are insignificant. If a large current flows though the device, the voltage drops across R_E and R_C are significant. V_D is the voltage drop across the collector drift region. The value of V_D is low due to conductivity modulation and it does not depend on collector current.

Example 1.7 A bipolar transistor as shown in Fig.1.15 has $\beta = 30$ and the load resistance $R_C = 15\Omega$. The dc supply voltage $V_{CC} = 200V$ and the input voltage to base is $V_B = 10V$. When $V_{CE(saturation)} = 1.2V$ and $V_{BE(saturation)} = 1.5V$, determine (a) the value of resistance R_B so that transistor operates in saturation and (b) power loss in the transistor.

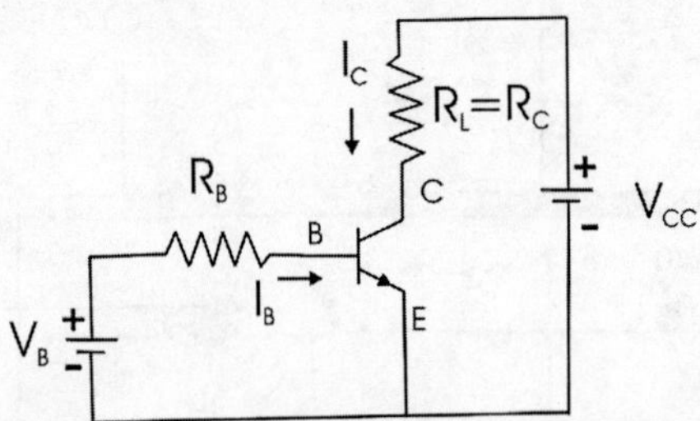

Fig.1.15: Bipolar transistor

Solution:

We know that $\beta_f = 30, R_C = 15\Omega, V_{CC} = 200V, V_B = 10V, V_{CE(saturation)} = 1.2V$, $V_{BE(saturation)} = 1.5V$

(a) When the transistor operates in saturation, load current is

$$I_{CS} = \frac{V_{CC} - V_{CE(saturation)}}{R_C} = \frac{200-1.2}{15} = 13.253\text{A}$$

The base current that drives the transistor in saturation is equal to

$$I_{BS} = \frac{I_{CS}}{\beta_f} = \frac{13.253}{30} = 0.4177\text{A}$$

The value of resistance R_B so that transistor operates in saturation is equal to

$$R_B = \frac{V_B - V_{BE(saturation)}}{I_{BS}} = \frac{10-1.5}{0.4177} = 20.349\ \Omega$$

(b) The power loss in transistor is

$$P_{Total-loss} = V_{BE}I_B + V_{CE}I_C = V_{BE(saturatiomm)}I_{BS} + V_{CE(saturation)}I_{CS}$$

$$= 1.5 \times 0.4177 + 1.2 \times 13.253 \text{ Watt} = 16.53 \text{ Watt}$$

Example 1.8 A bipolar transistor as depicted in Fig.1.14 has β in the range 10 to 40 and the load resistance $R_C = 10\Omega$. The dc supply voltage $V_{CC} = 150V$ and the input voltage to base is $V_B = 12V$. When $V_{CE(saturation)} = 1.0V$ and $V_{BE(saturation)} = 1.25V$, determine (a) the value of resistance R_B so that transistor operates in saturation with $ODF=4$ (b) forced current gain (c) power loss in the transistor.

Solution:

Given: $\beta_{\min}=10, \beta_{\max}=40,\ R_C=10\Omega, V_{CC}=150V,$

$$V_B=12V, V_{CE(saturation)}=1.0V, V_{BE(saturation)}=1.25V$$

(a) When the transistor operates in saturation, load current is

$$I_{CS}=\frac{V_{CC}-V_{CE(saturation)}}{R_C}=\frac{150-1.0}{10}=14.9\text{A}$$

The base current that drives the transistor in saturation is equal to

$$I_{BS}=\frac{I_{CS}}{\beta_{\min}}=\frac{14.9}{10}=1.49\text{A}$$

The base current with $ODF=4$ is

$$I_B=ODF\times I_{BS}=4\times1.49\text{A}=5.96\text{A}$$

The value of resistance R_B so that transistor operates in saturation with $ODF=4$ is equal to

$$R_B=\frac{V_B-V_{BE(saturation)}}{I_{BS}}=\frac{10-1.5}{5.96}=1.426\,\Omega$$

(b) Forced current gain is $\beta_{forced}=\frac{I_{CS}}{I_B}=\frac{14.9}{5.96}=2.5$

(c)The power loss in transistor is

$$P_{Total-loss}=V_{BE}I_B+V_{CE}I_C$$

$$=1.5\times5.96+1.0\times14.9 \text{ Watt} = 23.84 \text{ Watt}$$

1.4.6 SAFE OPERATING AREA (SOA) OF POWER BJT

The safe operating area (SOA) indicates the maximum values of collector current I_C and collector emitter voltage V_{CE} at which the power BJT can be operated safely. According to manufacture datasheet related to the detail specification of power BJT, there are two types of safe operating areas such "Forward bias safe operating area (FBSOA)" and "Reverse bias safe operating area (RBSOA)".

1.4.6.1 Forward bias safe operating area (FBSOA)

The SOA corresponding to the turn-on of power BJT is called as the Forward bias safe operating area (FBSOA). Fig.1.16 shows the FBSOA of power transistor where collector current I_C and collector emitter voltage V_{CE} are denoted by log scale. During forward biased state, the base-emitter junction is forward biased and base current flows from power supply to base.

The boundary of the FBSOA is limited by four factors such as maximum collector current I_{Cmax}, V_{CEo}, junction temperature T_{jmax} and second breakdown. The FBSOA depends on the operation in continuous dc signal and a pulse signal at different frequency. When the power transistor operates with dc signal, the boundary of the FBSOA is A-B-C-D-E. Boundary A-B indicates the maximum limit of collector current I_{Cmax} with V_{CE} is less than 100V. Whereas the collector emitter voltage V_{CE} is more than 100V, the collector current I_C has been reduced as represented by boundary B-C. Subsequently the operating junction temperature of power BJT should be less than the maximum operating junction temperature T_{jmax}. It is clear from Fig.1.16 that for high value of V_{CE}, the collector current is further reduced to avoid the second breakdown of power transistor and the operating boundary of power BJT is denoted by CD. The Boundary D-E denotes the maximum voltage capability of power transistor, V_{CEo}.

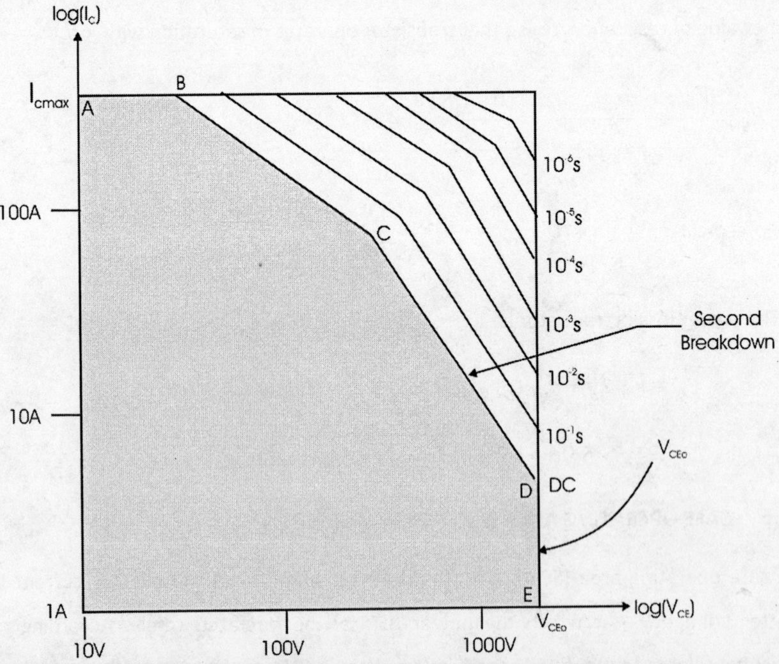

Fig.1.16 Forward bias safe operating area (FBSOA) of power BJT

If the power BJT operates as a switch and it is driven by pulse signal at different frequency, the boundaries of the FBSOA of the device is extended as depicted in Fig.1.16. Hence, the extension of the FBSOA is only possible for switch mode operation. In fact, power transistor has a specified thermal capacitance due to use of specified silicon wafer and its proper packaging. Therefore, power transistor has an ability to absorb a limited amount of energy without increasing the junction temperature excessively.

While the power BJT turns on for a few microseconds, the device can absorb very small amount of energy and the junction temperature rise is low and the boundary of FBSOA will be increased. When the pulse width is 10^{-1}s, the FBSOA is more than the boundary A-B-C-D-E. When the pulse width of signal is more reduced and its approximate value is 10^{-6} s, the FBSOA is greater than the boundary with pulse with 10^{-1}s. Therefore, the FBSOA increases with the decrease of pulse-width of signal.

1.4.6.2 Reverse bias safe operating area (RBSOA)

The safe operating area corresponding to turn off of power BJT is called as reverse bias safe operating area (RBSOA). In reverse biased condition, the base-emitter junction is reverse biased and base current flows in reverse direction. During turn-off of power BJT, the base–emitter junction is extremely reverse biased and the transistor should able to withstand at high current. The RBSOA is a plot of collector current I_C with respect to collector emitter voltage V_{CE} as shown in Fig.1.17. If the base current I_B=0, power transistor can able to sustained at voltage V_{CE0}. When the base-emitter junction is reverse biased, the power transistor can able to withstand up to collector-base breakdown voltage V_{CB0} along with low collector current.

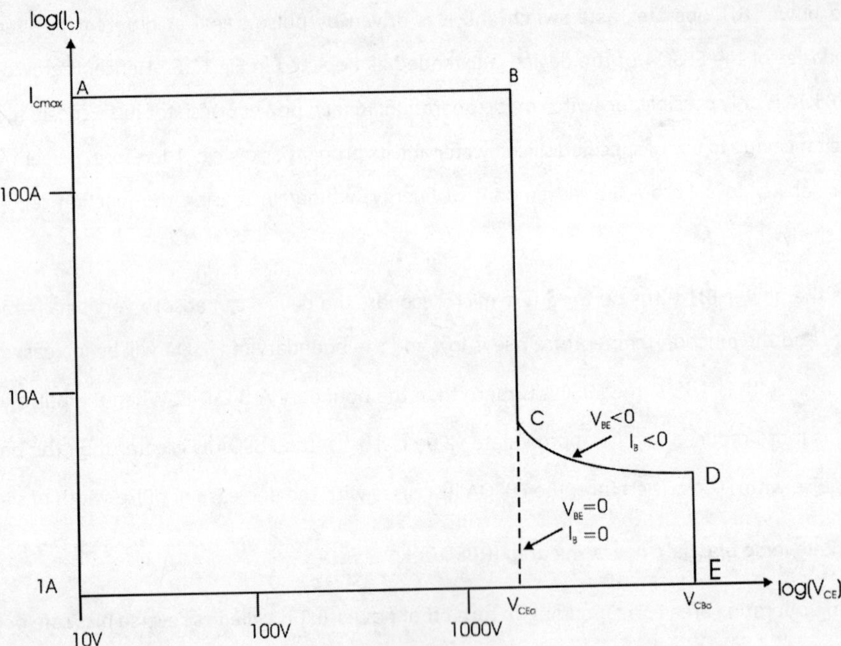

Fig.1.17 RBSOA of a power BJT

1.4.7 SERIES AND PARALLEL OPERATION of Power BJT

Power BJTs are connected in series to increase the voltage-handling capability. During series connection of transistors, the series-connected power transistors must be turned-on and turned-off simultaneously. If the series-connected transistors are not turned-on and turned-off simultaneously, the slowest transistor can be at turn-on state while the fastest transistor will be at turn-off state. Therefore, the turned-off transistor should be able to withstand at full voltage of the collector-emitter. When the transistor cannot able to withstand at high voltage, the device will be completely damaged. Consequently, the power BJTs should have same gain, on-state voltage, turn-on time, turn-off time, and base drive circuit during series connection of transistors.

Power BJTs are connected in parallel when one power transistor cannot able to handle the load current. For the equal current sharing between power transistors, the power BJTs should have same turn-on time, turn-off time, gain, saturation voltage, and transconductance. The parallel connections of two power BJTs is shown in Fig.1.18. Here, total current I_T is shared by transistors T_1 and T_2. Actually current I_{E1} flows through transistor T_1 and current I_{E2} flows through transistor T_2.

From Fig.1.18, we can write

$$I_T = I_{E1} + I_{E2} \tag{1.24}$$

and $V_{CE1} + I_{E1}R_{E1} = V_{CE2} + I_{E2}R_{E2}$ (1.25)

After substitute the equation (1.24) in equation (1.25) we obtain

$$V_{CE1} + I_{E1}R_{E1} = V_{CE2} + (I_T - I_{E1})R_{E2} \tag{1.26}$$

Therefore, $I_{E1} = \dfrac{V_{CE2} - V_{CE1} + I_T R_{E2}}{R_{E1} + R_{E2}}$ (1.27)

Almost 45% to 55% current can be shared by connecting resistance in series with transistors. In fact, resistances are very useful for current sharing under steady-state condition. Under dynamic condition, current can be shared by connecting coupled inductors in place of resistances.

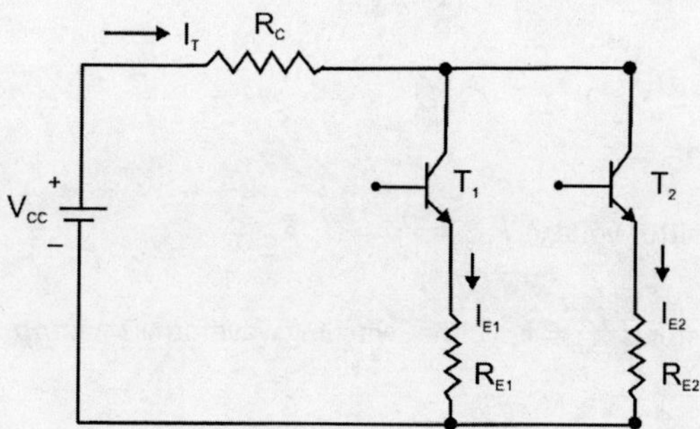

Fig.1.18 Parallel connection of transistors with current sharing at steady-state condition

Example 1.9 Two power bipolar junction transistors are connected in parallel to share a load current 30A. Calculate the emitter current of each transistor and the difference of current sharing when the current sharing series resistance are (i) $R_{E1} = 0.2\ \Omega$ and $R_{E2} = 0.3\ \Omega$ (ii) $R_{E1} = 0.5\ \Omega$ and $R_{E2} = 0.5\ \Omega$. Assume that the collector to emitter voltage of T_1 and T_2 are 1.5V and 1.6V respectively.

Solution:

We know that $I_T = 30A$, $V_{CE1} = 1.5V$, $V_{CE2} = 1.6V$,

(i) When $R_{E1} = 0.2\ \Omega$ and $R_{E2} = 0.3\ \Omega$

$$I_{E1} = \frac{V_{CE2} - V_{CE1} + I_T R_{E2}}{R_{E1} + R_{E2}} = \frac{1.6 - 1.5 + 30 \times 0.3}{0.2 + 0.3} = 18.2A$$

$I_{E2} = I_T - I_{E1}$=30-18.2=11.8A

$\Delta I = I_{E1} - I_{E2} = (18.2 - 11.8)A = 6.4A$

(ii) When $R_{E1} = 0.5\ \Omega$ and $R_{E2} = 0.5\ \Omega$

$$I_{E1} = \frac{V_{CE2} - V_{CE1} + I_T R_{E2}}{R_{E1} + R_{E2}} = \frac{1.6 - 1.5 + 30 \times 0.5}{0.5 + 0.5} = 15.1A$$

$I_{E2} = I_T - I_{E1}$=30-15.1=14.9 A

$\Delta I = I_{E1} - I_{E2} = (15.1 - 14.9)A = 0.2A$

1.4.8 Switching Characteristics of Power BJT

The npn Power BJT can be used as a switch as shown in Fig.1.19 whenever a square pulse input voltage is applied to the base of transistor, the base current is equal to

$$I_B = \frac{V_i - V_{BE}}{R_B} \tag{1.28}$$

and the collector emitter voltage $V_{CE} = V_{CC} - I_C R_C$

The output voltage of transistor is $V_o = V_{CE}$. The switching waveforms of an *npn* transistor are given in Fig.1.20.

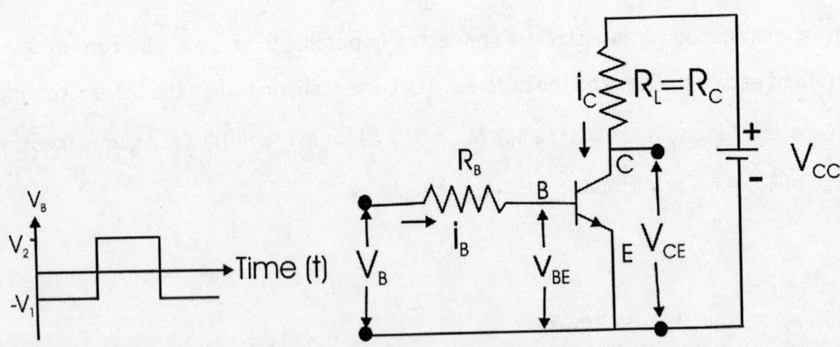

Fig.1.19 n-p-n Power BJT as a switch

Assume that the input voltage V_i varies between $-V_1$ to $+V_2$. At time $t = t_0$, $-V_1$ voltage is applied as input voltage at the base of power transistor and emitter-base junction is reverse biased.

Subsequently transistor operates in OFF state since base current $I_B = 0$ and the collector current I_c =0 and the output voltage is $V_o = V_{CE} = V_{CC}$.

At $t = t_1$, voltage increases slowly from $-V_1$ to $+V_2$ and the collector current I_C increases gradually from zero and the collector-emitter voltage V_{CE} starts to fall from V_{CC}. Consequently, the transistor changes its state from cut-off to saturation progressively and the maximum steady state current $I_C = \frac{V_C}{R_C}$ flows through transistor.

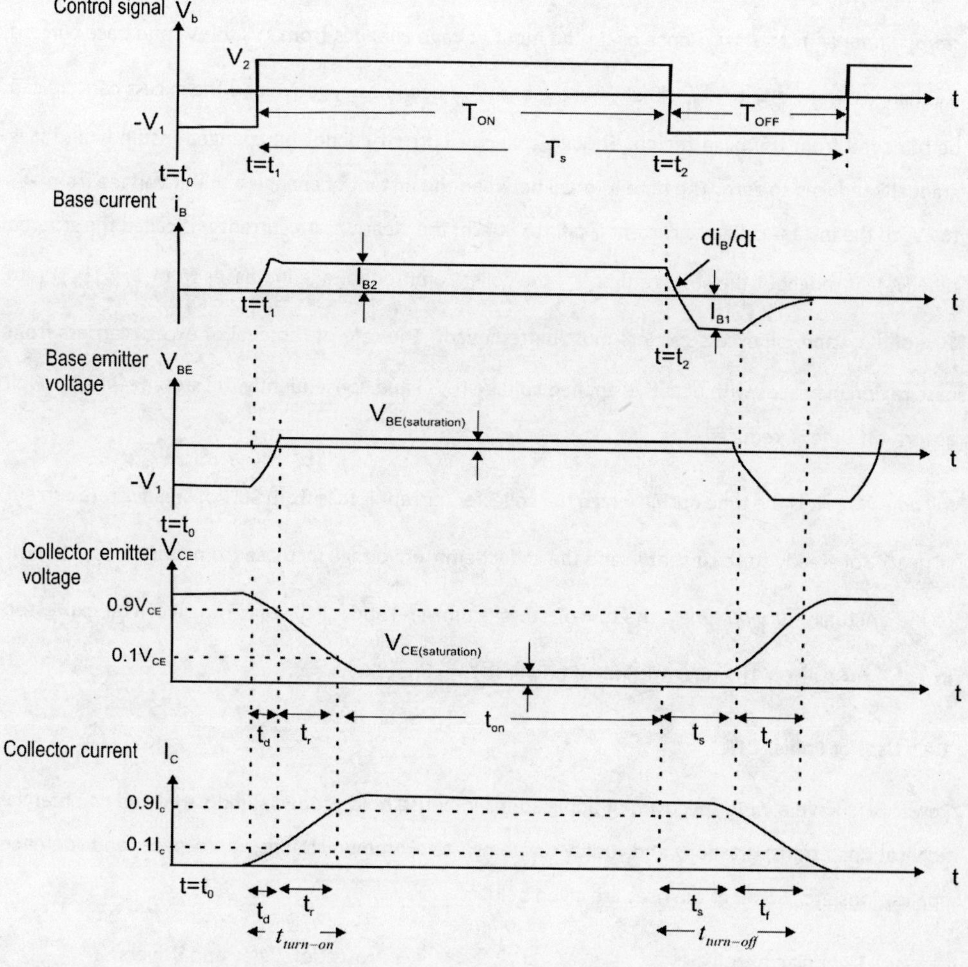

Fig.1.20 The switching waveforms of an *npn* transistor

delay time (t_d) : The *delay time* (t_d) is the time required for collector current i_C to reach 10% of the steady state current I_C and the collector emitter voltage v_{CE} falls from V_{CC} to 90% of V_{CC}.

rise time (t_r): The *rise time* (t_r) is the time required for collector current i_C to increase from 10% of the steady state current I_C to 90% of I_C and the collector emitter voltage v_{CE} falls from 90% of V_{CC} to 10% of V_{CC}.

turn on time ($t_{turn-on}$): The *turn on time* ($t_{turn-on}$) is the sum of the delay time (t_d) and the rise time (t_r) and it can be represented by $t_{turn-on} = t_d + t_r$. The turn-on time of power BJTs is about 30 to 300 ns. While the power transistor is ON, it operates in saturation region providing input voltage is equal to V_2.

storage time (t_s): At $t = t_2$, once again the input voltage changes from $+V_2$ to $-V_1$ and base current I_B changes from I_{B2} to I_{B1} slowly. As the base I_{B1} becomes negative and the excess carriers can be removed from the base region. Though collector current cannot be changed instantly and it is gradually reduced to zero. The time interval between the instant of change of input voltage from $+V_2$ to $-V_1$ to the instant collector current i_C falls to 90% of the steady state current I_C is called the *storage time* (t_s). Throughout the storage time t_s, the collect emitter voltage increases from $V_{CE(saturation)}$ to 10% of V_{CC} and all excess carriers must be removed. The rate of removal of excess carriers from base region increases with negative applied voltage ($-V_1$) and consequently the storage time as well as turn-off time is reduced.

fall time (t_f); It is the time during which the collector current i_C falls from 90% of steady state current I_C to 10% of steady state current I_C and the collect emitter voltage increase from 10% of V_{CC} to 90% of V_{CC}. Actually *turn off time* is the sum of storage time (t_s) and fall time (t_f) and it can be expressed as $t_{turn-off} = t_s + t_f$. The turn-off time of power BJT is about 60 ns.

1.4.9 Uses of Power BJT

Power BJTs have a wide spectrum of applications in house hold products, industrial and commercial applications, transportation and utility systems, telecommunications, aerospace and defence applications etc., Some of the uses are given below:

- Lighting control
- Micro-oven
- Protection Relay and Drivers
- DC to AC inverter

- Switched Mode Power Supply (SMPS)
- DC and AC motor speed controller
- Power Amplifier
- audio amplifier in the stereo system
- Battery chargers
- Washing machines

Example 1.10 Fig 1.21 shows the switching waveforms of a power BJT where $V_{CC} = 220V$, $V_{CE(saturation)} = 1.5V$, $I_{CS} = 100A$, $t_d = 0.25\mu s$, $t_r = 1.25\mu s$, $t_{on} = 40\mu s$, $t_s = 2\mu s$, $t_f = 1.5\mu s$. Calculate (a) energy loss during delay time (b) energy loss during rise time (c) energy loss during conduction time t_{on} (d) average power loss of power transistor during turn-on when switching frequency is 10kHz and $I_{EO} = 5mA$.

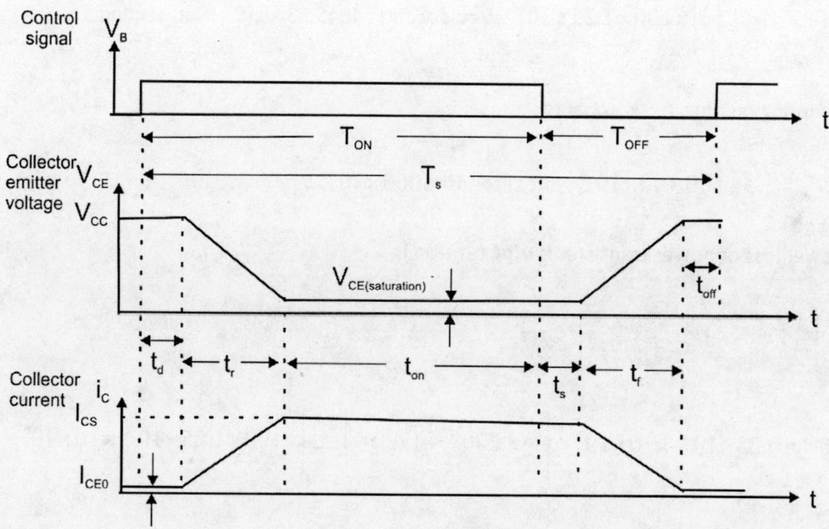

Fig.1.21 Switching waveforms of a power BJT

Solution:

We know that $V_{CC} = 220V$, $V_{CE(saturation)} = 1.5V$, $I_{CS} = 100A$, $t_d = 0.25\mu s$, $t_r = 1.25\mu s$, $t_{on} = 40\mu s$, $t_s = 2\mu s$, $t_f = 1.5\mu s$, $f_s =$ 10kHz and $I_{EO} = 5mA$

(a) Energy loss during delay time t_d is equal to

$$W_{delay} = V_{CC} I_{CEO} t_d = 220 \times 5 \times 10^{-3} \times 0.25 \times 10^{-6} \text{ Watt second} = 0.275 \times 10^{-6} \text{ Watt second}$$

(b) For the period of rise time t_r the voltage v and current i waveforms are expressed by

$$v_{CE}(t) = [V_{CC} - (V_{CC} - V_{CES})\frac{t}{t_r}] \text{ and } i_C(t) = I_{CS}\frac{t}{t_r} \text{ as } I_{CEO} << I_{CS}$$

So, the instantaneous power is $p = v_{CE}(t)i_C(t) = [V_{CC} - (V_{CC} - V_{CES})\frac{t}{t_r}]\, I_{CS}\frac{t}{t_r}$

The energy loss is $W_{rise} = \int_0^{t_r} p\,dt = \int_0^{t_r} [V_{CC} - (V_{CC} - V_{CES})\frac{t}{t_r}] I_{CS}\frac{t}{t_r})dt$

$$= \left[\frac{1}{2}V_{CC} - \frac{1}{3}(V_{CC} - V_{CES})\right] I_{CS} t_r$$

$= \left[\frac{1}{2}\times 220 - \frac{1}{3}(220 - 1.5)\right]\times 100 \times 1.25 \times 10^{-6}$ Watt second $= 4645.83 \times 10^{-6}$ Watt second

(c) Energy loss during on time t_{on} is equal to

$W_{on} = V_{CES} I_{CS} t_{on} = 1.5 \times 100 \times 40 \times 10^{-6}$ Watt second $= 6000 \times 10^{-6}$ Watt second

(d) Average power loss of power transistor during turn-on is

$$V_{CC} I_{CEO} t_d f_s + \left[\frac{1}{2}V_{CC} - \frac{1}{3}(V_{CC} - V_{CES})\right] I_{CS} t_r f_s$$

$$= 220 \times 5 \times 10^{-3} \times 0.25 \times 10^{-6} \times 10 \times 10^3 + \left[\frac{1}{2}\times 220 - \frac{1}{3}(220 - 1.5)\right]\times 100 \times 1.25 \times 10^{-6} \times 10 \times 10^3$$

$= 46.46$ Watt

1.5 POWER MOSFET

Power BJT is a minority carrier device and it is also a current controlled device. When a base current (I_B) flows, the transistor operates in the ON state. If the base current (I_B) is equal to zero, the power BJT is in the OFF state. Since collector current (I_C) depends on base current, the current gain of power BJT depends on the junction temperature. Power BJT is most commonly used as switching devices in power electronics circuits.

Another popular power transistor is power MOSFET. The MOSFET operation depends on the control of current by an applied electric field. Consequently, MOSFET is a voltage controlled device and it requires a very small input current. Unlike the power BJT, power MOSFET is a majority carrier device and this device has three terminals such as gate (G), drain (D) and source (S). The MOSFET has many advantages over BJT such as

i. It requires very small area on an integrated circuit (IC)
ii. Digital circuits can be designed using MOSFET only
iii. High density VLSI circuits such as microprocessors, microcontrollers and memory ICs can be manufactured using MOSFET
iv. MOSFET can be used in analog circuits as switching device and amplifier

In MOSFET, field effect principle is used for proper operation of this device. In this device, the current is controlled by an applied electric field perpendicular to both the semiconductor surface and to the current direction. Generally, there are two types of MOSFET such as depletion-type MOSFET and enhancement-type MOSFET. Again each type MOSFET is classified as n-channel and p-channel MOSFET. The basic structure, output characteristics and transfer characteristics of depletion-type and enhancement-type MOSFET is illustrated in Table 1.3.

Table 1.3 Enhancement and depletion type MOSFETs

Type	***Cross-section***	***Drain or output characteristics***	***Transfer characteristics***
n-channel enhancement type MOSTFET (Normally OFF)	+ G, +, I_D, D, n^+, p, n^+	I_D, $V_G = 4V$, 3, 2, 1, 0, V_D	I_D, –, 0, V_{Tn}, +
p-channel enhancement type MOSTFET (Normally OFF)	– G, –, I_D, D, p^+, n, p^+	$-V_D$, 0, –1, –2, –3, $V_G = -4V$, I_D	V_G, –, V_{Tp}, 0, +, I_D
n-channel depletion type MOSTFET (Normally ON)	± G, +, I_D, D, n^+, p, n^+, n-Channel	I_D, $V_G = 1V$, 0, –1, –2, 0, V_D	I_D, V_{Tn}, –, 0, +, V_G

Type	***Cross-section***	***Drain or output characteristics***	***Transfer characteristics***
p-channel enhancement and depletion type MOSTFET (Normally ON)	$\pm G$, I_D, D, p^+, n, p^+ p-Channel	$-V_D$, 0, 1, 2, 0, $V_G = -1V$, I_D	V_G, −, 0, +, V_{Tp}, I_D

1.5.1 Construction & Working Principle of Power MOSFET

The construction details of a vertically oriented enhancement mode n-channel power MOSFET is depicted in Fig.1.22. The power MOSFET is a four-layer $n^+ p\, n^- n^+$ semiconductor structure. In reality, a power MOSFET consists of a parallel connection of many MOSFET cells in a single chip of silicon i.e. integrated circuit (IC).

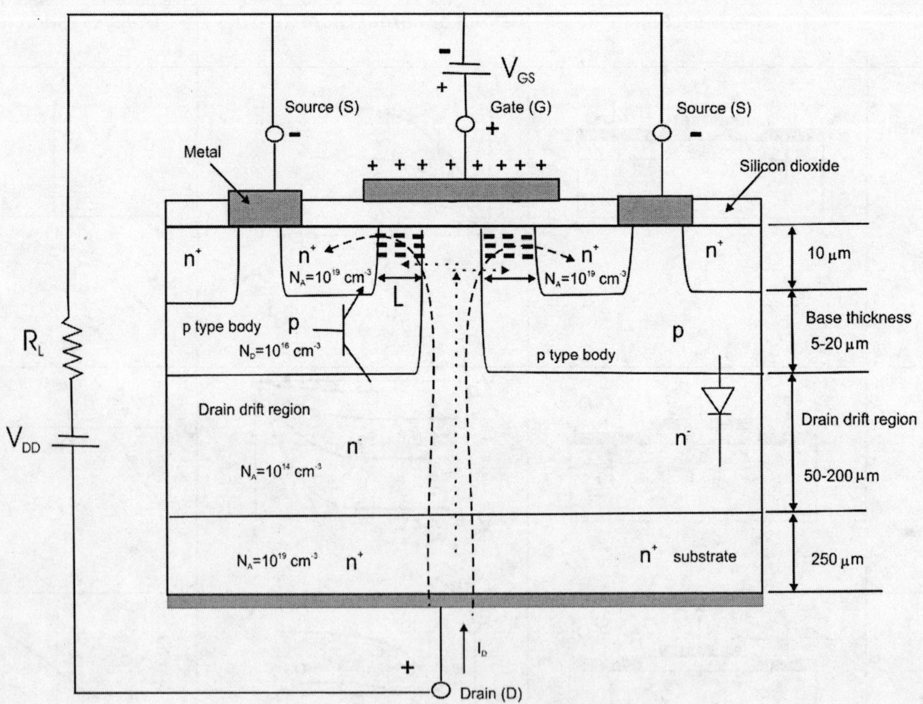

Fig.1.22 Structure of a n channel power MOSFET

Three are two n^+ end layers in a power MOSFET. The doping density in the n^+ end layers of power MOSFET is quite large and its value is about $10^{19} cm^{-3}$. One end is called as source (S) and the other end n^+ substrate is known as drain (D). The thickness of n^+ substrate is about 100μm and on the n^+ substrate, the n^- layer is grown epitaxially. After that p type semiconductor is diffused in the epitaxially grown n^- layer and subsequently p region developed. Then n^+ semiconductor is diffused in the p region and the n^+ region is developed.

The n^- layer is known as *drain drift region* and the doping density in n^- drain drift region layer is low and its value is about $10^{14} cm^{-3}$ to $10^{15} cm^{-3}$. The thickness of n^- drift region determines the breakdown voltage of the device. Usually the thickness of n^- drift region is about 60 to 70μm. It is clear from Fig.1.22 that the p layer is the region where the channel is established between source and drain. As a result, p region is called as the body of power MOSFET. The doping density of p region is about $10^{16} cm^{-3}$ and the thickness of p-type body is about 5 to 5.5μm. This structure is called as vertical diffused metal oxide semiconductor (VDMOS) power MOSFET. In the same way, a four-layer p-channel power MOSFET can also be manufactured while the doping density of different region of p-channel MOSFET is the opposite doping density of n-channel MOSFET.

Two pn junctions such as drain to body ($n^- p$) junction and body to source (pn^+) junction are developed with in the structure. If gate to source voltage $V_{GS} = 0$ and V_{DD} is present, the drain current (I_D) does not flow from drain to source of power MOSFET because $n^- p$ junction is reverse biased by the input voltage V_{DD}. The gate is isolated from the p type body region by a layer of silicon dioxide SiO_2. The thickness of silicon dioxide layer is about 1000 A^o. As silicon dioxide is used as very good insulator, the minority carriers cannot be injected into p region through gate terminal.

When gate is positive with respect to source, an electric field is developed and an n-channel is formed in the p region as shown in Fig.1.22. Subsequently source is connected with drain through *n* channel and drain current (I_D) flows from drain to source. As length of n-channel can be controlled, on-resistance can be low by using n-channel of short length. If the gate to source voltage is further increased, the drain current (I_D) increases. Drain current I_D depends on the thickness of silicon dioxide, width of gate and number of parallel connected gate and source regions with in the single chip.

It revels from Fig.1.22 that a parasitic *n-p-n* bipolar junction transistor exists between the source and drain. The p-type body region acts as the base, n^+ region as emitter and n^- drain drift region as collector of parasitic BJT. Source is connected with base and emitter of parasitic BJT as p-type body region is shorted with source region due to overlapping source metallization on the p-type body region. As a result, potential difference between base and emitter of parasitic BJT is zero and the parasitic BJT operates in cut-off region.

Fig.1.22 shows the parasitic diode also. In parasitic diode, source is acting as anode and drain acts as cathode. This inbuilt body diode can be used in half-bridge and full-bridge inverter circuits. The n-channel power MOSFET has three terminals such as gate(G), drain (D) and source (S). The circuit symbol of n channel power MOSFET and p channel power MOSFET are illustrated in Fig.1.23.

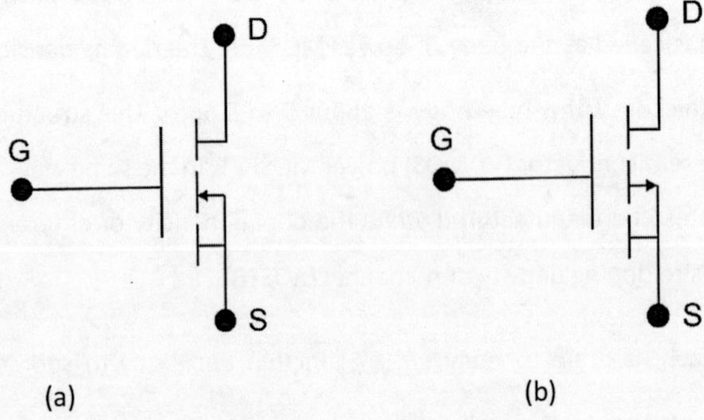

Fig.1.23 (a) Symbol of n-channel power MOSFET (b) Symbol of p-channel power MOSFET

1.5.2 V-I CHARACTERISTICS of Power MOSFET

The circuit diagram of a n-channel power MOSFET is shown in Fig.1.24 where V_{GS} is applied voltage across gate to source and V_{DS} is the output signal obtained from drain. The drain current I_D flows from drain to source and it is controlled by gate signal.

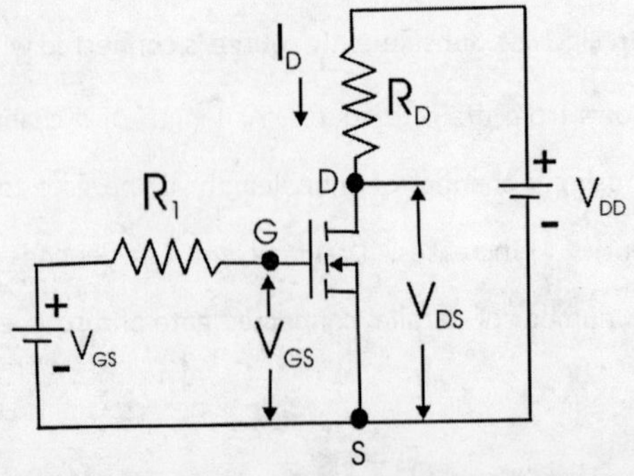

Fig.1.24 Circuit diagram of n channel power MOSFET

Transfer Characteristic: The transfer characteristic of an n-channel power MOSFET shows the variation of drain current I_D as a function of gate to source voltage V_{GS} (Fig.1.25). Threshold voltage $V_{GS(th)}$ is an important parameter of power MOSFET. The value of $V_{GS(th)}$ for power MOSFET is about 2 to 3V. If gate to source voltage is less than threshold voltage $V_{GS(th)}$, the current I_D flow from drain to source is zero. As $I_D = 0$, drain to source is open circuit and the device can able to with stand at supply voltage V_{DD}.

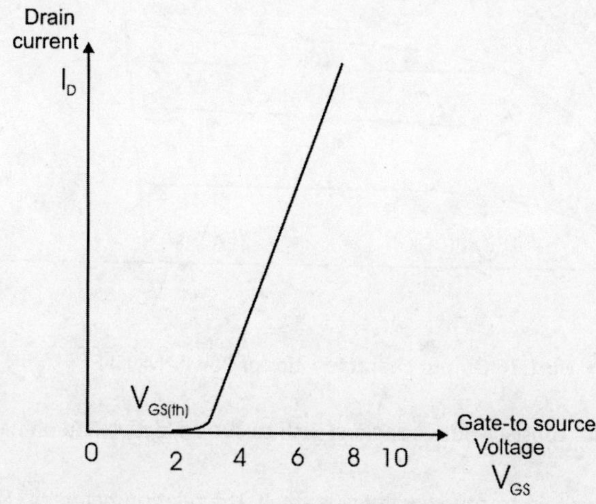

Fig.1.25 Transfer characteristics of power MOSFET

Output Characteristics:

Fig.1.26 shows the output characteristics of a power MOSFET. The drain current I_D is a function of drain to source voltage V_{DS} with constant gate to source voltage V_{GS}. For low values of V_{DS}, the relation between V_{DS} and I_D is almost linear and the device has a constant value of on-resistance $R_{DS}=V_{DS}/I_D$. When V_{DS} is increased, output characteristic is relatively flat means the drain current is nearly constant.

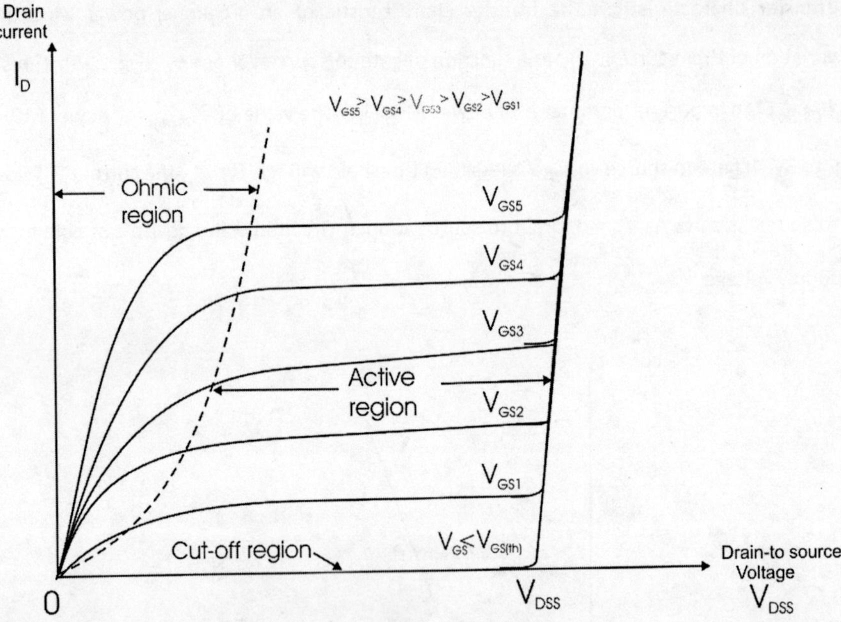

Fig.1.26 Output characteristics of Power MOSFET

The output characteristics consist of three regions such as cut-off, active and ohmic regions.

Ohmic region: While the drain to source voltage is small, the relation between I_D and V_{DS} is linear and power MOSFET operates in the ohmic region. The device operates in the ohmic region if $V_{GS} > V_{GS(th)}$ and $V_{DS} > 0$. The power dissipation should be within a limit and its value must be minimum with small drain to source voltage $V_{DS(on)}$ through large drain current I_D flows from drain to source.

If power MOSFET is used as a switch to control the flow of load current, I_D - V_{DS} characteristics must be traversed from the cut-off region to the ohmic region through the active region. The device is turned off in the cut-off region and turns on in the ohmic region.

Active region: During active region, drain current I_D does not depend on drain to source voltage V_{DS} although it varies with gate to source voltage V_{GS}. Since current I_D is called as saturated current, this region is known as saturation region. The drain current is equal to

$$I_D = K[V_{GS} - V_{GS(th)}]^2 \qquad (1.29)$$

Where, K is a constant and it depends on the device parameters.

At transition point between the active region and ohmic region, $[V_{GS} - V_{GS(th)}] = V_{DS}$. Therefore, the equation (1.29) can be expressed by

$$I_D = KV_{DS}^2 \tag{1.30}$$

Drain to source breakdown voltage V_{DSS} must be greater than V_{DD} to keep away from device breakdown. When drain to source voltage is larger than V_{DSS}, power MOSFET can be breakdown due to avalanche breakdown of drain to body $n^- p$ junction.

$V_{GS(max)}$: This is the maximum allowable gate to source voltage $V_{GS(max)}$. Whenever a voltage is applied across gate to source, an electric field can be developed. $V_{GS(max)}$ is the maximum voltage across gate to source when gate oxide cannot be broken due to large electric field. Typically silicon dioxide can break at electric field about $5-10$ Mega Volt/cm. If the thickness of silicon dioxide is about 1000 A^o, $V_{GS(max)}$ is equal to 20-30V.

V_{DSS} : This is the maximum allowable drain to source voltage at which power MOSFET can sustained without avalanche breakdown of the drain to body $n^- p$ junction. For high breakdown voltage, the n^- drift region must be lightly doped. The value of V_{DSS} is about few hundred volts.

The output characteristics for a p-channel power MOSFET should be same as the output characteristics for an n-channel power MOSFET, however voltage and current polarities are reversed.

1.5.3 ON STATE LOSS Power MOSFET

$R_{DS(on)}$ is the on-state resistance of power MOSFET and it is sum of drain resistance, n^- drift region resistance R_D, accumulation layer resistance, channel resistance and source region resistance as shown in Fig.1.27. When breakdown voltage is low, all resistance components contribute equally to determine on-state resistance of power MOSFET. If the breakdown voltage V_{DSS} is more than few hundred volts, the n^- drift region resistance R_D contributes maximum value of on-state resistance $R_{DS(on)}$.

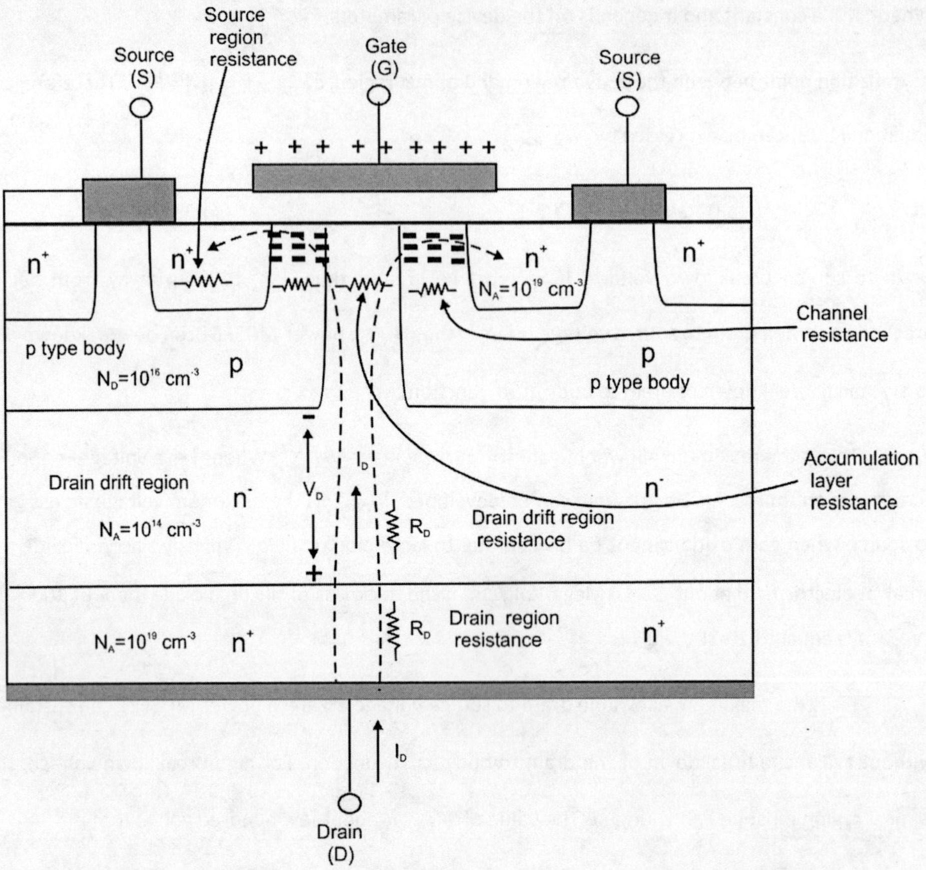

Fig.1.27 On-state resistances of n-channel enhancement mode MOSFET

As R_D depends on the breakdown voltage V_{DSS}, R_D increases with increasing the breakdown voltage for power MOSFET. The on-state resistance also depends on the junction temperature T_j. Actually, the on-state resistance of power MOSFET increases with increasing junction temperature.

When the device operates in on-state, the power dissipation is equal to

$$P_{on} = I_D R_{DS(on)} \tag{1.31}$$

where, I_D is current flow from drain to source and

$R_{DS(on)}$ is total on-state resistance and its value is about 9 to 100 mΩ.

1.5.4 SAFE OPERATING AREA OF POWER MOSFET

The safe operating area (SOA) is the locus of I_D-V_{DS} points represent the boundary between safe and unsafe operation. As per datasheet related to specification of power MOSFET, the safe operating area (SOA) depends on (i) maximum drain current $I_{D(\max)}$ (ii) internal junction temperature T_j (iii) breakdown voltage V_{DSS} and (iv) maximum power dissipation $P_{D(\max)}$.

The safe operating area (SOA) of power MOSFET is depicted in Fig.1.28 where drain current I_D and drain to source voltage V_{DS} are represented in log scale. If V_{DS} is low value, the maximum drain current $I_{D(\max)}$ is limited by the power dissipation. The maximum power dissipation $P_{D(\max)}$ can be limited by the maximum tolerable internal junction temperature T_j. At low drain current, the breakdown voltage V_{DSS} is limited to avoid the avalanche breakdown of $n^- p$ junction.

If power MOSFET operates with dc signal, the boundary of SOA is A-B-C-D as shown in Fig.1.28. As input signal is dc, continuous power dissipation takes place within device and the junction temperature increases significantly. The boundary A-B indicates the maximum limit of drain current I_{Dmax} with V_{DS} is less than 30V. While V_{DS} is more than 30V, the drain current decreases as per boundary B-C so that the operating junction temperature of power MOSFET is less than the maximum operating junction temperature T_j. The maximum voltage capability of a power MOSFET is represented by boundary C-D.

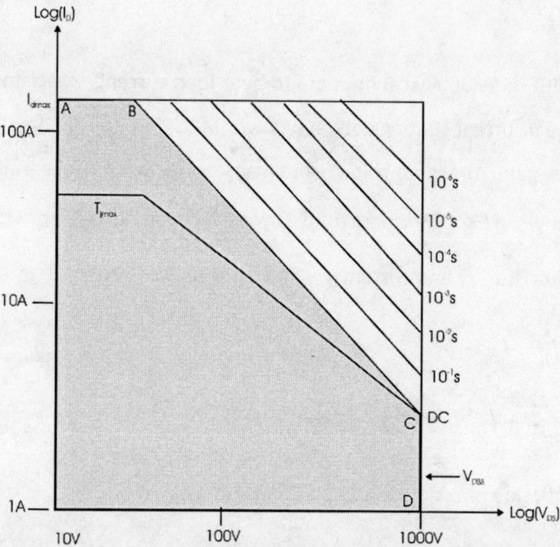

Fig.1.28 SOA of a power MOSFET

Whenever power MOSFET is driven by a pulse signal and operates as a switch, the boundaries of SOA are extended as shown in Fig.1.28. This is only possible when power MOSFET operates in switch mode operation. In fact, the silicon wafer and its proper packaging of power MOSFET has a specified thermal capacitance and it has an ability to absorb specified amount of heat energy without increasing junction temperature tremendously. If the device turns-on for a few microseconds, power dissipation will be low and junction temperature rise will also be low and subsequently boundary of SOA will be increased. The SOA increases with the decrease of pulse-width as depicted in Fig.1.28. When the pulse width is about 10^{-1} s, the SOA is more than the boundary A-B-C-D. While the pulse width is reduced to about some 10^{-6} s, SOA is larger than the boundary with pulse with 10^{-3} s.

1.5.5 SERIES AND PARALLEL OPERATION of POWER MOSFET

Power MOSFETs are connected in series to increase voltage-handling capability. It is most vital that series-connected power MOSFETs should be turned-on and turned-off simultaneously. If series-connected power transistors are not turned-on and turned-off at the same time, the slowest device can be operating at turn-on state but the fastest device operates at turn-off state. As a result, the turned-off power MOSFETs must be able to withstand at full drain-source voltage. When the device can not able to withstand at high voltage, the device can be damaged completely due to very high drain-source voltage. Therefore, power MOSFETs should have same gain, on-state voltage, transconductance, threshold voltage, turn-on time, turn-off time, and gate drive circuit for proper series operation of transistors.

When a power MOSFET individually cannot able to bear load current, transistors must be connected in parallel. For the equal current sharing between power MOSFETs, transistors should have same turn-on time, turn-off time, gain, threshold voltage and transconductance. The parallel connections of two power MOSFETs is shown in Fig.1.29 where total current I_T is shared by transistors T_1 and T_2. Current I_{D1} flows through transistor T_1 and current I_{D2} flows through transistor T_2.

The total current $I_T = I_{D1} + I_{D2}$ (1.32)

and $$V_{DS1} + I_{D1}R_{S1} = V_{DS2} + I_{D2}R_{S2} \quad (1.33)$$

After substituting the value of I_{D2} in equation (1.33), we get

$$V_{DS1} + I_{D1}R_{S1} = V_{DS2} + (I_T - I_{D1})R_{S2} \quad (1.34)$$

Therefore, $$I_{D1} = \frac{V_{DS2} - V_{DS1} + I_T R_{S2}}{R_{S1} + R_{S2}} \tag{1.35}$$

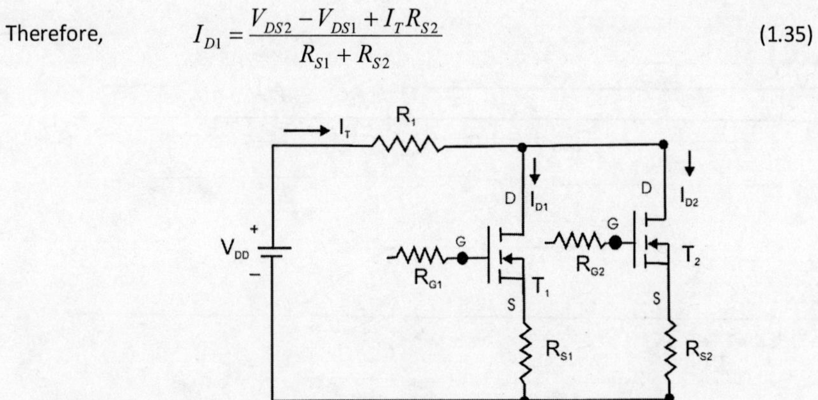

Fig.1.29 Parallel connection of power MOSFETs with current sharing at steady-state condition

Fig.1.29 shows that about 45% to 55% current can be shared by connecting resistance in series with power MOSFETs. In fact, resistances are used for current sharing under steady-state condition. Under dynamic condition, current can be shared by connecting coupled inductors.

1.5.6 SWITCHING CHARACTERISTICS OF POWER MOSFET

A circuit diagram of Power MOSFET acts as switch is shown in Fig.1.30. The switching characteristics of power MOSFET are affected by the internal capacitance of the device and the internal impedance of gate drive circuit. If a pulse input voltage is applied to the gate of power MOSFET, the device will be turn-on when gate to source voltage V_{GS} is greater than threshold voltage $V_{GS(th)}$. The switching waveforms of a Power MOSFET are depicted in Fig.1.31.

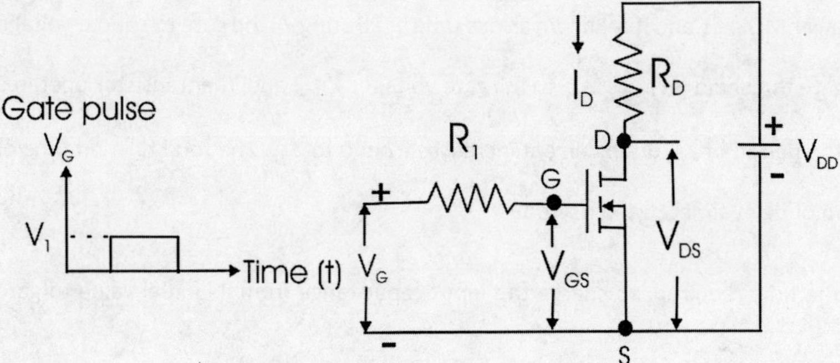

Fig.1.30 Power MOSFET as a switch

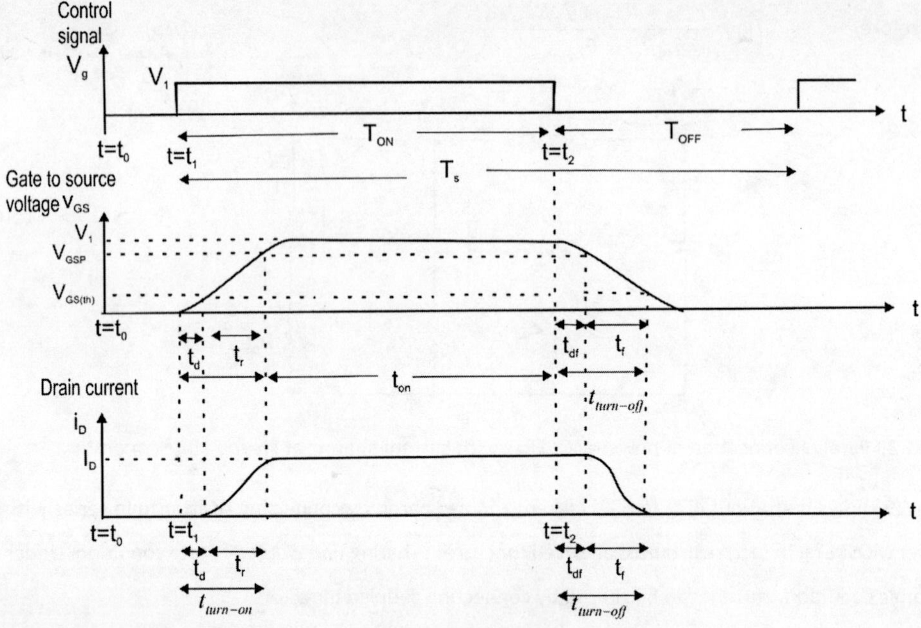

Fig.1.31 The switching waveforms of a Power MOSFET

At $t = t_0$, input voltage at the gate of power MOSFET is $V_g = 0$. Since the gate to source voltage V_{GS} is less than threshold voltage $V_{GS(th)}$, the device operates in OFF state and the drain current I_D is zero and the output voltage is $V_o = V_{DS} = V_{DD}$.

At $t = t_1$, voltage V_g increases from 0 to V_1 and the input capacitance C_{gs} starts to charge. During turn-on delay time t_d, capacitance C_{gs} is charged to gate threshold voltage $V_{GS(th)}$. There is a further delay in turn on of power MOSFET and it is known as rise time t_r. During t_r the gate to source voltage V_{GS} increases from gate threshold level $V_{GS(th)}$ to full gate voltage, V_{GSP} and the transistor operates in linear region. For the period of t_r, drain current increases from 0 to I_D. The total turn-on time of power MOSFET is sum of delay time t_d and rise time t_r.

delay time t_d : It is the time required to charge the input capacitance from its initial value to gate threshold voltage $V_{GS(th)}$.

rise time t_r : It is the time required to charge the input capacitance C_{gs} from gate threshold level $V_{GS(th)}$ to the full gate voltage, V_{GSP}.

turn on time $t_{turn-on}$: \The *turn on time* is the sum of the delay time t_d and the rise time t_r and it is $t_{turn-on} = t_d + t_r$. The turn on time of power MOSFET is about 1.6 μs .

Since power MOSFET is a majority carrier device, the turn-off process is initiated after removal of gate voltage V_g at $t = t_2$ and the input capacitance starts to discharge from gate voltage V_1 to V_{GSP}. V_{GS} must be decreased significantly so that V_{DS} starts to increase. The *turn off delay time* t_{df} is the time during which the input capacitance discharges from gate voltage V_1 to V_{GSP}. The *fall time* t_f is the time during which the input capacitance discharges from gate voltage V_{GSP} to $V_{GS(th)}$ and the drain current becomes zero. *Turn-off time* is the sum of the turn off delay time (t_{df}) and fall time (t_f) and it can be expressed by $t_{turn-off} = t_{df} + t_f$. Usually, turn off time of power MOSFET is about 30 to 300 ns.

Average power loss if the power MOSFET in off state is equal to

$$P_{off-state} = V_{DS} I_{DSS} t_{off} f_s \text{ Watt} \quad (1.36)$$

Average power loss during turn-on is

$$P_{turn-on} = \frac{1}{6} V_{DS} I_D t_r f_s \text{ Watt} \quad (1.37)$$

Average power loss during turn-off is expressed by

$$P_{turn-off} = \frac{1}{6} V_{DS} I_D t_f f_s \text{ Watt} \quad (1.38)$$

Average power loss in on state of device is equal to

$$P_{on-state} = I_D^2 R_{DS(on)} t_{on} f_s \text{ Watt} \quad (1.39)$$

1.5.7 COMPARISION BETWEEN POWER BJT and POWER MOSFET

The comparison between Power BJT and Power MOSFET is shown in Table 1.4

Table 1.4 Comparison between Power MOSFET and Power BJT

POWER BJT	POWER MOSFET
It is a bipolar device	It is a unipolar device
It is a current controlled device	It is a voltage controlled device

POWER BJT	POWER MOSFET
It is a majority and minority carrier device	It is a majority carrier device
Input impedance of Power BJT is low compared to Power MOSFET	Input impedance of Power MOSFET is very high compared to Power BJT
Secondary breakdown occurs in Power BJT	Secondary breakdown does not occur in power MOSFET
Operating frequency of Power BJT is low about 10kHz	Operating frequency of Power MOSFET is high about 100kHz
Power BJT has high switching loss but conduction loss is less.	Power MOSFET has low switching loss but conduction loss is more due to high on-state resistance
On state voltage of Power BJT is low compared to power MOSFETs	On state voltage of Power MOSFETs is high compared to power BJTs
Powers BJT are available with high current rating compared to Power MOSFET. Current rating of power BJT is about 800A	Powers MOSFET are available with low current rating compared to Power BJT. Current rating of power MOSFET is about 140A
Power BJTs are less sensitive to voltage spikes compared to power MOSFETs	Power MOSFETS are very sensitive to voltage spikes compared to power BJTs

1.5.8 Uses of Power MOSFET

The main applications of Power MOSFETs I are given below:

- DC to DC converters or Switched mode power supplies
- Uninterruptible Power Supplies (UPS)
- AC and DC drives
- Battery charger
- Office equipment
- VLF transmitters, radars
- Solar inverters
- Automotive applications
- Power amplifiers for public address systems
- Induction heating, Electric dryer
- Power supply of wireless communication
- Non-conventional energy sources

1.6 INSULATED GATE BIPOLAR TRANSISTOR (IGBT)

An insulated gate bipolar transistor (IGBT) has been developed by combining into all advantages of both power BJT and power MOSFET. Therefore, an IGBT has high input impedance like a power MOSFET and has low on-state power loss as in a power BJT. IGBT is free from second breakdown which is present in power BJT. It is also called as metal oxide insulated gate transistor (MOSIGT), insulated gate transistor (IGT) or conductivity modulated field effect transistor (COMFET). Now a-days, IGBTs are extensively used in industrial power electronics circuits due to its different advantages.

1.6.1 Construction and Working Principle of IGBT

IGBT is a multi-layer semiconductor structure with p-type and n-type doping. Fig.1.32 shows a vertically oriented n-channel IGBT which consists of a $p^+ n^+ n^- p\ n^+$ structure. Firstly n^+ layer is epitaxially grown on the p^+ substrate and a p^+n^+ junction J_1 is developed. In this device, p^+ substrate is used as drain and minority carriers are injected through drain region. As a result, p^+ is called *injecting layer*. The doping density in the p^+ and n^+ layers is about $10^{19} cm^{-3}$. The thickness of p^+ injecting layer is about 100µm. Highly doped p^+ and n^+ regions are the two end layers of a vertically oriented IGBT. n^+ region is used as source and p^+ region is used as drain as shown in Fig.1.32.

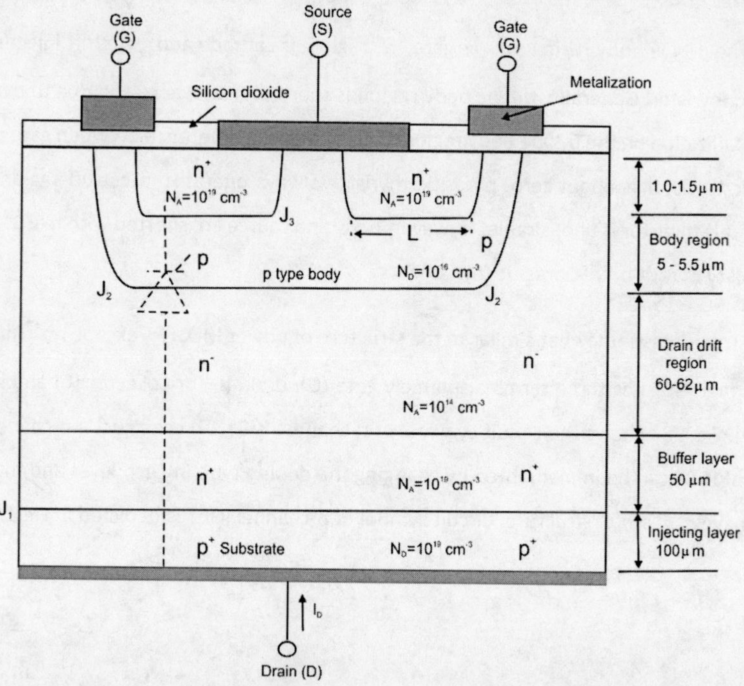

Fig.1.32 Structure of a n channel IGBT

The n^- layer is also epitaxially grown on the n^+ substrate. Subsequently p-type semiconductor is diffused in the epitaxially grown n^- layer and the *p* region is formed. Lastly n^+ type semiconductor is diffused in the *p* region and the n^+ region is created.

The n^+ region which is developed above p^+ substrate is called as *buffer layer*. The thickness of n^+ buffer layer is about 50µm. The n^- layer is known as *drain drift region* and the doping density of n^- layer is about $10^{14} cm^{-3}$ to $10^{15} cm^{-3}$. The thickness of n^- drain drift region is about 60-62 µm which determines the breakdown voltage of IGBT. The *p* type semiconductor layer is the region where the channel is established between source and drain. Therefore, *p* region is called as *body* of an IGBT. The doping density of *p* region is about $10^{16} cm^{-3}$ to $10^{17} cm^{-3}$ and the thickness of body region is about 5-5.5 µm.

Fig.1.32 shows that there are three p-n junctions such as drain to buffer layer $p^+ n^+$ junction J_1, drain drift region to body $n^- p$ junction J_2 and body to source pn^+ junction J_3. When n^+ buffer layer exist in between p^+ injecting layer and n^- drift layer and doping density and thickness of n^+ buffer layer are selected appropriately, the operating performance of the IGBT can be improved significantly.

It is clear from Fig.1.32 that a parasitic $pnpn$ thyristor is developed between the source and drain contacts. The p-type body region acts as gate, n^+ region as cathode and p^+ drift injecting as anode of parasitic thyristor. Generally, p-type body region is shorted to source region due to overlapping of source metallization on the p-type body region. Since potential difference between gate and cathode of parasitic thyristor is about zero, parasitic thyristor always operates in cut-off region. Essentially turn-on of this thyristor is undesirable. However, body and source are shorted within IGBT to minimize the possibility of turn-on of parasitic thyristor.

The IGBT structure is somewhat similar to the structure of power MOSFET except p^+ injecting layer. The n-channel IGBT has three terminals namely gate (G), drain (D) or collector (C) and source (S) or emitter (E). Fig.1.33 shows the circuit symbols of n channel IGBT. To construct a p-channel IGBT, the structure of IGBT can be implemented by changing the doping type in each layer and thus p channel IGBT will have $n^+ p^+ p^- n\, p^+$ structure. Circuit symbol of p-channel IGBT is depicted in Fig.1.34.

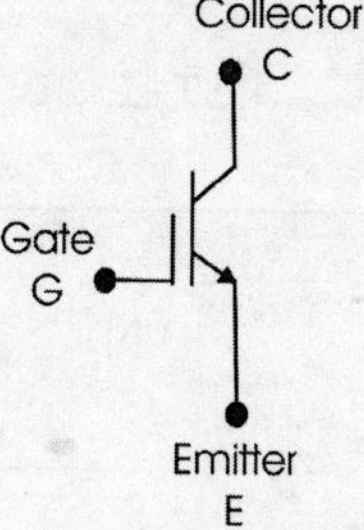

Fig.1.33 Circuit symbol of n-channel IGBT

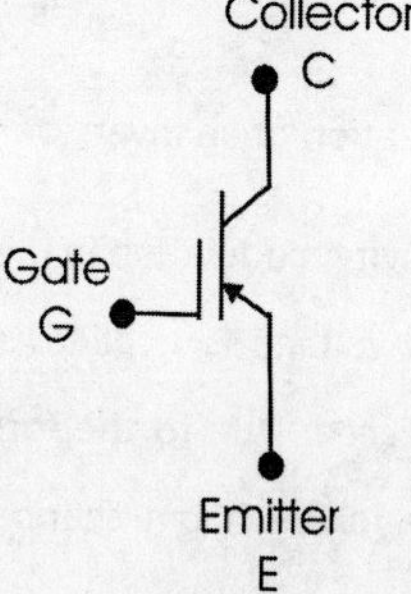

Fig.1.34 Circuit symbol of p-channel IGBT

When the gate is positive with respect to source/emitter by voltage V_G, the electric field across the dielectric develops negative charge and accumulates near gate on the p-type body region. It behaves like a space charge capacitance. This n channel can develop short circuit the n^- and n^+ regions near the dielectric. Due to biasing of the drain/collector terminal, n^- drift region is positively biased. Therefore, current I_D flows from drain to source. Holes from the p^+ substrate moves into n^- drift region and some of the holes can recombine with the electron exist in n^- drift region.

The principle of operation of IGBT can be divided into parts such as (i) creation of inversion layer and (ii) conductivity modulation.

Creation of inversion layer: Fig.1.35 shows the creation of an inversion layer in IGBT. When the gate to source voltage V_{GS} is greater than threshold voltage $V_{GS(th)}$, *n*-type inversion layer is created under the layer of silicon dioxide. Due to formation of n type inversion layer in the p type body, an $n^+ \, n \, n^-$ channel is formed and a current flows from drain to source through this channel.

V_{DD}
- +
R_D
Source (S)
P
N^+
N^-
N^+
P^+
Drain (D)
N
Electrons flow
Inversion layer
Gate (G)
Silicon dioxide
- +
V_{GS}

Fig.1.35 Creation of an inversion layer in IGBT

Conductivity modulation: The conductivity modulation in the n^- drift layer of IGBT is shown in Fig.1.35 and Fig.1.36. When the forward bias voltage is applied in between drain (collector) and source (emitter), the junction J_1 is forward biased. Due to the formation of inversion layer, electrons are injected from source into the n^- drift region via $n^+ p\ n^-$ channel. Since J_1 is forward biased, it will inject holes in the n^+ buffer region. The electrons injected in the n^- drift region create a space charge which can attract hole from the n^+ buffer region which is injected from p^+ region. Thus double injection takes place into the n^- drift region. Consequently, the conductivity of drift region increases and as resistance becomes minimum, the conductivity modulation can reduce the on state voltage drop across IGBT.

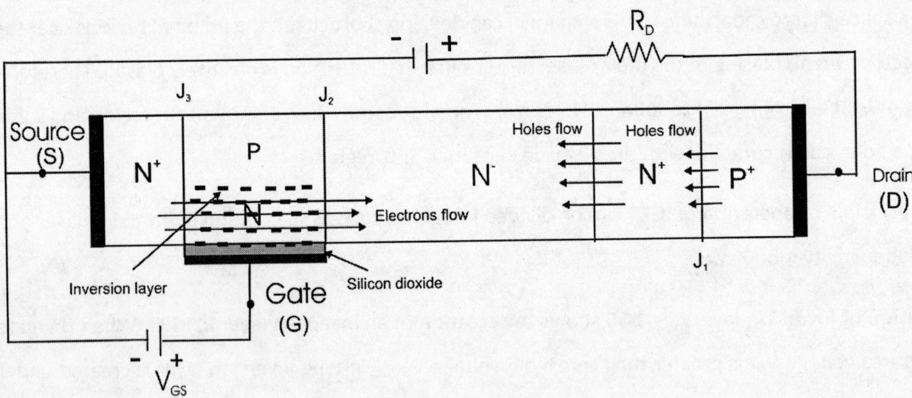

Fig.1.36 Conductivity modulation in IGBT

1.6.2 V-I CHARACTERISTICS of IGBT

The circuit diagram of n-channel IGBT is shown in Fig.1.37 where the input signal V_{GE} is applied across gate to emitter and the output signal V_{CE} is obtained from the collector terminal. The emitter terminal is common between the input and output of an IGBT. The current I_C flows from collector to emitter and it is controlled by gate signal V_{GE}.

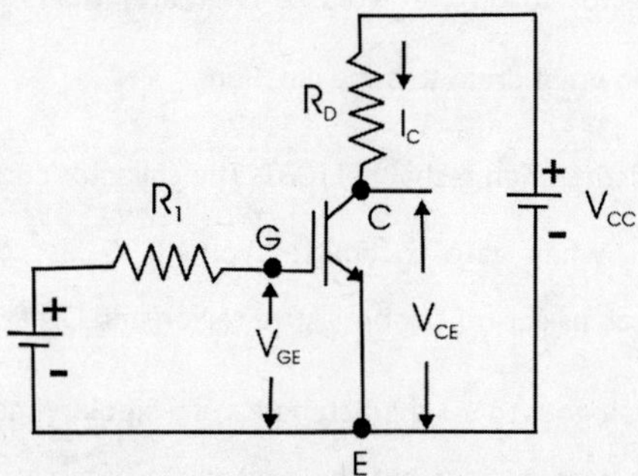

Fig.1.37 n-channel IGBT circuit

The transfer characteristic of an n-channel IGBT is a plot of V_{GE} vs I_C as shown in Fig.1.38. The collector current I_C is a function of gate to emitter voltage V_{GE} and the curve is almost linear in most of the collector current I_C when gate to emitter voltage is greater than threshold voltage $V_{GE(th)}$. When the collector current I_C is low and the gate to emitter voltage is approaching the threshold voltage $V_{GE(th)}$, the nature of curve is nonlinear. If gate to emitter voltage is less than threshold voltage $V_{GE(th)}$, current flows from collector to emitter is about zero. The approximate value of $V_{GE(th)}$ is about 2 to 3 volts. When I_C=0, the collector to emitter is open circuit and IGBT operates in off-state and it can able to hold the supply voltage V_{CC}.

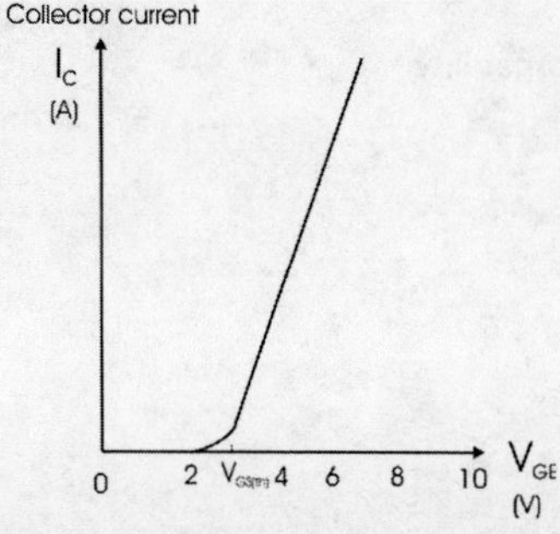

Fig.1.38 Transfer characteristic of an n-channel IGBT

Whenever gate to emitter voltage is less than threshold voltage $V_{GE(th)}$, inversion layer is not developed to connect the collector to the emitter. As a result, collector to emitter is open circuit. Subsequently applied voltage is dropped across the junction J_2 and very small leakage current flows. Therefore, IGBT operates in off-sate.

The collector to emitter breakdown voltage $V_{CE(\max)}$ should be greater than V_{CC} to keep away from the device breakdown. When collector to emitter voltage is greater than $V_{CE(\max)}$, IGBT will be breakdown due to avalanche breakdown of drain to body junction.

Fig.1.39 shows the output characteristics of an n channel IGBT. The collector current I_C is a function of collector to emitter voltage V_{CE} when gate to emitter voltage V_{GE} is constant. The output characteristics have three regions such as cut-off region, active region and ON regions.

The V-I characteristics of IGBT are analogous to V-I characteristics of a bipolar junction transistor (BJT). In BJT, the controlling parameter is base current but the controlling parameter of IGBT is gate to emitter voltage V_{GE}. The V-I characteristics of p-channel IGBT is same V-I characteristics of n-channel IGBT however the polarities of voltages and currents must be reversed.

If IGBT operates in off-state, junction J_2 blocks the forward voltages. The IGBT has also reverse voltage blocking capability. The reverse blocking voltage of IGBT is equal to the forward blocking voltage when IGBT is manufactured without n^+ buffer layer. In reverse bias condition, junction J_1 behaves as reverse blocking junction. When n^+ buffer layer is present within the device, the breakdown voltage of J_1 junction can be reduced drastically and it is about few tens of volts due to presence of highly doped semiconductor material on both sides of J_1 junction. If the applied reverse voltage is greater than the breakdown voltage of J_1 junction, IGBT lost its reverse blocking capability. V_{BR} is the reverse breakdown voltage of IGBT as shown in Fig.1.39. IGBT can be used in ac circuits due to reverse voltage blocking capability.

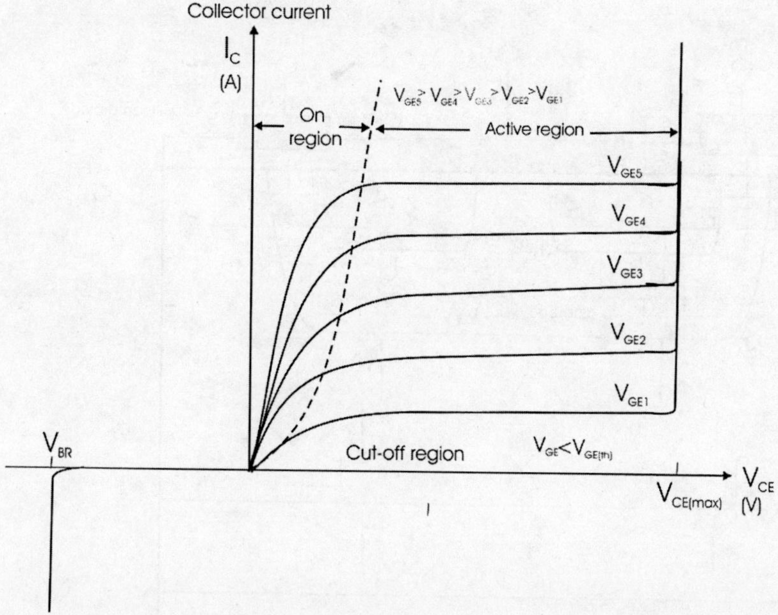

Fig.1.39 Output characteristic of an n-channel IGBT

1.6.3 ON STATE VOLTAGE DROP in IGBT

An inversion layer is created just below the gate of the IGBT when the applied gate to source voltage V_{GS} is greater than the threshold voltage $V_{GS(th)}$. The developed inversion layer can able to interconnect n^-drift region and n^+ drift region and then electron current flows through this inversion layer. In this instant, holes are injected from p^+ drain region to n^-drift region. The injected holes come across the n^-drift region using drift and diffusion methods and can able to reach to p-type body region. When holes enter in the p-type body region, these holes can attract electrons from the source and excess holes are recombined with electrons.

Fig.1.40 shows the formation of MOSFET, *pnp* and *npn* transistors with in the structure of IGBT. The equivalent circuit of IGBT is depicted in Fig.1.41. The IGBT can be represented by a Darlington circuit with the *pnp* transistor (main transistor) and the MOSFET as the driver device. An approximate equivalent circuit of Fig.1.41 for normal operating condition is shown in Fig.1.42. The resistance of the n^-drift region is the resistance between *p-n-p* base and MOSFET drain and it is denoted by R_{drift} .

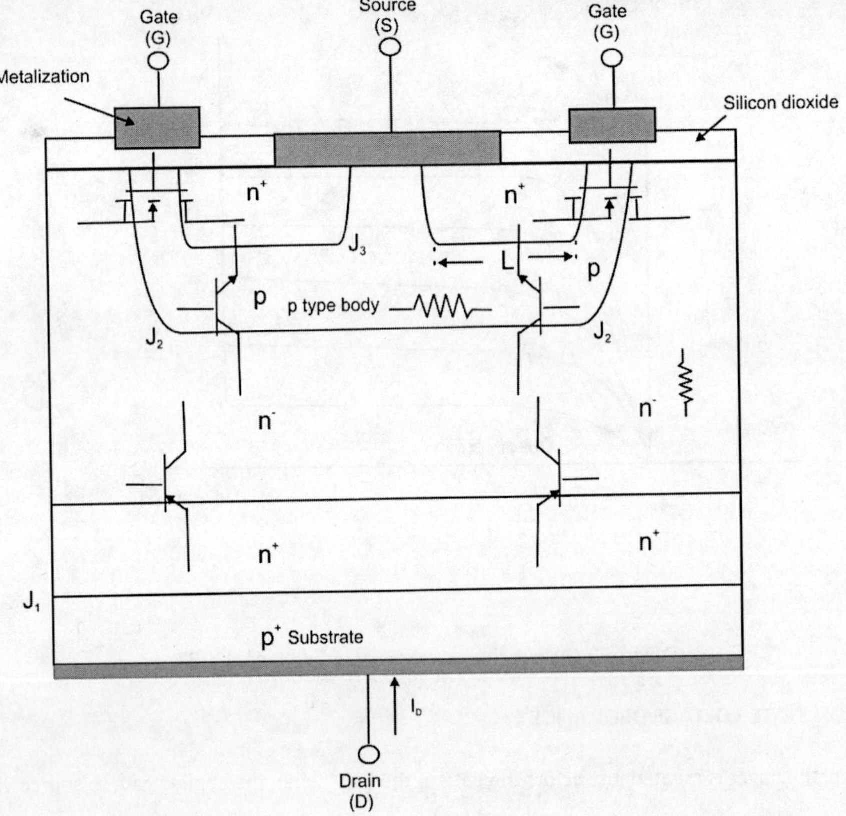

Fig.1.40 Formation of MOSFET, *pnp* and *npn* transistors within structure of IGBT

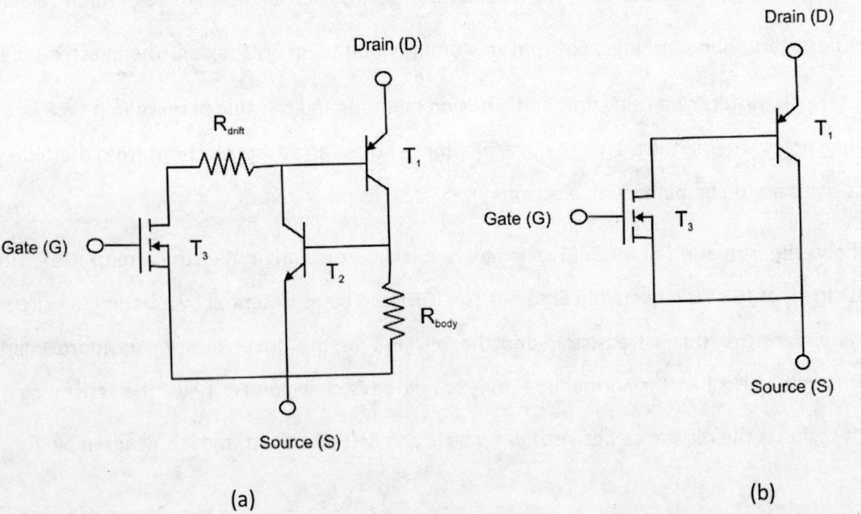

Fig.1.41 (a) Equivalent circuit diagram of an IGBT (b) Darlington representation of MOSFET and p-n-p transistor

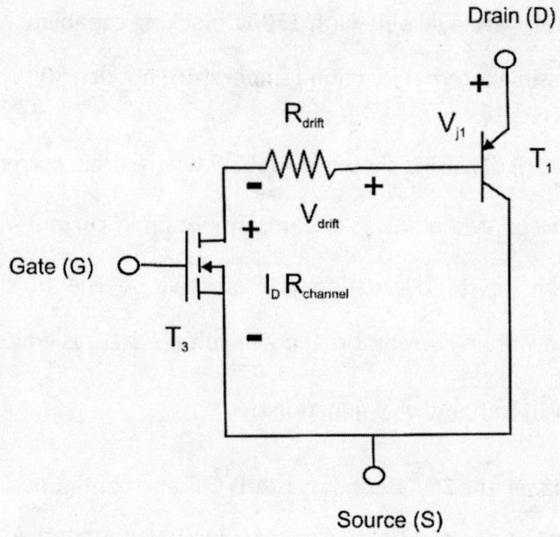

Fig.1.42 Approximate equivalent circuit diagram of an IGBT

The *on state voltage* across drain to source is $V_{DS(on)}$ and it is equal to

$$V_{DS(on)} = V_{J1} + V_{drift} + I_D R_{Channel} \tag{1.40}$$

Where, V_{J1} is the forward biased voltage drop across the junction J_1 and its value is about 0.7V to 1.0V; V_{drift} is the voltage drop across the drift region; $I_D R_{Channel}$ is voltage drop across the channel due to the channel resistance. The value of V_{drift} voltage is less in the IGBT than in the MOSFET due to the conductivity modulation of the drift region.

1.6.4 SAFE OPERATING AREA OF IGBT

The safe operating area (SOA) of IGBT states the maximum values of voltage and current at which the device can withstand safely. According to manufacturer datasheet related to the specification of IGBT, the safe operating area (SOA) depends on (i) maximum drain current $I_{D(\max)}$ or maximum collector current $I_{C(\max)}$ (ii) maximum permissible internal junction temperature $T_{j(\max)}$ (iii) forward blocking voltage in FBSOA V_{DSS} or $V_{CE(\max)}$ (iv) Reverse breakdown voltage in RBSOA V_{BR} (v) maximum power dissipation $P_{D(\max)}$ and (vi) maximum gate voltage $V_{GE(\max)}$.

$I_{C(\max)}$ is the maximum permissible collector current of IGBT and its value is about 200A to 400A. $V_{GE(\max)}$ is the maximum gate voltage and it is determined by the breakdown of the silicon dioxide (SiO_2) layer. $V_{CE(\max)}$ is the maximum collector to emitter voltage which is represented by breakdown voltage of transistor T_2. IGBT are available with 1700V blocking capability for industrial applications. $T_{j(\max)}$ is maximum permissible internal junction temperature about $150°$ C.

The safe operating area (SOA) of IGBT is shown in Fig.1.43 where drain current I_D and drain to source voltage V_{DS} are in log scale. When V_{DS} is low, maximum drain current $I_{D(\max)}$ is limited by power dissipation. The maximum power dissipation is restricted by the maximum tolerable junction temperature $T_{j(\max)}$. At low drain current, breakdown voltage V_{DSS} is limited to avoid the avalanche breakdown of the drain drift to body $n^- p$ junction.

If IGBT operates with dc signal, the SOA boundary is A-B-C-D and continuously power will be dissipated within device and junction temperature increases considerably. Boundary A-B indicated the maximum limit of drain current I_{DM} when V_{DS} is less than 100V. If V_{DS} is more than 100V, drain current decreases as shown in boundary B-C when the operating junction temperature of IGBT is less than the maximum operating junction temperature. The maximum voltage capability of IGBT is denoted by boundary C-D.

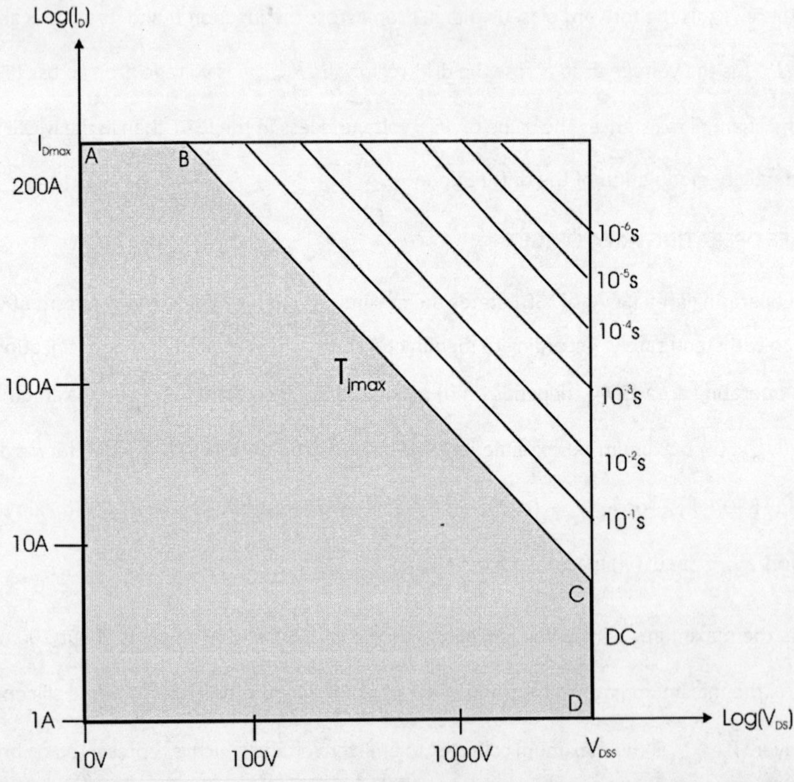

Fig.1.43 FBSOA of a IGBT

When IGBT operates as a switch and it is driven by pulse signal, SOA can be extended as depicted in Fig.1.43. The extension of SOA is possible during switch mode operation only.

If IGBT turned-on for a few microseconds, the device can absorb very small amount of energy and junction temperature rise will be low and SOA will be increased. When pulse width is about 10^{-2} s, SOA is greater than the boundary A-B-C-D. If pulse width of signal is further reduced to about some 10^{-6} s, SOA is greater than the boundary with pulse width 10^{-2} s. Thus SOA can be increased with decrease of pulse-width of signal.

Reverse bias safe operating area (RBSOA) of IGBT is dissimilar from FBSOA of IGBT. With the increase of the rate of change of drain to source voltage dV_{DS}/dt, the upper right hand corner of RBSOA of IGBT is cut out gradually and RBSOA of IGBT decrease significantly as shown in Fig.1.44.

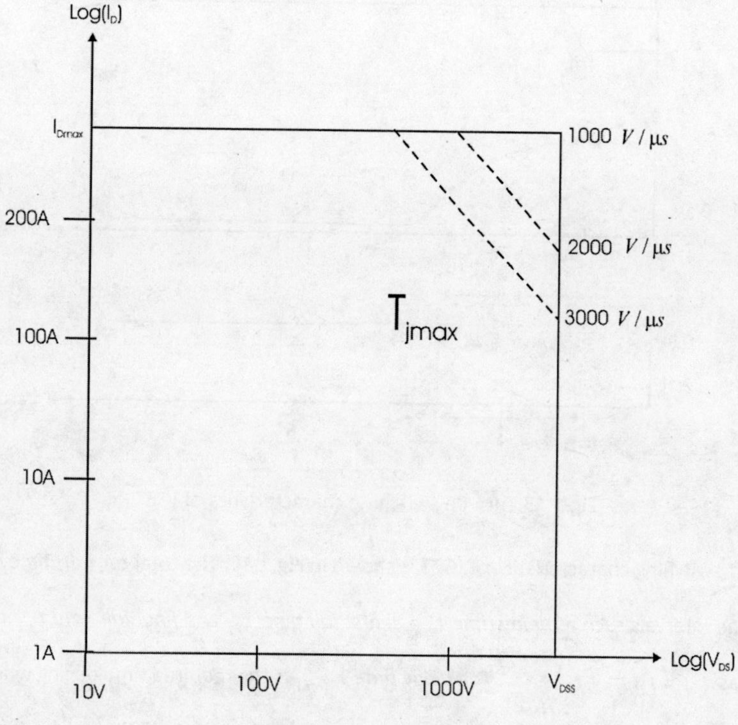

Fig.1.44 RBSOA of a IGBT

1.6.5 SWITCHING CHARACTERISTICS of IGBT

There are two types switching characteristics of IGBT such as turn-on and turn-off. Fig.1.45 shows the turn-on switching characteristics of IGBT. Turn-on *delay time* $t_{d(on)}$ is the required time to fall the collector-emitter voltage from V_{CE} to $0.9V_{CE}$. $t_{d(on)}$ is also the time required for collector current to increase from its initial value to 10% of rated collector current, $0.1I_C$. The *rise time* t_r is the required time to increase collector current from 10% of rated collector current $0.1I_C$ to rated current I_C and collector-emitter voltage falls from 90% of V_{CE} to 10% of V_{CE}. Therefore total *turn-on time* t_{on} is sum of delay time and rise time and it is equal to $t_{on} = t_{d(on)} + t_r$. Just after turn-on, the collector-emitter voltage of IGBT falls to small value which is called on-state voltage drop V_{CES}.

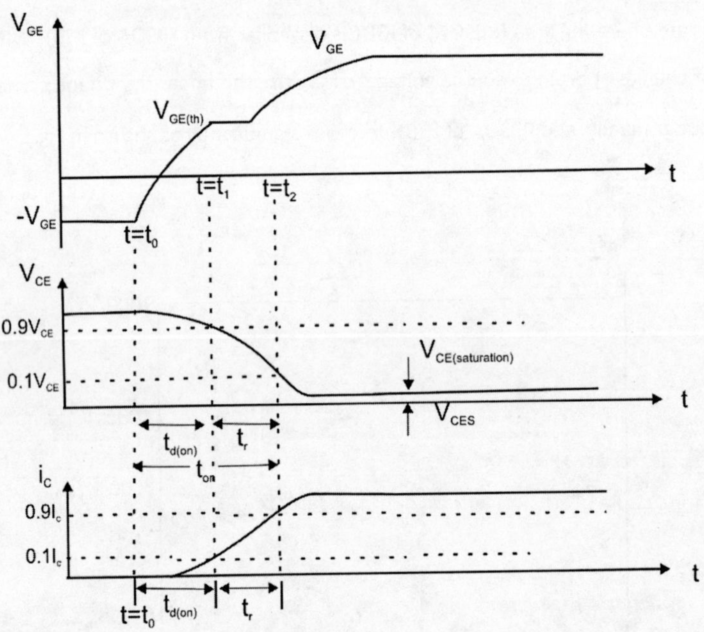

Fig.1.45 Turn on switching characteristics of IGBT

The turn off switching characteristics of IGBT is shown in Fig.1.46. The total *turn-off time* t_{off} consists of three time intervals such as *delay time* $t_{d(off)}$, *first fall time* t_{f1} and *final fall time* t_{f2} and it can be expressed as $t_{off} = t_{d(off)} + t_{f1} + t_{f2}$. The *delay time* $t_{d(off)}$ is the required time during which the gate-emitter voltage falls from V_{GE} to the threshold voltage $V_{GE(th)}$. As gate-emitter voltage decreases to $V_{GE(th)}$ during $t_{d(off)}$, the collector current decreases falls from rated current I_C to 90% of rated collector current $(0.9I_C)$. At the end of delay time $t_{d(off)}$, the collector-emitter voltage starts to increase.

The *first fall time* t_{f1} is the time interval during which the collector current falls from the 90% of rated collector current $(0.9I_C)$ to 20% of rated collector current $(0.2I_C)$ and it is also the interval time during which the collector emitter voltage rises from V_{CES} to 10% of V_{CE}. The *final fall time* t_{f2} is the time interval during which the collector current falls from 20% of rated collector current $(0.2I_C)$ to 10% of rated collector current $(0.1I_C)$ or time interval during which the collector emitter voltage rises from 10% of V_{CE} to final value of V_{CE}.

The switching characteristics of IGBT are similar to that of a Power MOSFET but the major difference is that IGBT has a tailing collector current due to stored charge in n- drift region. The tail current increases the turn-off loss. To reduce switching losses, IGBT will be turned off by applying a negative voltage about -15V at gate.

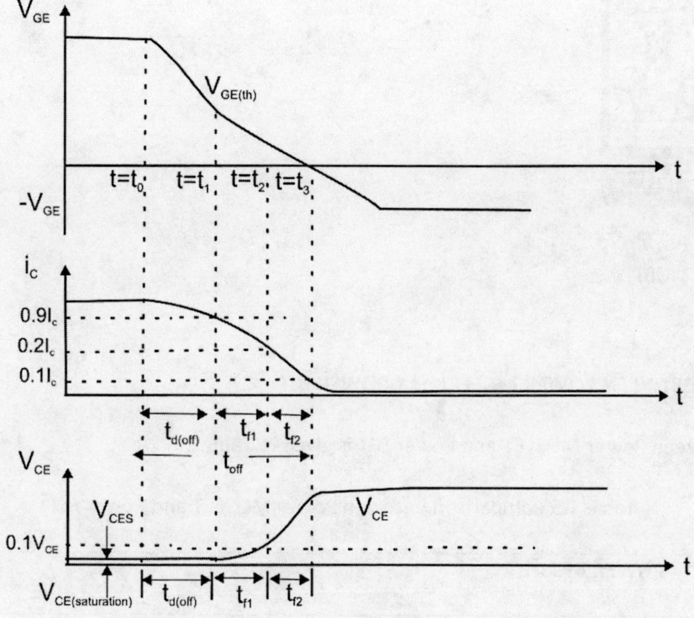

Fig.1.46 Turn off switching characteristics of IGBT

During turn-on process of IGBT, the energy loss in the device is

$$W_{on} = \frac{1}{6} V_{CE\max} I_{C\max} t_{on} \tag{1.41}$$

The average power loss during turn-on process of IGBT is

$$P_{on(average)} = W_{on} f_s = \frac{1}{6} V_{CE\max} I_{C\max} t_{on} f_s \tag{1.42}$$

where, f_s is the switching frequency

During turn-off process of IGBT, the energy loss with in the device is

$$W_{off} = \frac{1}{6} V_{CE\max} I_{C\max} t_{off} \tag{1.43}$$

Then the average power loss during turn-off process of IGBT is

$$P_{off(average)} = W_{off} f_s = \frac{1}{6} V_{CE\max} I_{C\max} t_{off} f_s \tag{1.44}$$

For more on Power BJT. MOSFET, IGBT

1.6.6 COMPARISION BETWEEN POWER MOSFET and POWER BJT

The comparison between Power MOSFET and Power BJT is given in Table 1.5

Table 1.5 Comparison between Power MOSFET and Power BJT

POWER MOSFET	IGBT
It is a voltage controlled device.	It is a voltage controlled device.
It has three terminals namely gate(G), drain(D) and source(S).	It has three terminals namely gate(G), drain(D) or emitter(E) and source(S) or collector (C).
Input impedance of Power MOSFET is very high.	Input impedance of IGBT is very high.

POWER MOSFET	IGBT
Power MOSFET has positive temperature coefficient. With increase in temperature, on-sate resistance increases compared to IGBT.	IGBT has positive temperature coefficient. With increase in temperature, on-state resistance increases but rate of increment is less than increase in MOSFET.
On-state voltage drop of MOSFET is large compared to on-state voltage drop of IGBT.	IGBT has a very low on-state voltage drop due to conductivity modulation. So smaller chip size is possible and the cost can be reduced.
The on-state voltage drop of MOSFET increases by 3 times for temperature rise from room temperature to $200°C$.	The increment of on-state voltage drop of IGBT is very small.
Current sharing between parallel connected MOSFETs is poor compared to IGBTs.	Current sharing between parallel connected IGBTs is better compared to MOSFETs.
At high ambient temperature, maximum current rating reduces.	At high ambient temperature, IGBT is well suited.
Switching speed of Power MOSFET is superior to that of a IGBT.	Switching speed of IGBT is inferior compared to Power MOSFET.
FBSOA is wide.	Wide FBSOA and RBSOA. It also has excellent forward and reverse blocking capability.
Power MOSFET has a *parasitic BJT* as an integral part of its structure.	IGBT has a *parasitic thyristor* as an integral part of its structure.

1.6.7 Uses of IGBT

IGBTs have a wide spectrum of applications in house hold equipment, industrial applications, transportation systems, telecommunications, aerospace and defence applications, and Space technology. The some of the uses of IGBT are given below:

- Inverter circuit
- Aircraft power system
- Air conditioners
- Washing machines
- Industrial Robot and Elevator
- Static VAR compensation
- Refrigerators
- Power quality control
- Industrial drives
- Traffic-signal control
- Automotive electronics
- DC power supply and UPS

1.7 CONCEPT OF SINGLE ELECTRON TRANSISTOR (SET)

Single-electron transistor (SET) is a very sensitive electronic device which works based on the coulomb blockade effect. In this device, electrons flow through the tunnel junction between source to a quantum dot (conductive island) and the tunnel junction between drain to a quantum dot. Actually, electrical potential of the island can be tuned by a third electrode, called as gate, which is capacitive coupled to the island. Fig.1.47 shows the construction of SET. The conductive island is sandwiched between two tunnel junctions as shown in Fig.1.47 and tunnel junctions can be modeled by a capacitor C_D and C_S and a resistor R_D and R_S in parallel (Fig.1.48). The first single-electron transistor (SET) based on the Coulomb blockade was developed by K. K. Likharev and D. V. Averin in 1986.

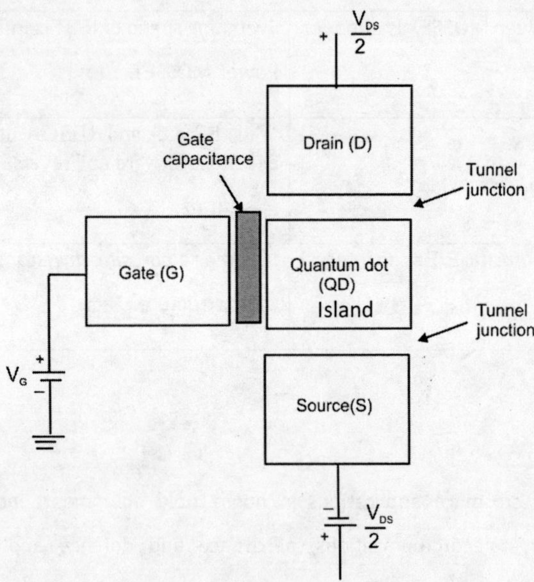

Fig.1.47 Construction of a basic SET

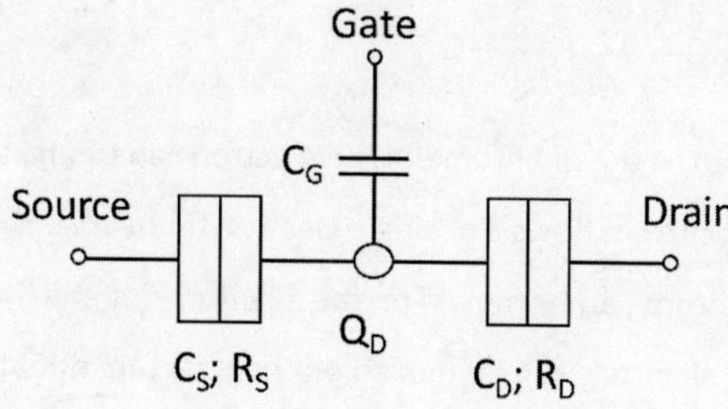

Fig.1.48 Schematic diagram of a basic SET

Basically SET is a three terminal (gate, source and drain) device which can transfer electrons from source to drain one by one. The purpose of SET is to control the tunnelling of electron into or from the quantum dot. Structure of SET is similar to field effect transistor (FET) excluding (i) SET has tunnelling junction instead of p-n junction and (ii) SET has quantum dot in the place of channel region of FET.

SET structure consist of (i) quantum dot, (ii) two tunnel junctions, (iii) a gate electrode and (iv) gate capacitor.

Operating Principle of SET:

To control tunnelling, voltage bias is applied to the gate electrode. A separate voltage bias is applied between source and drain electrode so that the current flows from source to drain through tunnel junctions. The gate bias voltage must be significantly large to overcome coulomb blockade energy and the current will start to flow. Single electron flow in SET is given in Fig.1.49.

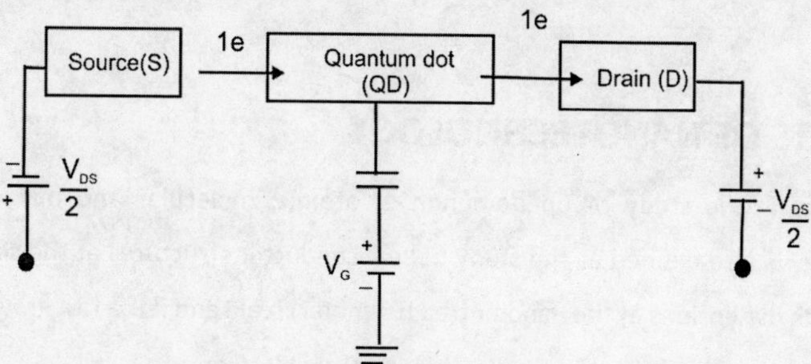

Fig.1.49 Single electron flow in SET

The energy need to move charge Q inside quantum dot is equal to

E=VQ where, Q=e

$$V = \frac{E}{e} = \frac{W_c}{e} \quad \text{as } E = W_c$$

Since $Wc = e^2 / 2C$, $V = \frac{e^2 / 2C}{e} = \frac{e}{2C}$ (1.45)

When the voltage V applied to the quantum dot, an electron can tunnel through quantum dot. If the gate voltage Vg is zero, no current flows. For single electron tunnelling, V_g=$V_{coulomb}$. When gate voltage V_g=C_{oulomb} + e/2C, two electrons can be moved on the quantum dot at a time. If gate voltage V_g=$V_{coulomb}$ + e/2C + e/2C, then three electrons can be moved on the quantum dot at time. Therefore, number of electrons in the quantum dot is controlled using the gate voltage. A single electron transistor is similar to a normal transistor except (i) the channel is replaced by a small dot and (i) the dot is separated from source and drain by thin insulators. Here an electron tunnels in two steps such as (i) source to quantum dot and (ii) quantum dot to drain. SET are used in supersensitive electrometer, single-electron spectroscopy, DC current standards, temperature standards, detection of infrared radiation, voltage state logics, charge state logics, and programmable single electron transistor logic etc.

For more on single electron transistor

1.8 ASPECTS OF NANOTECHNOLOGY

Nanotechnology is the study of phenomenon at atomic, molecular and macromolecular scale. Nanotechnology is also defined as the study of semiconductor structures at the size of 1 nm to 100 nm. It deals with dimensions at the nanometre (10^{-9} meter) scale and it is a fast growing scientific field with applications in many different areas including power electronics.

Nanotechnology is used to develop new circuit materials, new processors, new memory devices for storing information and telecommunication technology for transferring information in new way. This technology can improve the capability of electronic components such as reducing the size of transistor used in an IC and increasing memory density to about one tetra-byte (TB) per square inch.

A nano-transistor is a transistor where the semiconductor components act as an electronic signal switch or amplifier in nano-scale. It has a gate of controlled electrode about 70 nm width and a gate oxide which separates the control electrode from the current carrying channels, thin about 1 nm. Usually semiconductor manufacturers manufactures logic circuits that incorporate more than 40 million MOSFETs in a single IC but now a days they develop logic circuits with nearly a half billion nanometre scale MOSFET, packing about 5-10 nanotransistors per micrometre square which improves the reliability, reduce the cost and the power consumption. Nanotransistors are billion times faster, pairs of metal contacts lie on a top printed carbon nanotubes, forming transistors visible only in the electron microscope image.

Field Effect Transistors, the most commonly used transistors, when made with carbon nanotubes (CNT), have shown a significant speed improvement compared to same-sized silicon-based transistors. Also, they can be fabricated at near-room temperatures, while silicon-based transistors need 450 to 500°C during the fabrication process. This means that CNT FET (carbon nanotube-based field-effect transistor) based circuits can be built, on top of a previously fabricated one. For example, logic circuits can be built on top of memory circuits. These cannot be done currently with the silicon-based process because, at 450 to 500 °C, the top layer will melt the bottom layer.

Flexible and printable electronics are another aspect of nanotechnology. Conductive ink made of nanomaterial, such as silver nanowires (AGNW), is used to print circuits using home printers on flexible surfaces. Devices such as smart phones and laptops don't need to be rigid anymore.

Carbon nanotube (CNT) based Nanotube Emissive Displays (NED) provide higher colour contrast, better resolution and better brightness than LED. The viewing angles of this type of display are unrestricted. They are also low-cost, lightweight and thin.

Sensors made out of nanomaterials such as Palladium, Carbon Nanotubes and Zinc coated nanowires are much more sensitive than currently used sensors. Their extremely-high surface area per unit volume allows them to be very sensitive. Gas sensors or chemical sensors made of nanomaterials can detect a single atom of gas or chemical.

Photovoltaic cells made of nanomaterials such as carbon nanowires cost less and are more efficient.

Nanotechnology is used to make paper battery which is flexible, ultra-thin energy storage and production device formed by two things- carbon nanotubes and nanocomposite paper. Nanocomposite paper is a hybrid energy storage device made of cellulose. It has the high energy storage capacity and high energy density of the super capacitor. E-Tattoos made of nanomaterials can monitor your health remotely and send the data to your phone wirelessly.

E-Textiles made of graphene-coated polypropylene can sense your body temperature. They are also washable and show good mechanical stability. E-Textiles made of nanomaterials can check your blood pressure and sugar levels and monitor your heartbeat. They can also change colour according to your mood and the environment. In this way, nanotechnology can amplify the positive aspects of electronics.

EXERCISE

MULTIPLE CHOICE QUESTIONS

1.1 Match the following devices on the left hand with circuit symbols on the right hand side

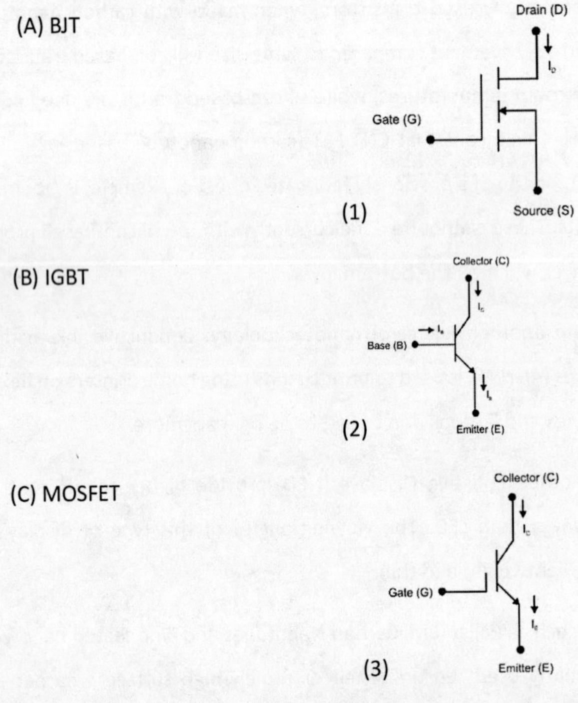

(a) $A\rightarrow1, B\rightarrow2, C\rightarrow3$ (b) $A\rightarrow2, B\rightarrow1, C\rightarrow3$

(c) $A\rightarrow2, B\rightarrow3, C\rightarrow1$ (d) $A\rightarrow3, B\rightarrow2, C\rightarrow1$

1.2 The V-I characteristics of ____________ is shown in Fig.1.50

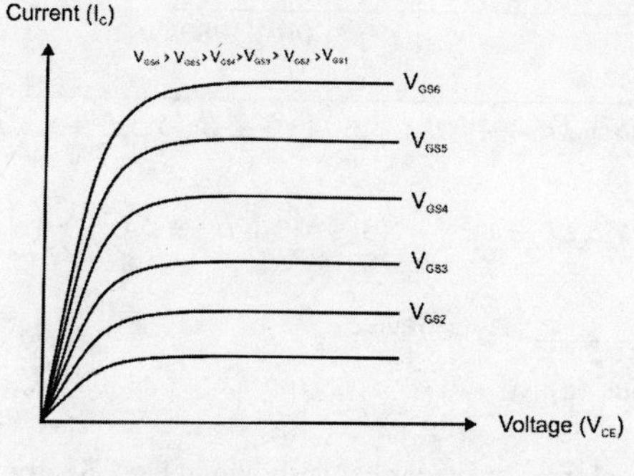

Fig.1.50

(a) SCR (b) BJT (c) MOSFET (d) IGBT

1.3 IGBT has three terminals namely

(a) Gate, collector and emitter (b) Gate, drain and source

(c) Base, collector and emitter (d) Base, drain and source

1.4 Match the following devices on the left hand side with the switching frequency on the right side

(A) Thyristors	(1) 100kHz
(B) Power MOSFET	(2) 1kHz
(C) Power BJT	(3) 50kHz
(D) IGBT	(4) 10kHz

(a) $A \to 2, B \to 1, C \to 4, D \to 3$ (b) $A \to 1, B \to 2, C \to 3, D \to 4$

(c) $A \to 4, B \to 3, C \to 2, D \to 1$ (d) $A \to 1, B \to 2, C \to 4, D \to 3$

1.5 Match the following device on the left hand side with the Voltage and current rating on the right side

(A) Thyristors	(1) 3.3kV/2.5kA
(B) Power MOSFET	(2) 1400V/400A
(C) Power BJT	(3) 10kV/5kA

(D) IGBT	(4) 1kV/50A

(a) $A \rightarrow 1, B \rightarrow 2, C \rightarrow 3, D \rightarrow 4$ (b) $A \rightarrow 4, B \rightarrow 3, C \rightarrow 2, D \rightarrow 1$

(c) $A \rightarrow 3, B \rightarrow 4, C \rightarrow 2, D \rightarrow 1$ (d) $A \rightarrow 1, B \rightarrow 2, C \rightarrow 4, D \rightarrow 3$

1.6 A power transistor is a _______ layer device

(a) two (b) three (c) four (d)five

1.7 When the switching waveforms of power BJT is shown in Fig.1.51, the average switching loss during turn on is ______

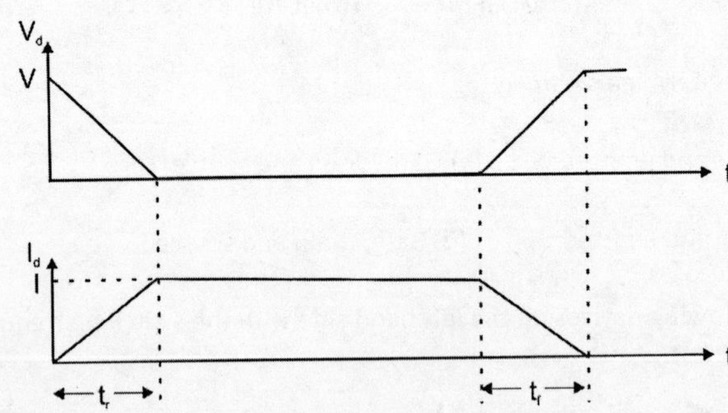

Fig.1.51

a) $\frac{1}{2}VIt_r f_s$ (b) $\frac{1}{4}VIt_r f_s$ (c) $\frac{1}{6}VIt_r f_s$ (d) $\frac{1}{8}VIt_r f_s$

1.8 High frequency operation of a electronics circuit is limited by

(a) switching losses (turn on loss and turn off loss) in the power semiconductor devices

(b) on-state loss in the power semiconductor devices

(c) off-state loss in the power semiconductor devices

(d) all of these

1.9 Which of the following statements are correct

(i) BJT is a unipolar and a majority carrier device

(ii) MOSFET is a majority carrier device

(iii) IGBT is a bipolar device

(iv) TRIAC is an unipolar device

(a) (ii) and (iii) (b) (i) and (iv) (c) (iii) and (i) (d) (ii) and (iv)

1.10 The power electronics based system has very high efficiency as

(a) the power semiconductor devices operates at high frequency, traverse active region at high speed and stay at the two states ON and OFF

(b) the power semiconductor devices always operate in active region

(c) the power semiconductor devices never operate in active region

(d) Heat sink is used to maintain device temperature.

1.11 A power MOSFET has ______________terminals

(a) Base, Emitter and Collector (b) Gate, drain and Source

(c) Base, drain and source (d) Gate, Emitter and Collector

1.12 Which of the following statement is correct

(a) BJT has higher switching losses and lower conduction loss with respect to MOSFET

(b) BJT has higher switching losses and higher conduction loss with respect to MOSFET

(c) BJT has lower switching losses and lower conduction loss with respect to MOSFET

(d) BJT has lower switching losses and higher conduction loss with respect to MOSFET

1.13 *npn* power transistor is a _______ layer device

(a) $n^+pn^-n^+$ (b) $p^+n^+n^-n^+$ (a) $n^-n^+p^-p^+$ (a) $n^+p^-p^+n^-$

1.14 If β_A is gain of auxiliary transistor and β_M is gain of main transistor, the gain of Darlington pair BJT is equal to

(a) $\beta = \beta_M\beta_A - \beta_M - \beta_A$ (b) $\beta = \beta_M\beta_A + \beta_A$

(c) $\beta = \beta_M\beta_A + \beta_A$ (d) $\beta = \beta_M\beta_A + \beta_M + \beta_A$

1.15 Overdrive factor (ODF) is equal to

(a) $\frac{I_E}{I_{BS}}$ (b) $\frac{I_C}{I_{BS}}$ (c) $\frac{I_{BS}}{I_B}$ (d) $\frac{I_B}{I_{BS}}$

1.16 Secondary breakdown occurred in

(a) BJT (b) MOSFET (c) Both BJT and MOSFET (d) none of these

1.17 The relation between α and β is

(a) $\beta = \dfrac{\alpha}{1+\alpha}$ (b) $\alpha = \dfrac{\beta}{\beta - 1}$ (c) $\alpha = \dfrac{\beta}{\beta + 1}$ (d) $\beta = \dfrac{\alpha}{1-\alpha}$

1.18 BJT is____________ controlled device whereas MOSFET is____________ controlled devices.

(a) current, current (b) current, voltage (c) voltage, current (d) voltage, voltage

1.19 Power MOSFET has _____ input impedance.

(a) low (b) medium (c) high (d) zero

1.20 During turn-off, the_______ base current must be applied to quickly remove the stored charges in the transistor.

(a) positive (b) negative (c) zero

Answer of MCQ

1.1 (c) 1.2 (d) 1.3 (a) 1.4 (a) 1.5 (c) 1.6 (c) 1.7 (c 1.8 (a) 1.9 (a) 1.10 (a)

1.11 (b 1.12 (a) 1.13 (a) 1.14 (d) 1.15 (d) 1.16 (a) 1.17 (d) 1.18 (b) 1.19 (c) 1.20(b)

Short and Long Answer Type Questions

Category I Short Answer Type Questions

1.21 Define is power electronics? Give a list of applications of power electronics

1.22 Draw the symbol of the following devices

(a) SCR (b) IGBT (c) TRIAC (d) DIAC

1.23 What is converter? What are the different types of converters?

1.24 Write the name of converters which are used for ac to dc and dc to ac conversion?

1.25 What is power BJT?

1.26 Write the difference between *npn* and *pnp* transistors

1.27 What is primary breakdown of a power BJT

1.28 What is secondary breakdown of a power Transistor

1.29 What is the on-state loss of a power BJT

1.30 What is power MOSFET? What are the types of power MOSFET?

1.31 What is IGBT?

1.32 State single electron transistor (SET)

1.33 Give a list of applications of nano-technology

Category II Long Answer Type Questions

1.34 Draw the block diagram of power electronics based system and explains briefly

1.35 Give a list of power electronics devices

1.36 Write the difference between general purpose BJT and power BJT

1.37 Draw the structure of a power BJT and explain its operating principle

1.38 Explain V-I characteristics of a power BJT

1.39 Describe quasi saturation and hard saturation of a power BJT

1.40 Draw the FBSOA and RBSOA of power and explain briefly

1.41 Sketch the switching characteristics of power BJT. State delay time, rise time, turn-on time, storage time, fall time and turn-off time.

1.42 Explain operating principle of power MOSFET with suitable diagram.

1.43 Compare power MOSFET and power BJT

1.44 Draw the structure of an IGBT and describe its operating principle briefly

1.45 What are advantages of IGBT over power BJT and power MOSFET?

1.46 Draw the V-I characteristics of a IGBT showing different operating regions

1.47 Write short note on (a) SET and (b) Nano-technology

Numerical Problems

1.48 If α of transistor is 0.97, determine the value of β

1.49 When β of transistor is 59, compute the value of α

1.50 If collector current of a transistor is 95 mA and it's β is about 65, determine the value of base current and emitter current

1.51 Determine the value of base current and collector current of a transistor when β=95 and the emitter current is 90 mA.

1.52 A bipolar transistor has $\beta_f = 30$ and load resistance $R_C = 15\Omega$. The dc supply voltage $V_{CC} = 210V$ and the input voltage to base is $V_B = 12V$. If $V_{CE(saturation)} = 1.2V$ and

$V_{BE(saturation)} = 1.7V$, Calculate (a) R_B when transistor operates in saturation and (b) power loss in the transistor.

1.53 Two power BJTs are connected in parallel to share the total current 50A. The collector to emitter voltage of T_1 and T_2 are 1.65V and 1.75V correspondingly. Calculate emitter current of each transistors and the difference of current sharing when the current sharing series resistance are (a) $R_{E1} = 0.35\ \Omega$ and $R_{E2} = 0.45\ \Omega$ (b) $R_{E1} = 0.5\ \Omega$ and $R_{E2} = 0.5\ \Omega$

1.54 Switching waveforms of a power BJT is shown in Fig.1.52 where $V_{CC} = 210V$, $V_{CE(saturation)} = 2.1V$, $I_{CS} = 125A$, $t_d = 0.25\mu s$, $t_r = 1.25\mu s$, $t_{on} = 40\mu s$, $t_s = 2\mu s$, $t_f = 1.5\mu s$. Determine (a) energy loss during delay time (b) energy loss during rise time (c) energy loss during conduction time t_{on} (d) average power loss of power transistor during turn-on if switching frequency is 10 kHz and emitter leakage current is $I_{EO} = 10mA$.

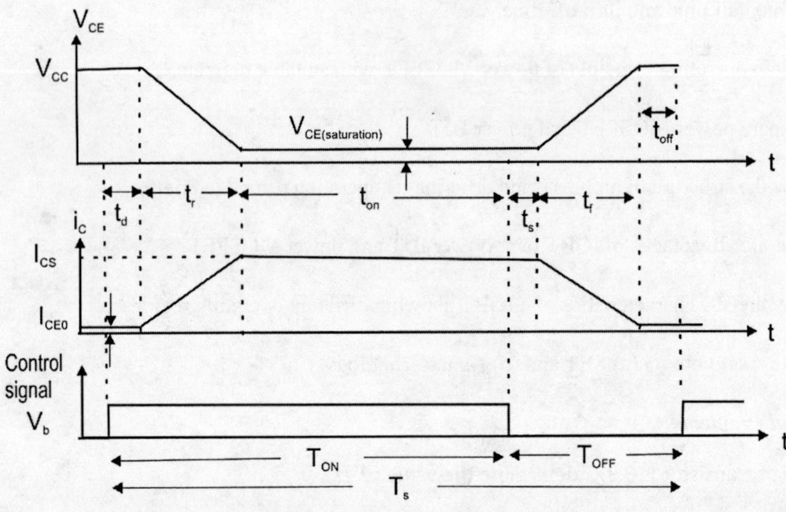

Fig.1.52

PRACTICAL

1. Experiment on Test the proper functioning of power transistor

Aim:

Investigate the input and output characteristics of a NPN power bipolar junction transistor in common emitter configuration

Apparatus:

Regulated dc power supply 0-30V; Digital multi-meter 0-200V DC, 0-1A/10A; Power Transistor 2N3055, Resistance R_C=100Ω,5W; Resistance R_B=3.3 kΩ,1W

Circuit Diagram:

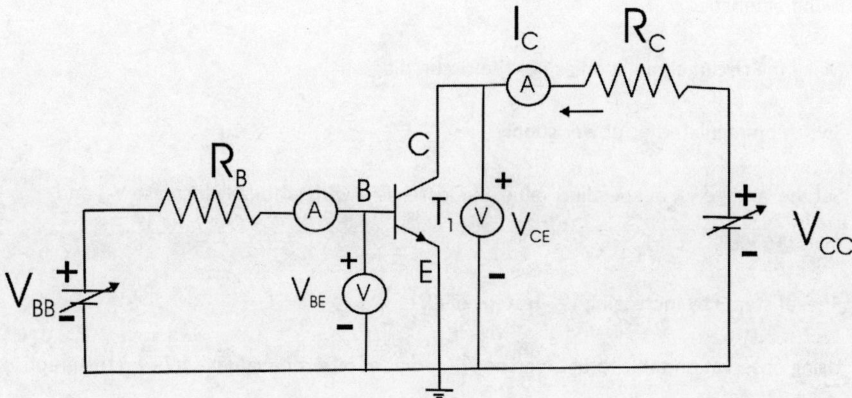

Fig.1.53 Circuit diagram for test the proper functioning of power transistor

Theory:

Power BJT is the most commonly used semiconductor device which has full control over its turn-on and turn-off operation. Power BJTs are available in NPN, PNP and Darlington pair forms. A power BJT is vertically oriented four layer structure device as explained in section 1.4.3. This device has low on-state resistance and power loss is minimum. It's size is large compared to conventional BJT and it has high current handling capability. Usually the collector of power BJT is connected to a metal base that acts as a heat sink to dissipate excess power. Power rating of Power BJT is about 10-300W, maximum collector current 1A-100A, frequency of operation about 100 Hz-100kHz. The input and output characteristics of a NPN power bipolar junction transistor is explained in section 1.4.4. Power transistors are used in high-power application such as power amplifiers, switch mode power supplies, chopper and inverter circuits.

Precautions:

1. Initially all the knobs of the power supplies are at zero value.
2. Do not increase the base current more than its rated value.
3. The applied voltage and current should not be more than the maximum rating of Power BJT.
4. Use appropriate heat sink for proper cooling of Power BJT.
5. Reading of observation should be error free.

Procedure:

A) Input Characteristics

1. Make the circuit connection as per the circuit diagram
2. Switch on regulated dc power supply
3. Set the voltage V_{CE} at specified value and increase I_B with gradually increase V_{BB} and measure V_{BE}
4. Repeat step 3 by increasing V_{CE} in steps of 1V.
5. Using observation table, plot I_B versus V_{BE} curves at different values of V_{CE} on the graph paper

B) Output Characteristics

1. Make the circuit connection as per the circuit diagram
2. Switch on regulated dc power supply
3. Increase V_{BB} gradually to increase base current I_B to set a set value
4. Keep base current I_B constant when V_{CC} increases in steps of 1Volt and take observation reading of collector current I_C and collector emitter voltage V_{CE}.
5. Repeat step 3 and 4 by increasing base current I_B in steps of 10 mA until I_C becomes rated value.
6. Using observation table, plot I_C versus V_{CE} curves at different values of I_B on the graph paper

Observation Table:

Table 1.6 Observation table for input characteristics for Power BJT

Sr.No.	V_{CE} =		V_{CE} =		V_{CE} =	
	I_B in mA	V_{BE} in Volts	I_B in mA	V_{BE} in Volts	I_B in mA	V_{BE} in Volts

Table 1.7 Observation table for output characteristics for Power BJT

Sr.No.	I_B=		I_B=		I_B=	
	I_C in mA	V_{CE} in Volts	I_C in mA	V_{CE} in Volts	I_C in mA	V_{CE} in Volts

VIVA Questions Related to power BJT

1. Write specifications of power BJT from manufacturer data sheet
2. Describe the V-I characteristics of Power BJT base on your observations
3. What is the voltage across the collector to emitter terminal when the power BJT operates in (i) saturation (ii) cut-off (iii) active region?

2. Experiment on Test the proper functioning of IGBT

Aim:

Investigate the transfer and output characteristics of an n channel Insulated gate bipolar transistor.

Apparatus:

Regulated dc power supply 0-50V; Digital voltmeter/multi-meter 0-50V DC, Digital ammeter/multi-meter 0-1A; IGBT IC T_1-IRG4BC20U, Resistance R_D=220Ω, 5W; Resistance R_B=1MΩ,1W

Circuit Diagram:

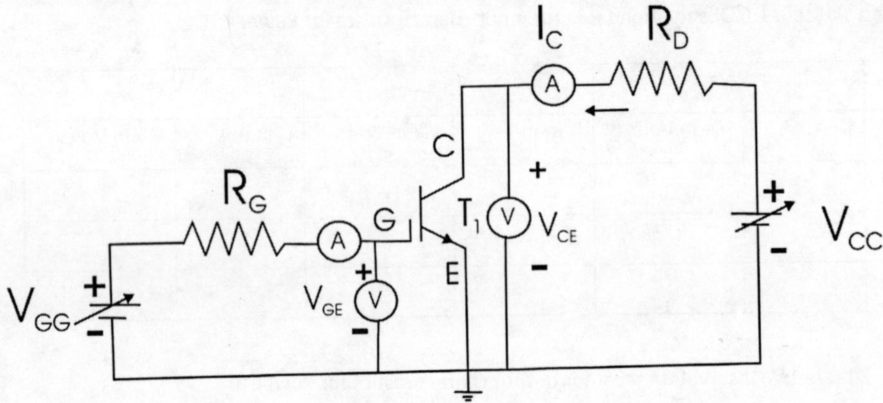

Fig.1.54 Circuit diagram for test the proper functioning of IGBT

Theory:

IGBT is a hybrid MOS gated turn-on and turn-off bipolar transistor. An IGBT has been developed by combining into all advantages of both power BJT and power MOSFET. It has high input impedance and low on-state conduction loss. IGBT is a multi-layer vertically oriented semiconductor structure with p type and n type doping as s discussed in section 1.6.1. This device has no second down problem like BJT and it has simple drive circuit, wide safe operating area, peak current capability, ruggedness and bipolar voltage blocking capability. IGBT IC-T_1 "IRG4BC20U" is optimized for high operating frequency 8-40 kHz, 600V and 6.5A. The transfer and output characteristics of IGBT is explained in section 1.6.2. IGBT are used in high-power applications such as UPS, SMPS, induction heating, power amplifiers, chopper and inverter circuits.

Precautions:

1. At first all the knobs of the power supplies are at zero value.
2. Identify gate (G), collector(C) and emitter (E) terminals of the IGBT
3. The applied voltage and current should not be more than the maximum rating of the IGBT
4. Use appropriate heat sink for proper cooling of IGBT.
5. Reading of observation should be error free.

Procedure:

A) Transfer Characteristics

1. Make the circuit connection as per the circuit diagram
2. Switch on regulated dc power supply
3. Set the voltage V_{CE} at specified value and gradually increase V_{GG} in steps of 1 Volts and measure I_C and V_{GE}
4. Using observation table, plot I_C versus V_{GE} curves at specified values of V_{CE} on the graph paper
5. Find the threshold voltage or the minimum gate voltage V_{GE} ($V_{GE(th)}$) which is required for conduction of IGBT

B) Output Characteristics

1. Make the circuit connection as per the circuit diagram
2. Switch on regulated dc power supply V_{GG} and set $V_{GE} \geq V_{GE(th)}$
3. Increase V_{DD} gradually in steps of 2 Volts and measure I_C and V_{CE}.
4. Repeat step 3 for different values V_{GE}
5. Using observation table, plot I_C versus V_{CE} curves at different values of V_{GE} on the graph paper

Observation Table:

Table 1.8 Observation table for transfer characteristics for IGBT

Sr.No.	V_{CE} = in Volts	
	I_C in mA	V_{GE} in Volts

Table 1.9 Observation table for output characteristics for Power BJT

Sr.No.	V_{GE}= in Volts		V_{GE}= in Volts		V_{GE}= in Volts	
	I_C in mA	V_{CE} in Volts	I_C in mA	V_{CE} in Volts	I_C in mA	V_{CE} in Volts

VIVA Questions Related to IGBT

1. Write specifications of IGBT from manufacturer data sheet
2. Describe the transfer and output characteristics of IGBT base on your observations
3. Calculate trans-conductance and output resistance

KNOW MORE

In this module, the construction, working principle, and V-I characteristics, switching characteristics, safe operating area (SOA) and series and parallel operation of power electronics devices namely Power bipolar junction transistor (Power BJT), Power metal oxide field effect transistor (Power MOSFET), and Insulated gate bipolar transistor(IGBT) are described in detail. Power BJT, Power MOSFET and IGBT are most commonly used in different converter and inverter circuits. To know more on applications of power semiconductor devices, please scan the following QR codes:

For more on Power BJT in DC to DC converter

For more on Power MOSFET Applications

For more on IGBT Module Applications

References:

1. Ramamoorty M., An Introduction to Thyristors and their applications, East-West Press Pvt. Ltd., New Delhi.

2. Sugandhi, Rajendra Kumar and Sugandhi, Krishna Kumar, Thyristors: Theory and

Applications, New Age International (P) ltd. Publishers, New Delhi.

3. Bhattacharya, S.K., Fundamentals of Power Electronics, Vikas Publishing House Pvt. Ltd. Noida.

4. Jain & Alok, Power Electronics and its Applications, Penram International Publishing (India) Pvt. Ltd, Mumbai.

5. Rashid, Muhammad, Power Electronics Circuits Devices and Applications, Pearson Education India, Noida.

6. Singh, M. D. and Khanchandani, K.B., Power Electronics, Tata McGraw Hill Publishing Co. Ltd, New Delhi.

7. Zbar, Paul B., Industrial Electronics: A Text –Lab Manual, McGraw Hill Publishing Co. Ltd., New Delhi.

8. Grafham D.R., SCR Manual, General Electric Co.

Dynamic QR Code for Further Reading

-QR codes embedded in the unit.

2 Thyristor Family Devices

UNIT SPECIFIES

Through this unit we have discussed the following aspects:

- SCR: construction, two transistor analogy, types, working and characteristics
- SCR mounting and cooling
- Types of Thyristors: SCR, LASCR, SCS, GTO, UJT, PUT, DIAC and TRIAC
- Thyristor family devices: symbol, construction, operating principle and V-I Characteristics
- Protection circuits: over-voltage, over-current, Snubber, and Crowbar.

The practical applications of the above topics are discussed for generating further curiosity and creativity as well as improving problem solving capacity of learners.

Not only a large number of multiple-choice questions are incorporated in this unit but also questions of short and long answer types are included in two categories following lower and higher order of Bloom's taxonomy, and assignments through a number of numerical problems. A list of references and suggested readings are illustrated in this unit so that learners can go through them for practice and get detailed knowledge. It is also be noted that some QR codes have been incorporated in different sections of this unit for getting more information on various topics of interest.

After discussion the content related to theory, there is a "Know More" section at end of this unit which is related to laboratory experiments. This section has been carefully designed so that the supplementary information provided in this section becomes beneficial for the learners. Usually, this section highlights the initial activity, examples of some interesting facts, analogy, and history of the development of thyristor family devices to focusing the salient observations and finding, timelines starting from the development of the concerned topics up to the recent time, applications of thyristor in our day-to-day life, and industrial applications of thyristor in power electronics converter circuits.

RATIONALE

This unit on thyristor family devices can be able to provide detail information regarding construction, two transistor analogy, types, working, V-I characteristics, mounting and cooling

of silicon-controlled rectifier (SCR) devices which are widely use in modern industry for the control and conversion of electrical power. This unit can help students to get knowledge about types of thyristors such as SCR, LASCR, SCS, GTO, UJT, PUT, DIAC and TRIAC, thyristor family devices: symbol, construction, operating principle and V-I characteristics. Turn-on and turn-off characteristics of thyristor are also discussed. The comparison between SCR, GTO, DIAC and TRIAC is incorporated in tabular manner so that students can be able to differentiate the SCR, GTO, DIAC and TRIAC very easily. Protection circuits such as over-voltage, over-current, snubber and crowbar for thyristor are explained in detail.

PRE-REQUISITES

Semiconductor physics

Analog and digital electronics

Electrical circuit theory

Electrical machines and power system

UNIT OUTCOMES

List of outcomes of this unit is as follows:

U2-O1: To explain construction and two transistor analogy of SCR

U2-O2: To describe working principle and V-I characteristics, mounting and cooling of SCR

U2-O3: To list types of thyristors

U2-O4: To discuss about construction, working principle, V-I characteristics, symbol of thyristor family devices

U2-O5: To select thyristor family devices for specific applications

U2-O6: To provide proper protection of converter circuits from over-voltage and over-current

U2-O7: To understand the operation of snubber and crowbar circuits

U2-O8: To troubleshoot turn-on and turn-off circuits of thyristors

U2-O9: To determine the latching current and holding current of SCR

Unit-2 Outcomes	EXPECTED MAPPING WITH COURSE OUTCOMES (*1-Weak Correlation; 2- Medium Correlation; 3-Strong Correlation*)				
	CO-1	**CO-2**	**CO-3**	**CO-4**	**CO-5**
U2-O1	3	-	-	-	-
U2-O2	3	2	-	3	-
U2-O3	3	-	-	-	-
U2-O4	3	2	-	3	-
U2-O5	3	-	-	-	3
U2-O6	2	3	3	-	-
U2-O7	2	3	3	-	-
U2-O8	2	3	3	-	-
U2-O9	-	3	-	-	-

2.1 INTRUDUCTION TO THYRISTOR FAMILY

The term "Thyristor" is a generic term which is applied to a family of power semiconductor devices such as silicon controlled rectifier (SCR), light activated SCR (LASCR), silicon controlled switch (SCS), gate turn off thyristors (GTO), UJT, PUT, DIAC and TRIAC etc. Thyristors are used extensively in power electronics circuits. These devices act as a bistable switch and operate in conducting and non-conducting states. The thyristor operates in conducting state when it is forward biased and current is supplied to the gate terminal. Among the above thyristor family devices, SCR is the most simplest in structure and most commonly used in power electronics converter circuits. SCRs are unidirectional devices as they can conduct current in one direction and they can be turned on by applying control signal and turned-off by natural commutation and forced commutation. GTOs are deigned to have both turn-on and turn-off facility. Triac and DIAC are bidirectional device. Thyristor was first developed in Bell Laboratories in 1957 and subsequently it was manufactured by General Electric Company, USA. SCRs are available for the rating of few volts (V) to several kilovolts (kV) and few amperes (A) to several kilo amperes (kA).

In this unit, construction of SCR, operating principle, V-I characteristics, two transistor analogy of SCR, switching characteristics, gate characteristics, ratings, and protection circuits are explained elaborately. The basic structure, operating principle and I-V characteristics of LASCR, SCS, GTO, UJT, PUT DIAC and TRIAC are also included in this unit.

2.2 CONSTRUCTION AND OPERATING PRINCIPLE OF SLILICON CONTROLLED RECTIFIER (SCR)

SCR is a four-layer semiconductor device. The basic structure of SCR and vertically oriented structure of SCR are shown in Fig.2.1 and Fig.2.2 respectively. It consists of four layer P-N-P-N structure of semiconductor materials. There are three junctions namely J_1, J_2 and J_3 and three external terminals such as anode (A), cathode (K) and gate (G).

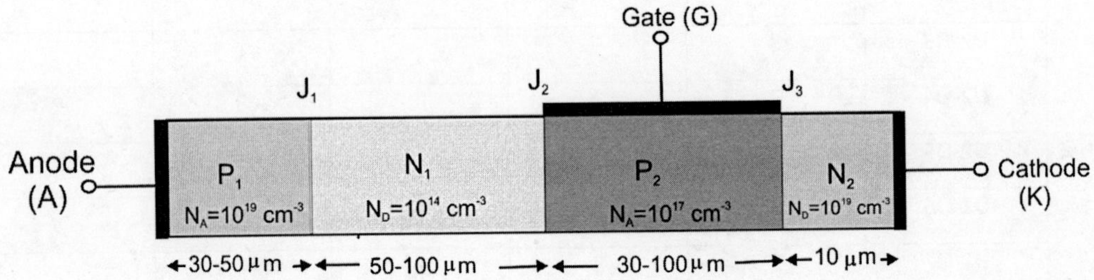

Fig.2.1 Basic structure of SCR

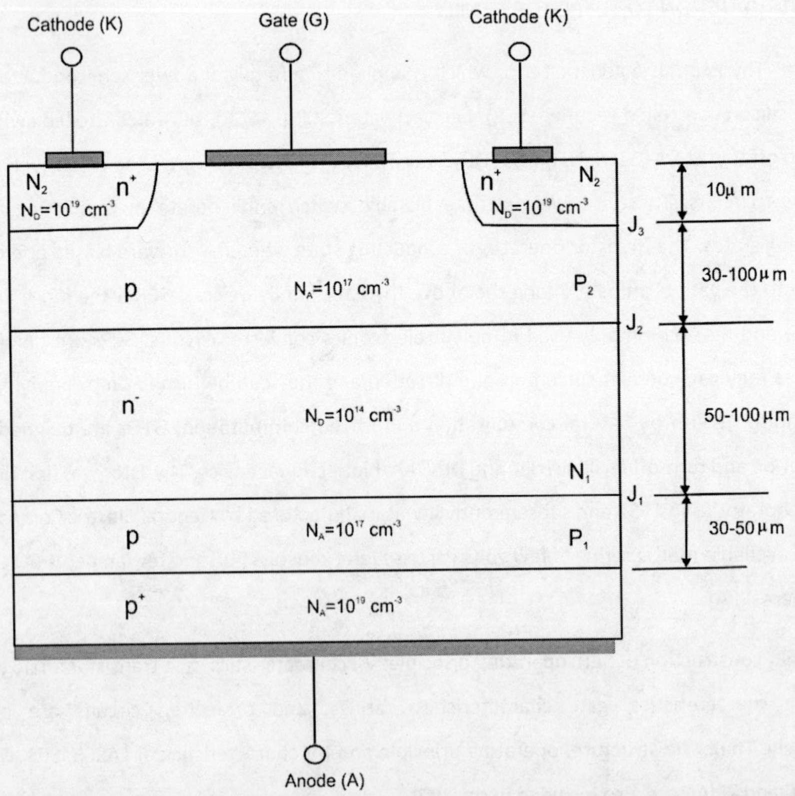

Fig.2.2 Vertically oriented structure of SCR

The width of each semiconductor layer and doping density are depicted in Fig.2.1. The P_1 and N_2 semiconductor layers are heavily doped. The width of P_1 *layer* is about 30-50μm and its doping density is about $N_A = 10^{17} cm^{-3}$. The width of N_2 is very small about 10μm and its doping density is $N_D = 10^{19} cm^{-3}$. The width of N_1 is large about 50 to 100μm but it is most lightly doped about $N_D = 10^{14} cm^{-3}$. The width of P_2 is about 30 to 100μm and its doping density is about $N_A = 10^{17} cm^{-3}$. The symbol of SCR is shown in Fig.2.3.

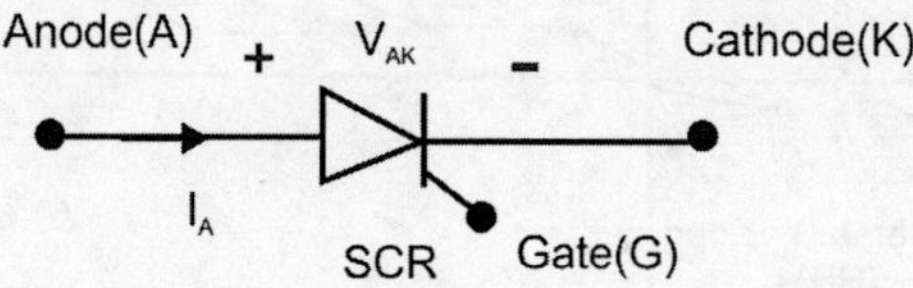

Fig.2.3 Symbol of SCR

When anode voltage is positive with respect to cathode, junctions J_1 and J_3 are forward biased and junction J_2 is reverse biased. Consequently, a small leakage current flows from anode to cathode and SCR operates in off-state. If anode to cathode voltage is increased to a large value, the reverse biased junction J_2 breaks down. As junctions J_1 and J_3 are forward biased, there are free movement carriers (electrons and holes) across all three junctions. Then SCR will be turned on and operates in conduction state, a large amount of forward anode current flows from anode to cathode.

For more on Thyristor

2.3 V-I characteristics of SCR

Fig.2.4 shows the V-I characteristics of a SCR. It is clear from Fig.2.4 that SCR has three operating regions such as *forward conduction state*, *forward blocking state* and *reverse blocking state.*

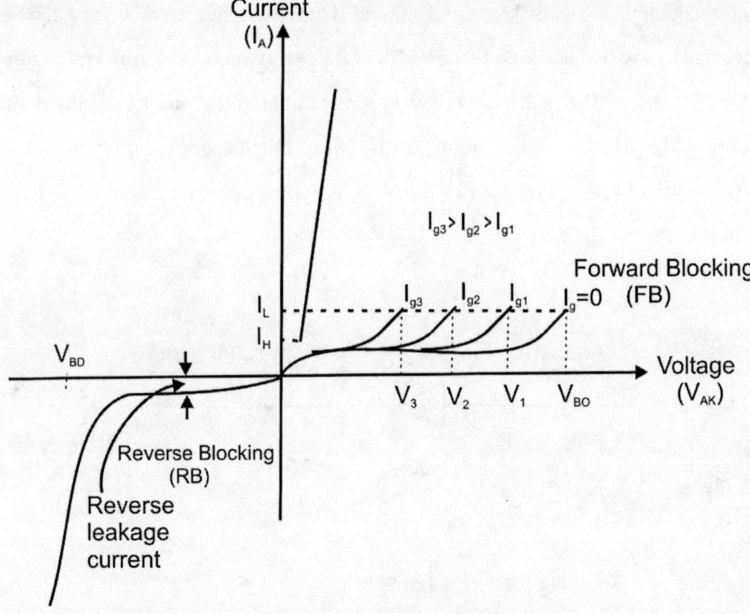

Fig.2.4 V-I characteristics of SCR

2.3.1 Forward Blocking (FB) State

In FB state, anode (A) is positive with respect to cathode (C). During this state, two junctions J_1 and J_3 are forward biased, however junction J_2 is reverse biased.

The forward biased junction depletion layer in J_1 and J_3 and the reverse biased junction depletion layer in J_2 changes with doping density and bias voltage. The width of forward biased junction depletion layer is thin but the width of reverse biased junction depletion layer is thick. If voltage applied across anode to cathode is positive, junction J_1 and J_3 are forward biased and junction J_2 block the input voltage.

The forward blocking state of SCR is shown in Fig.2.5. When the switch S is opened, current will not flows through gate and gate current $i_g = 0$. If anode to cathode voltage is increased progressively, depletion layer across J_2 will be increased. As junction J_2 is reverse biased, forward leakage current flows through from anode (A) to cathode (K). The amplitude of current is few mA. This current flows due to thermally generated minority carriers. When anode to cathode voltage V_{AK} increased to the forward break over voltage V_{BO}, junction J_2 breaks down. The depletion layer of N_1 increases and it can contact to opposite end of the depletion layer at J_1. Hence breakdown is possible through punch-through and this break down is called as *avalanche breakdown.* Then anode current I_A increases sharply to a very high value and the voltage across anode to cathode reduce sharply to a low value called as *on-state voltage*. The switching of SCR from turned-off sate to conduction state is done within a short time and this state is known as *transition state*.

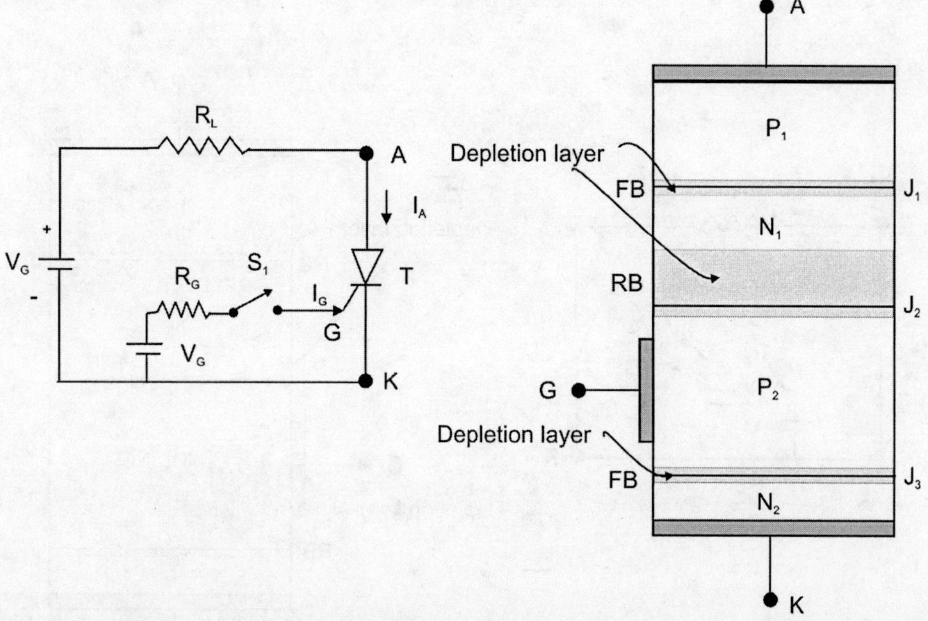

Fig.2.5 Forward blocking state of SCR with forward bias of J_1 and J_3 and reverse bias of J_2

2.3.2 Forward Conduction (FC) State

If SCR is forward biased and the gate current is applied through gate terminal (G), the junction J_2 breaks down. Subsequently all four P-N-P-N layers are filled with charge carriers. Then SCR give very low impedance between A and K and anode current I_A flows through anode to cathode. Amplitude of I_A depends on load resistance R_L. Usually gate voltage V_G varies with in few volts. This state is called as *forward conduction state*. When SCR operates in conduction state, the on-state voltage drop across anode to cathode is equal to about 1V.

2.3.3 Reverse Blocking (RB) State

When anode terminal is negative with respect to cathode, SCR is reverse biased. Then junctions J_1 and J_3 are reverse biased and junction J_2 is forward biased as shown in Fig. 2.6. Since N_1 layer is lightly doped, the reverse biased depletion layer at junction J_2 is large but the depletion layer width of J_2 and J_3 are thin. Consequently, the reverse leakage current flows from cathode (K) to anode (A). If the applied voltage becomes more than break down voltage V_{BD}, the junction breaks and large current flows in reverse direction. This state is known as *reverse blocking state*. The operation in reverse breakdown region should be avoided to protect the device from damage permanently.

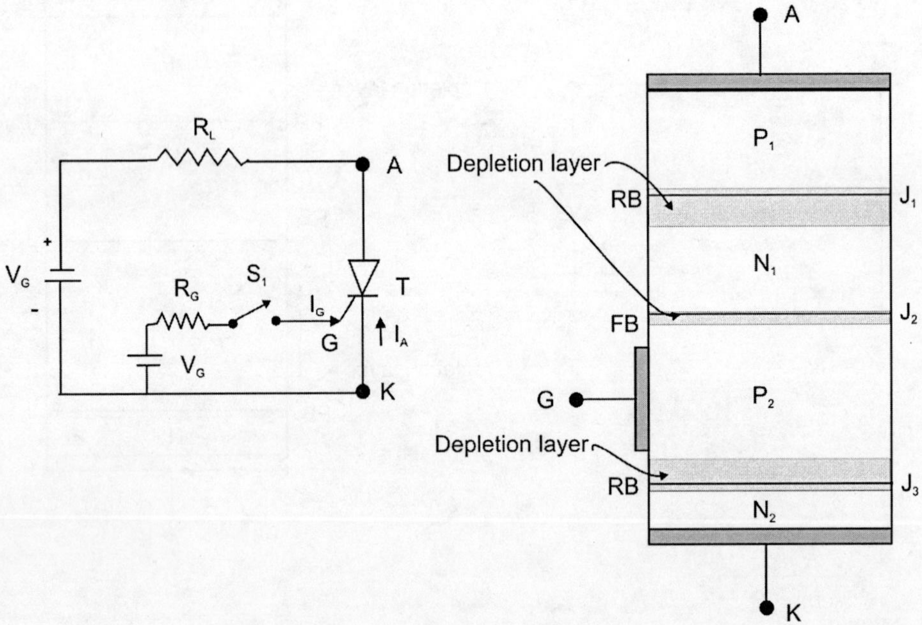

Fig.2.6 Reverse blocking state of SCR with reverse bias of J_1 and J_3 and forward bias of J_2

2.4 TWO TRANSISTOR ANALOGY OF SCR

The PNPN structure of SCR can be represented by two transistors T_1 and T_2 as depicted in Fig.2.7. Here, T_1 is a PNP transistor and T_2 is a NPN transistor. The collector terminal of T_1 is connected with base of T_2. When the gate current flows through gate terminal (G), base current flows through the base of T_2. After that the collector current of T_2 is applied to base of T_1 which produce very large collector current I_{C1}. Subsequently the sum of gate current (I_G) and collector current (I_{C1}) flow through the base of T_2. It means that the amplified collector current I_{C1} is applied to the base of T_2. Therefore, the regenerative operation is performed and SCR operates in *on-state* (conduction state) with in few μs. The connection of *p-n-p* and *n-p-n* transistors to model a SCR is illustrated in Fig.2.8.

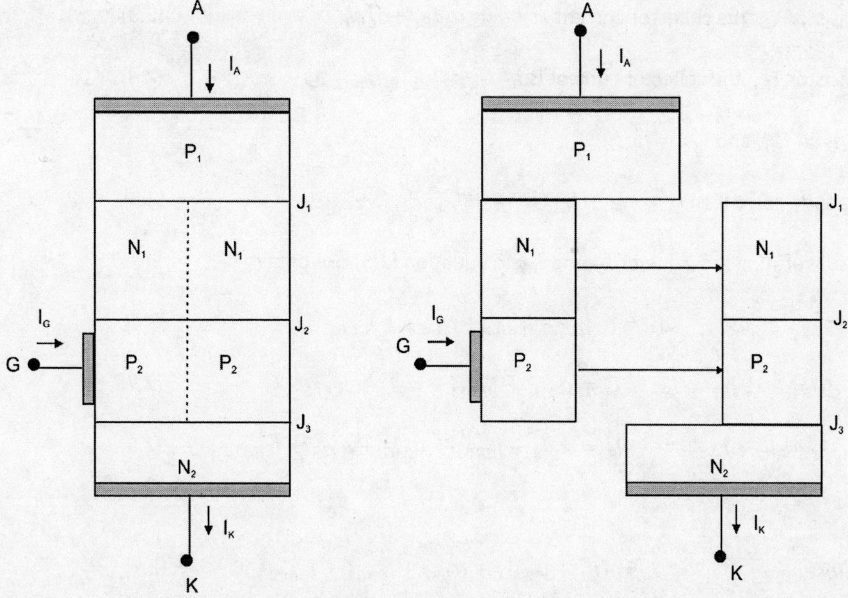

Fig.2.7 Two transistor model of SCR

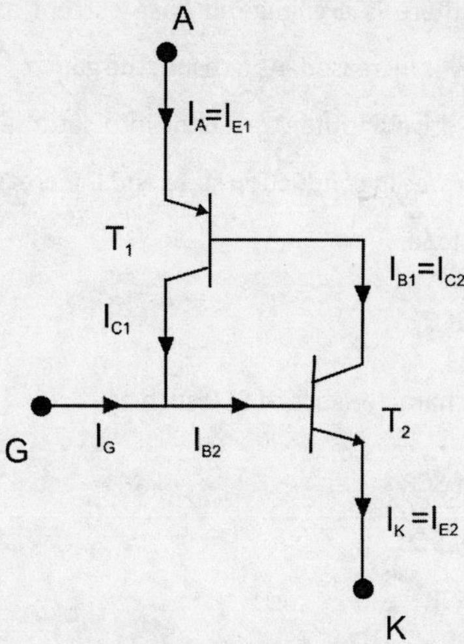

Fig.2.8 Connection of *p-n-p* and *n-p-n* transistors to model a SCR

In a power BJT, the collector current I_C is related with emitter current I_E, the leakage collector current of collector-base junction I_{CBO} and the dc current gain α as follows:

$$I_C = \alpha I_E + I_{CBO} \tag{2.1}$$

The base current of T_2 is, $I_{B2} = I_{C1} + I_G$ (2.2)

In transistor T_1, the collector current is $I_{C1} = \alpha_1 I_{E1} + I_{CBO1}$ (2.3)

In transistor T_2, the collector current is $I_{C2} = \alpha_2 I_{E2} + I_{CBO2}$ (2.4)

Here, $I_A = I_{E1}$, and $I_{E2} = I_K$

Then anode current is $I_A = I_{B1} + I_{C1} = I_{C2} + I_{C1}$ (2.5)

After substituting the value of I_{C1} and I_{C2} in equation (2.5), we obtain

$$I_A = \alpha_2 I_{E2} + I_{CBO2} + \alpha_1 I_{E1} + I_{CBO1} \quad (2.6)$$

As $I_A = I_{E1}$ and $I_{E2} = I_K$, $\quad I_A = \alpha_2 I_K + I_{CBO2} + \alpha_1 I_A + I_{CBO1}$ (2.7)

Since $I_K = I_G + I_A$, $\quad I_A = \alpha_1 I_A + I_{CBO1} + \alpha_2 I_G + \alpha_2 I_A + I_{CBO2}$ (2.8)

Therefore, $\quad I_A = (\alpha_1 + \alpha_2) I_A + \alpha_2 I_G + I_{CBO1} + I_{CBO2}$ (2.9)

Finally anode current is equal to $\quad I_A = \frac{\alpha_2 I_G + I_{CBO1} + I_{CBO2}}{1 - (\alpha_1 + \alpha_2)}$ (2.10)

If a small gate current I_G is supplied, I_{E2} increases and current gain α_2 increases. Since α_2 is increased, I_{C2} increases. In the same way, there is a change in base current as $I_{B1} = I_{C2}$. Subsequently, I_{E1} increases as $I_{E1} = I_{C1} + I_{B1}$ and α_1 is increased. As a result, the gain $\alpha_1 + \alpha_2$ becomes about 1 (*unity*). Consequently, one transistor can drive another transistor into saturation. Therefore, both T_1 and T_2 operate in saturation and SCR operates in conduction state and I_A increases extensively. The amplitude of anode current is limited by the load.

2.5 Switching Characteristics of SCR

There are two types of switching characteristics of SCR such as

- Turn-on characteristics of SCR
- Turn-off characteristics of SCR

2.5.1 Turn- on characteristics of SCR

The turn on characteristics of SCR is depicted in Fig.2.9 which shows the waveform of anode current with respect to gate signals (gate voltage V_G and gate current I_G) . When SCR is forward biased and a positive gate pulse is applied between gate and cathode, it will be turned on but there is a finite delay for transition from forward off-state to forward on-state. This finite delay time is called as *turn-on time* $t_{turn-on}$. The $t_{turn-on}$ time is sum of the *delay time* t_d, *rise time* t_r and *spread time* t_p and it is equal to $t_{turn-on} = t_d + t_r + t_p$. The *total turn on time* of SCR is equal to about 1 to $4\mu s$.

Delay time t_d: The *delay time* t_d is the time interval from initial anode current (forward leakage current) to attain 10% of final value of anode current $(0.1I_A)$. The delay time t_d can also be measured from the instant of the gate current reaches $0.9I_G$ to the instant where anode current reaches to $0.1I_A$.

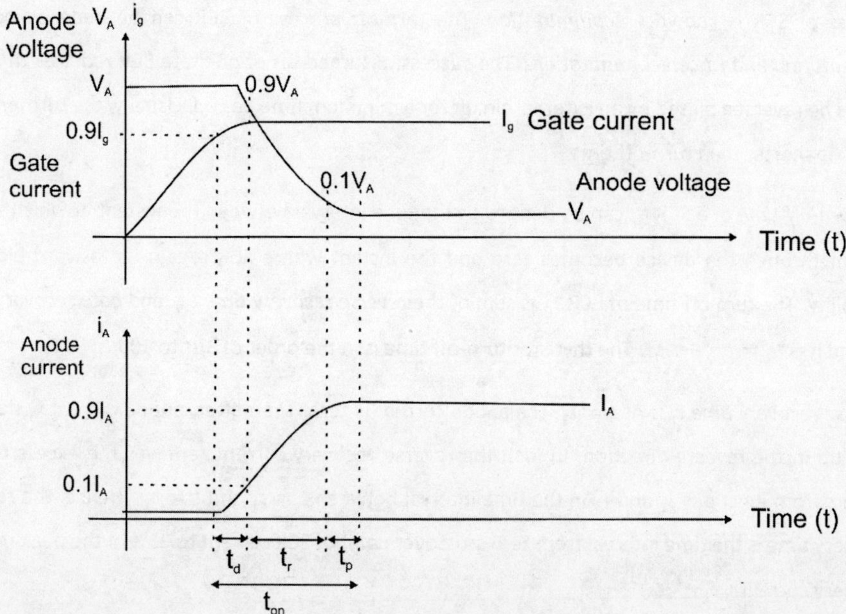

Fig.2.9 Turn-on characteristics of SCR

Rise time t_r: The *rise time* t_r is the time interval during which anode current I_A increases from 10% of on-state current ($0.1I_A$) to 90% of on-state current ($0.9I_A$). t_r is the time interval during which forward blocking off-state voltage decreases from the instant of 90% of forward blocking off-state voltage $(0.9V_A)$ to the instant of 10% forward blocking off-state voltage $(0.1V_A)$. It can be measured from the instant of the gate current reaches $0.9I_G$ to the instant of final value of gate current (I_G). The phenomenon of decreasing anode voltage and increasing anode current is completely dependent upon the type of the load

Spread time t_p: The *spread time* t_p is the time interval during which the forward blocking voltage falls from 10% of its value $(0.1V_A)$ to the on-state voltage drop about 1V. It can be defined as the time interval between the anode current rise from $0.9I_A$to I_A. In this time interval, the conduction spreads over the entire cross section of SCR cathode. t_p depends on the area of cathode as well as gate structure of SCR.

2.5.2 Turn-off characteristics of SCR

The turn off characteristics of SCR is shown in Fig.2.10. When a SCR is in conducting state and it can be turned off if the anode current is reduced gradually less than holding current I_H. This turn-off process of SCR is known as *commutation*. The turn-off process of SCR can be done by *natural commutation* and *forced commutation*. The successful turned-off of on-state SCR requires that SCR should be reversed biased by an external circuit for a minimum time period. Usually, the turn-on time of SCR is shorter than turn-off time.

The *turn-off time* of a SCR t_q can be defined as the time interval between the instant at which anode current through the device becomes zero and the instant where SCR regain its forward blocking capability. The turn off time of SCR t_q is sum of the *reverse recovery time* t_{rr} and *gate recovery time* t_{gr} and it is $t_q = t_{rr} + t_{gr}$. The thyristor turn-off time is in the order of $3\mu s$ to $100\mu s$.

Reverse recovery time t_{rr}: At $t = t_1$, the anode current is zero. Thereafter, anode current I_A starts to build up in the reverse direction. In fact, the reverse recovery current removes the excess charge carriers from junctions J_1 and J_3 in the time interval between $t = t_1$ and $t = t_3$. Hence, the reverse recovery time is the time interval from zero crossover point of SCR current to 25% of the peak reverse recovery current I_{rr}.

In the beginning, the rate of decrease of reverse recovery current is very fast however thereafter it is gradually reduced. SCR can be damaged permanently due to fast decay of reverse recovery current. This problem can be resolved by using snubber circuit.

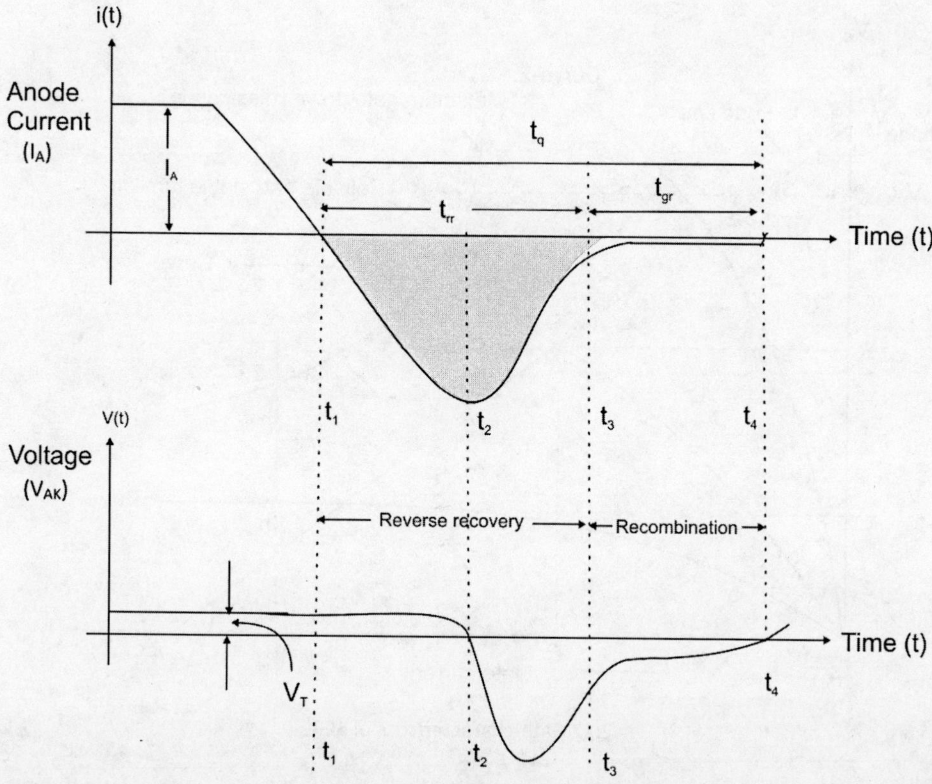

Fig.2.11 Turn-off characteristics of SCR

Gate recovery time t_{gr}: It is the time required for the recombination of excess charges in J_2 junction. When a negative reverse voltage is applied across SCR, *gate recovery time* t_{gr} can be reduced significantly.

2.6 GATE CHARACTERISTICS

Fig.2.12 shows the gate triggering characteristics which is used to design the gate triggering circuit for SCR and to estimate amplitude of gate current which will be applied to turn on SCR effectively. Curve 1 and curve 2 are the gate characteristics of SCR at minimum and maximum temperature respectively.

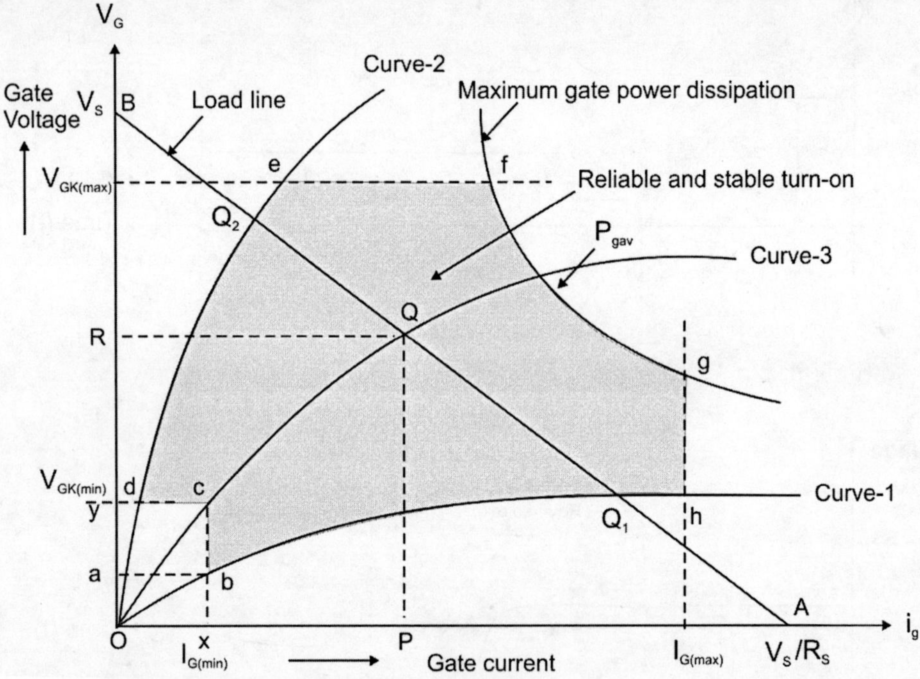

Fig.2.12 Gate-characteristics of SCR

The gate characteristics are widely spread between curve-1 and curve-2 due to inadvertent difference in doping density of P_2 and N_2 layers. The curve-1 represents the minimum gate voltage values which are applied to gate-cathode of SCR for turn-on adequately. In the same way, the curve-2 corresponds to the maximum possible gate voltage values which are also applied to gate-cathode of SCR for turn-on satisfactorily.

Fig.2.13 shows a triggering circuit of an SCR where V_s is gate source voltage, R_s is resistance of source voltage, V_g is gate to cathode voltage, and i_g is gate current. Here, $V_s = i_g R_s + V_G$. The value of internal resistance R_s must be such that the current $\frac{V_s}{R_s}$ should not be very high so gate drive circuit operates safely during turn-on of SCR. When R_s is very small, an external resistance R_1 must be connected across gate-cathode terminals of SCR.

If minimum gate voltage V_{gmin} is applied between gate and cathode, minimum gate current I_{gmin} flows from gate to cathode. The current flows through resistance R_1 is equal to V_{gmin}/R_1 and the source voltage is equal to

$$V_s = V_{gmin} + \left(i_g + \frac{V_{gmin}}{R_1}\right) R_s \qquad \text{as } V_G = V_{gmin} \qquad (2.11)$$

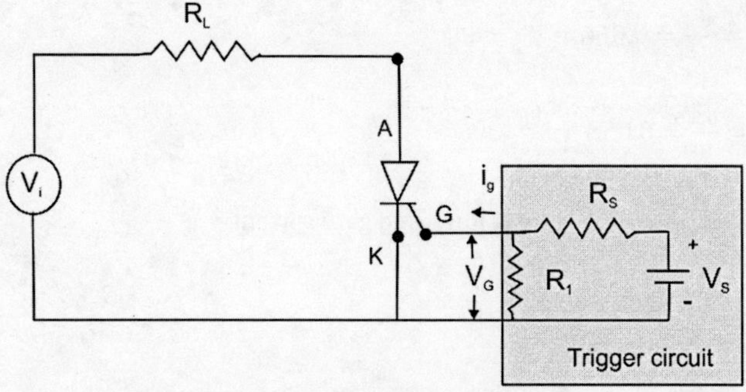

Fig.2.13 Triggering circuit of SCR

A-B is the load line of gate-cathode circuit. OB is source voltage V_S and OA is current $\frac{V_S}{R_S}$ while triggering circuit is short circuit. Consider that SCR operates at point Q which is the intersection of load line and curve-3. At point Q, OR=QP is gate voltage and the gate current is OP. The intersection between load line and curve-1 and curve-2 are Q_1 and Q_2 respectively.

SCR can operate within operating points Q_1 and Q_2 as operating point must lie within the limit curve-1 and curve-2. The slope of load line AB is $\frac{OB}{OA} = R_S$. The maximum gate to cathode voltage is V_{GKmax} and maximum gate current is I_{Gmax}. During design of gate drive circuit, we should not exceed these values to save the gate junction due to excessive power dissipation. In the same way, the minimum gate to cathode voltage is $V_{GKmin}$and minimum gate current is I_{Gmin} for successful turn on of SCR.

Fig.2.12 shows the variation of maximum gate power dissipation. The operating point should not be outside the maximum power dissipation curve. The shaded region in Fig.2.12 shows the safe operating region and reliable to turn-on of SCR. The operating point of gate drive circuit should be in this region.

Example 2.1 The average gate power dissipation of a SCR is $P_{gav} = 0.5\ Watt$. When the gate voltage varies from 2.5 V to 10 V, plot the curve where gate voltage is a function of gate current.

Solution:

Given: $P_{gav} = 0.5\ Watt,$ and V_g varies from 2.5 V to 10 V

Assume four gate voltages $V_{g1} = 2.5V, V_{g2} = 5V, V_{g3} = 7.5V$ and $V_{g4} = 10V$

The corresponding gate currents are

$$I_{g1} = \frac{P_{gav}}{V_{g1}} = \frac{0.5}{2.5} = 0.2A \qquad \text{as } V_{g1}I_{g1} = P_{gav}$$

$$I_{g2} = \frac{P_{gav}}{V_{g2}} = \frac{0.5}{5} = 0.1A$$

$$I_{g3} = \frac{P_{gav}}{V_{g3}} = \frac{0.5}{7.5} = 0.066A$$

$$I_{g4} = \frac{P_{gav}}{V_{g4}} = \frac{0.5}{10} = 0.05A$$

Fig. 2.14 shows the plot between gate voltage and gate current

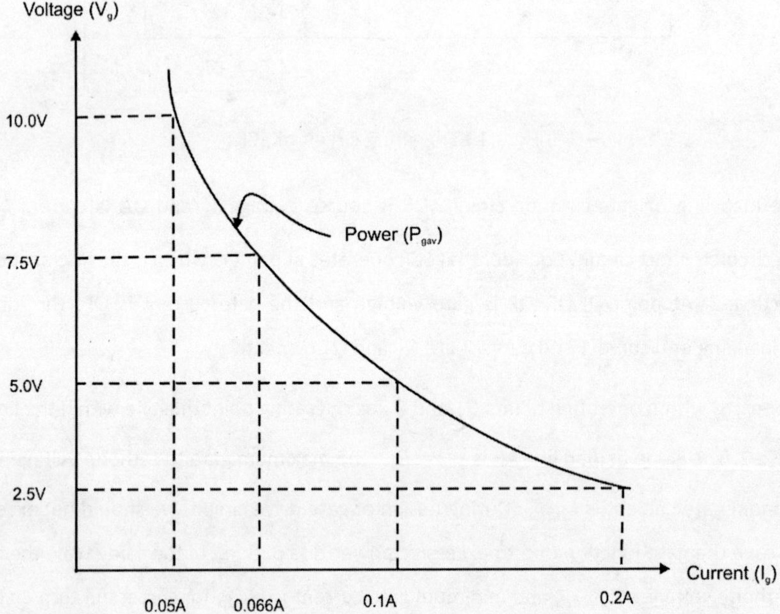

Fig.2.14 Gate voltage V_g vs gate current I_g

Example 2.2 Assume that slope of the gate-cathode characteristics of a thyristor is about 125 and average power dissipation is 0.8 Watt. If the gate to source voltage is about 12V, determine the gate source resistance.

Solution: Given: $\frac{V_g}{i_g} = 125,\ P_{gav} = 0.8\ Watt,\ V_s = 12\ V$

As $V_g I_g = P_{gav}$, $125\ i_g^2 = 0.8$

Therefore, $i_g = 0.08A$

Then gate voltage $V_g = 125\ i_g = 125 \times 0.08 = 10\ V$

The gate to source voltage is $V_s = i_g R_s + V_g$

Therefore, the value of gate to source resistance is equal to $R_s = \frac{V_s - V_g}{i_g} = (12 - 10)/0.08 = 25\ \Omega$

Example 2.3 The gate-cathode characteristics of a thyristor are spread by the following equations:

$$i_g = 2.3 \times 10^{-3} V_g^2 \ \ and \ i_g = 1.5 \times 10^{-3} V_g^{1.5}$$

When the gate source voltage is about 15 V and $R_s = 100\,\Omega,$ determine the triggering voltage and triggering current. Assume the gate power dissipation is 0.5Watt.

Solution: Given $P_{gav} = 0.5\,Watt, V_s = 15\,V$

We know that $V_g i_g = P_{gav} = 0.5\,Watt$

Or, $V_g = 0.5/i_g$

The source voltage $V_S = i_G R_S + V_g$

Or, $15 = 100 i_g + \frac{0.5}{i_g}$

Or, $100 i_g^2 - 15 i_g + 0.5 = 0$

After solving the above equation, we get $i_g = 0.1\,A \ or \ 0.05\,A$

According to $i_g - v_g$ characteristics

$$i_g = 2.3 \times 10^{-3} V_g^2 \ = 2.3 \times 10^{-3} \frac{0.5^2}{i_g^2}$$

Therefore, $i_g = 83.15\,mA\,,\ v_g = \frac{0.5}{i_g} = \frac{0.5}{83.15\,x\,10^{-3}} = 6.01\,V$

As $i_g = 1.5 \times 10^{-3} V_g^{1.5} = 1.5 \times 10^{-3} \frac{0.5^{1.5}}{i_g^{1.5}}\ ,\ i_g = 48.95\,mA$ and $v_g = \frac{0.5}{i_g} = \frac{0.5}{48.95\,x\,10^{-3}} = 10.21\,V$

Since the gate current will be in between $i_g = 83.15\,mA$ and $i_g = 48.95\,mA,$ the gate current at operating point Q, $i_g = 50\,mA$

As $i_g = 50\,mA\,,\ v_g = \frac{0.5}{i_g} = \frac{0.5}{50\,x\,10^{-3}} = 10\,V$

Hence, the operating point Q is (50mA, 10V)

Example 2.4 The gate-cathode characteristics of a thyristor is a straight line passing through origin with a gradient of $\frac{V_g}{i_g} = 13\ V/A$. The maximum turn-on time is $100\mu s$ and the minimum gate current required is 80mA. If the gate to source voltage is 10V, (i) determine the value of gate-source resistance which is connected in series with gate drive circuit as shown in Fig.2.15 and (ii) compute the power dissipation.

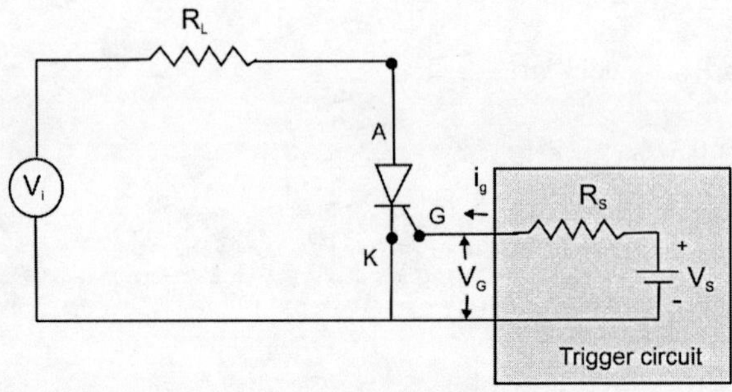

Fig.2.15 Tigger Circuit

Solution: Given: $\frac{V_g}{i_g} = 13\ V/A,\ i_{gmin} = 80\ mA,\ V_s = 10V$

The value of gate voltage is $V_g = 13 i_g$ as : $\frac{V_g}{i_g} = 13\ V/A$

Or, $V_g = 13\ x\ 80\ x\ 10^{-3} = 1.04\ V$

(i)The value of gate-source resistance which is connected in series with gate drive circuit is

$$R_s = (V_s - V_g)/I_g) = (10\text{-}1.04)/80\ x10^{-3} = 112\ \Omega$$

(ii) Power dissipation $P_g = V_g i_g = 1.04\ x\ 80\ x\ 10^{-3} = 83.2\ mWatt$

2.7 SCR Mounting and Cooling

Heat is generated within SCR due to on-state loss when it operates in conduction state and switching loss due to transition from OFF to ON and ON to OFF. This heat should be transferred from SCR to a cooling medium to maintain junction temperature with in limit. Transfer of heat is possible by conduction, convention, and radiation, natural cooling and forced air cooling.

SCR should be mounted on the heat sink which is used as a cooling medium. A wide variety of aluminium heat sinks are commercially available and sometimes cooling fins are used to increase the heat transfer capability. Fig.2.16 shows heat sinks which are used for mounting SCRs. In this section we explain about different methods of cooling of SCR such as natural cooling, forced air cooling, and forced liquid cooling.

Fig.2.16 Heat sinks are mounting with SCRs.

Natural Air Cooling: In this method, heat is transferred from heat sink to atmosphere. Usually heat sinks are mounted vertically for proper cooling. If heat sinks are mounted horizontally, it will be derated by about 20%.

Forced Air Cooling: The effectiveness of heat sink can be improved by using forced air cooling. When we use blower, the volume of air flowing over the heat sink increased significantly. Consequently, forced air can able to transfer more heat from device. The size of heat sink can be reduced due to forced air cooling.

Forced Liquid Cooling: SCR may be cooled by heat pipes which are partially filled with low pressure liquid. Actually, the SCR is mounted on one side of the pipe and other side is connected with heat sink for condensing mechanism. The heat generated by SCR vaporizes the liquid and subsequently vapour flows to the condensing end, where it condenses and the liquid returns to the heat source. Fig.2.17 shows the construction of heat pipes for cooling.

In high power applications, the forced liquid cooling system is highly efficient. Water and oil are generally used as cooling liquids. The water cooling is very efficient and more effective compared to the oil cooling. However, distilled water is used as coolant to minimize corrosion and an antifreeze agent to avoid freezing.

Since oil is flammable, the oil cooling is restricted to only very few applications. Oil cooling provides good insulation and eliminates the problems of corrosion and freezing. Heat pipes and liquid cooed heat sinks are available in market.

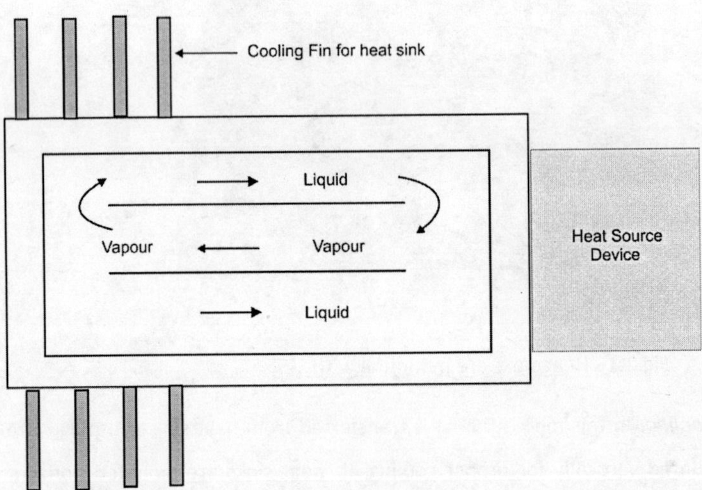

Fig.2.17 Construction of heat pipes for cooling

For more on Heat sink

2.8 RATINGS OF SCR

Power handling capability of SCR is limited by voltage and current ratings. In fact the power handling capability of SCR depends on the temperature withstand capacity of *p-n* junction at steady state and dynamic conditions. SCR should always be operating within the safe operating area (SOA) when it operates within voltage, current, power and temperature limit. Generally SCR ratings are specified in the manufacturer data sheet. In this section, some very useful specification of SCR voltage and current ratings are explained.

On State Voltage Drop (V_T): It is the voltage drop across anode and cathode with specified forward on state current and junction temperature. Its value is about 1V to 1.5V.

Peak Forward Blocking Voltage (V_{DWM}) : This is the maximum forward blocking voltage at which SCR can withstand during its working as depicted in Fig.2.18.

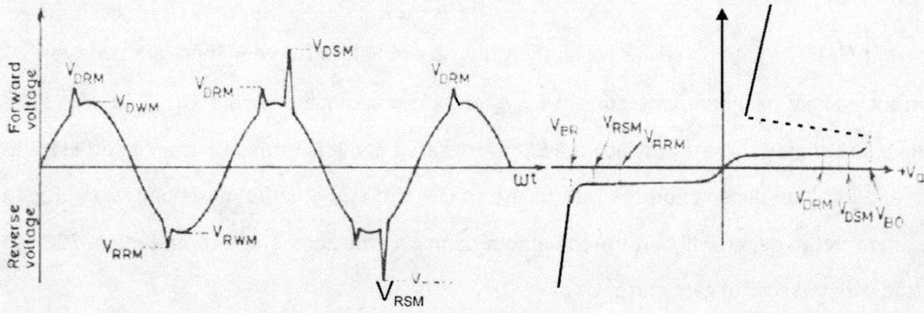

Fig.2.18 Voltage rating of SCR

Peak repetitive forward blocking voltage (V_{DRM}): This is the repetitive peak transient voltage at which SCR can withstand in it's forward blocking state. It is specified at a maximum allowable junction temperature when gate circuit is open or a specified biasing resistance is present between gate and cathode. V_{DRM} is obtained when a SCR is commutated or turned-off.

Peak reverse voltage (V_{RWM}): This is the maximum reverse voltage at which SCR can withstand repeatedly. It's value is equal to the peak negative value of ac voltage.

Repetitive peak reverse voltage (V_{RRM}): This is the repetitive maximum reverse voltage at which SCR can withstand at the allowable maximum junction temperature.

Voltage Safety Factor (V_{SF}): It is the ratio of peak repetitive reverse voltage (V_{RRM}) to maximum value of input voltage and it is equal to

$$V_{SF} = \frac{\text{Peak.repetitive. reverse.voltage.}(V_{RRM})}{\sqrt{2}.\text{RMS.value.of.operating.voltage}} \tag{2.12}$$

Generally, the voltage safety factor is about 2 to 3.

Latching Current (I_L): The latching current is the minimum value of anode current to turn on SCR from its off-state to on-state even after removal of trigger pulse. To trigger an SCR, the anode current must be build up to the latching current before the gate pulse is removed.

Holding Current (I_H): The holding current is the minimum value of anode current to hold the SCR in on-state. In turn-off process, the anode current must be below the holding current. Generally the value of holding current is few *mA*.

Gate Current (I_g): The gate current is required at the gate of SCR to turn on. There are two types of gate current namely minimum gate current i_{gmin} and maximum gate current i_{gmax}. i_{gmin} is the minimum value of gate current at which the SCR is triggered and it is turned on from its off-state to on-state and its value depends on the rate of rise of current. i_{gmax} is the maximum value of gate current at which the device will be turn-on without damaging the gate. The turn-on time of SCR can be reduced with increase of gate current.

Forward dv/dt rating: It is the maximum rate of rise of anode voltage at which thyristor will not be triggered when gate signal is not applied and anode to cathode voltage is less than forward break over voltage. In forward blocking mode, the applied voltage appears across the junction J_2 as junctions J_1 and J_3 are forward biased and junction J_2 is reverse biased. Then reverse biased junction J_2 behaves as capacitor. If a forward voltage is applied suddenly and dv/dt is very high, a charging current $i_c = C_j \frac{dv}{dt}$ starts to flow and the thyristor can be turned-on. Therefore, this type of unwanted triggering of thyristor must be avoided. Whenever dv/dt is less than forward dv/dt rating, thyristor must be remain in forward blocking state.

$\frac{di}{dt}$ **rating:** It is the maximum permissible rate of rise of anode to cathode current without any damage. The value of maximum rate of change of current for a particular thyristor device is always specified at its highest value of junction temperature that it can safely bear. If the rate of rise of anode current is very rapid compared to the spreading velocity of carriers across the junctions during the turn-on period, the local hot spots are created due to high current density in the junction regions. This increases the junction temperature beyond the safe limit and thyristor may be damaged. Typical values of $\frac{di}{dt}$ lies in the range of 50-800 ampere/microseconds ($A/\mu s$).

2.9 Types of Thyristors

Thyristors are power semiconductor switching devices which are made up of four layers of alternating p and n-type semiconductor materials. These devices are used in converters, inverters, cycloconverters, choppers, ac voltage controllers, ac and dc drives and ac power control applications. The most commonly used thyristor family devices are

- Silicon Controlled Rectifier (SCR)
- Light Activated Silicon-Controlled Rectifier (LASCR)
- Silicon-Controlled Switch (SCS)
- MOS Controlled Thyristor (MCT)
- MOS Turn-off Thyristor (MTO)
- Gate Turn off Thyristor (GTO)
- Unijunction Transistor (UJT)
- Programmable Unijunction Transistor (PUT)
- DIAC
- TRIAC
- Static Induction Thyristor (SITH)

In this unit, the symbol, construction, operating principle and V-I characteristics of Silicon Controlled Rectifier (SCR), Light Activated Silicon-Controlled Rectifier (LASCR), Silicon-Controlled Switch (SCS), Gate Turn off Thyristor (GTO), Unijunction Transistor (UJT), Programmable Unijunction Transistor (PUT), DIAC and TRIAC are explained in detail.

2.10 LASCR (Light Activated Silicon Controlled Rectifier)

2.10.1 Construction of LASCR: Fig.2.19 shows the construction of LASCR. It is a four-layer PNPN semiconductor device. It is also a semiconductor opto-electronic switch, and it has a lens that focuses light on its gate. The silicon pellet is used in the bottom of LASCR, and the light intensity dislodges electrons in the semiconductor crystal and contributes to conduction.

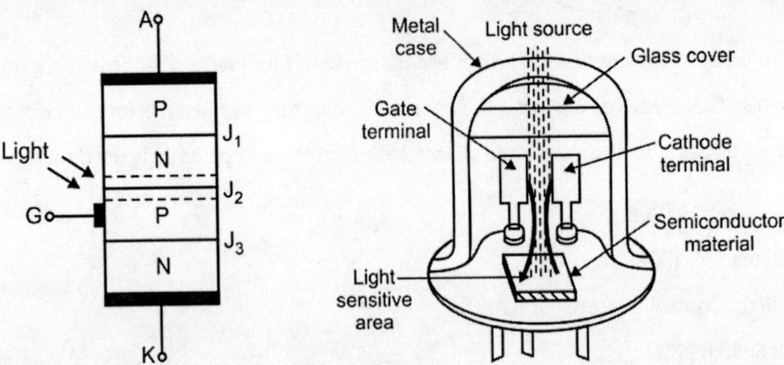

Fig.2.19 The basic construction of LASCR

It will be turns ON when it is exposed to light. This device is triggered into conduction state by light. When sufficient light does not fall on the LASCR, it remains in off-state. Therefore, LASCR is a type of thyristor that is triggered by photons present in the light rays.

LASCR is a three terminal device which consists of cathode, anode and gate terminal. The gate terminal is used while the triggering is provided to the LASCR. If a pulse of light with suitable wave length fall on the wafer of the SCR and the intensity of light exceeds a threshold value, electron-hole pairs are generated and a current flows anode to cathode. The advantage of triggering of SCR using light is prevention from electrical noise disturbances. Consequently, LASCR can be used as one of the useful devices in industrial applications.

2.10.2 Working Principle of LASCR: LASCR works on the principle of photoconduction as shown in Fig.2.20. Once light is focused on LASCR, the incident photons can generate electron-hole pairs. The number of optically generated electron-hole pairs is directly proportional to intensity of light. Therefore, a gate current flows into LASCR. The LASCR operates into conduction due to photon striking on the semiconductor surface. Fundamentally LASCR is a thyristor, and it is made up of semiconductor material. The light rays falling on LASCR are focused at one place to intensify it.

When the intensity of light increases, the more current flows through the LASCR. LASCR consists of two transistors in such a way that the collector of one transistor is connected to the base of another transistor. LASCR does not get turn off even when the supply of external light is stopped. To turn off the LASCR, we require to reverse the polarity of anode and cathode (A and K).

Though LASCR is a light-triggered device, the combination of light signal and the electrical signal may be used to operate device in conduction for two conditions. Firstly, when the lower strength electrical signal is applied to gate terminal, the intensity of light required to dislocate electrons from the semiconductor wafer will be high. Secondly, when the higher strength electrical signal is applied to gate terminal, the intensity of light required to displace electrons from the semiconductor crystal will be low.

LASCR is most sensitive to light while its gate terminal is open. Its sensitivity can be controlled by inserting a resistor between its gate and cathode terminals. Actually, it provides complete electrical isolation between the source of triggering and device. The forward breakdown voltage decreases with increase in light intensity.

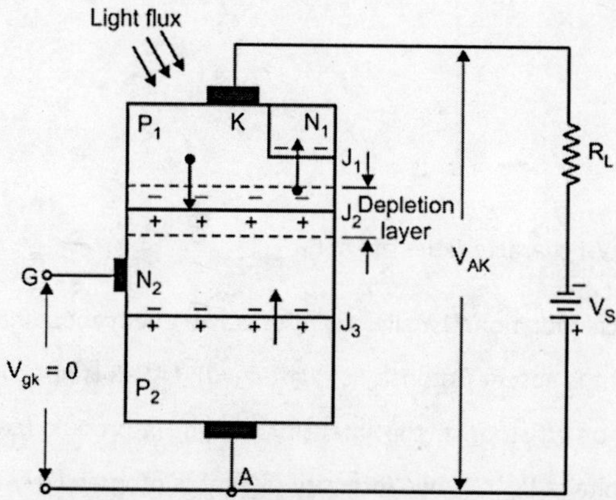

Fig.2.20 The basic operation of LASCR

The symbolic representation of LASCR is shown in Fig.2.21. The Light Activated SCR are used for low power applications, motor control, computer applications, solid state relay and optical light controls.

Fig.2.21 Symbol of LASCR

2.10.3 V-I Characteristics of LASCR: As the LASCR is photosensitive, it is provided with a glass top to allow the light input. It is also provided with a lens at the top to focus the light on the gate. It has three terminals such as anode, cathode and gate. It can be turned ON by direct radiation on the silicon wafer of gate with light. Fig.2.22 shows the V-I characteristics of LASCR.

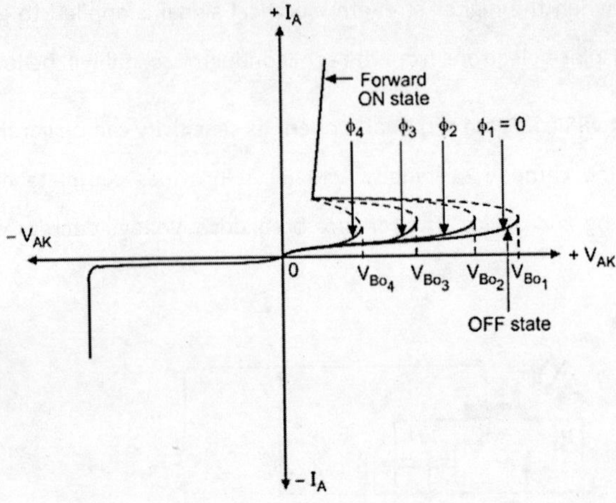

Fig.2.22 V-I characteristics of LASCR

LASCR can be triggered by optical radiation. Usually, a light of wave-length about 0.8 to 1.5 μm is used for triggering LASCR. When light is absent (intensity of light φ_1=0), LASCR is not conducting, and it is called *forward blocking state* or *off-state*. If the intensity of light is present (φ_2>0), LASCR starts conducting from brake-over voltage $V_{B_{O2}}$. While intensity of light is progressively increased from φ_2 (φ_4>φ_3>φ_2>0), LASCR starts conducting at lower break-over voltage ($V_{B_{O4}} < V_{B_{O3}} < V_{B_{O2}}$) as shown in Fig.2.22. It is clear from Fig.2.22 that larger in intensity of light on LASCR, the lower forward break down voltage is required to turn on the device. Usually a variable resistance is connected between gate and cathode to change the sensitivity of circuit due to light intensity.

2.11 SCS (Silicon Controlled Switch)

2.11.1 Construction of SCS: Silicon controlled switch (SCS) is a unilateral four-layer silicon device with four electrode such as anode, cathode, anode gate and cathode gate. It is a unidirectional device of thyristor family. The construction of SCS is shown in Fig.2.23. This device consists of four-layer PNPN semiconductors. It has three junctions like a SCR. The circuit symbol of SCS is shown in Fig.2.24 and it is available for industrial applications at low power ratings.

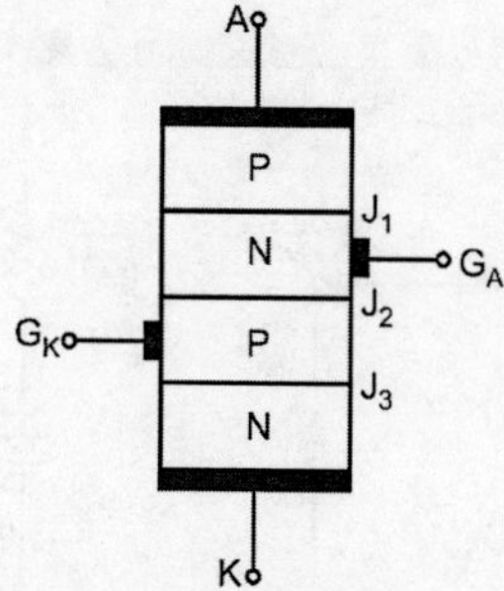

Fig.2.23 Basic construction of SCS

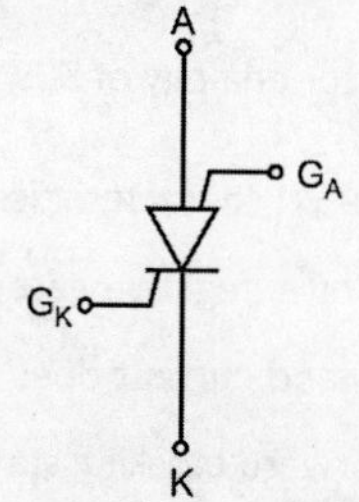

Fig.2.24 Symbol of SCS

2.11.2 Working Principle of SCS: Fig.2.25 shows the equivalent circuit of two transistor analogy of a SCS. SCS consists of a complementary pair of transistors with regenerative feedback. Like a SCR, this device will be turned on by applying a positive trigger pulse at G_K or a negative triggering pulse at gate G_A. The construction SCS is similar to that of a SCR.

The SCS is forward biased when a positive voltage is applied at the anode terminal with respect to cathode and there is no voltage at the gates G_K and G_A. Consequently, the gate current is zero, both transistors will be in off-sate. When a positive triggering pulse is applied at the terminal G_K, the gate current will flow though the base of transistor Q_1. As regenerative feedback action takes place, transistors Q_1 and Q_2 operate into saturation region and high current will flow from anode to cathode. This is called as *conduction* or *on-state* of SCS. SCS can be turned on by applying a negative pulse at G_A. Similarly, SCS can be turned-off by applying a positive pulse at G_A or a negative pulse at cathode gate terminal G_K. SCS can also be turned off just like SCR when anode current is less than the holding current I_H.

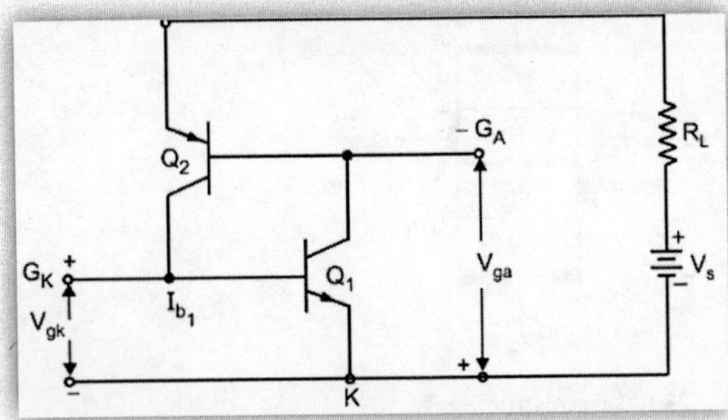

Fig.2.25 Equivalent circuit of two transistor analogy of SCS

2.11.3 V-I Characteristics of SCS: Fig.2.26 shows the V-I characteristics of SCS. During forward biased condition, a positive triggering pulse applied at G_K or a negative gate pulse voltage at the gate G_A is less than break over voltage, SCS will not conduct, and current does not flow. At that time leakage current flows through SCS. Hence SCS operates in forward blocking state. When SCS is forward biased, a positive triggering pulse voltage at G_K or a negative pulse voltage at the G_A is more than break over voltage, SCS will operate in *on-state* and current flows from anode to cathode. Subsequently the voltage across SCS drops to about 1V though anode current increases rapidly. This state of device is known as *forward conduction state* of SCS.

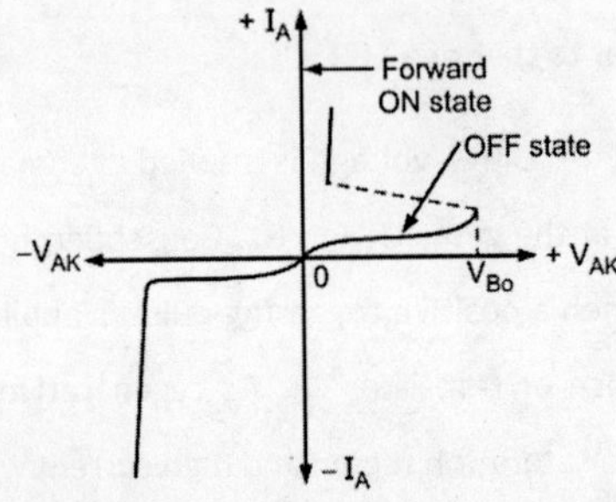

Fig.2.26 V-I characteristics of SCS

Advantages of SCS are (i) it has fast turn-off and (ii) it can be turned-off with positive or negative pulse at either gate.

The applications of SCS are (i) it is used in low power circuits, (ii) it is used in timers, registers and counters, (iii) it is used in digital logic circuits, (iv) it is used in pulse generators and (v) it is used in oscillators.

2.12 GTO (Gate Turn-off Thyristor)

SCR can be used as ideal switches in power electronics. When SCRs are forward biased and a gate pulse is applied between anode to cathode, these devices will be turned-on. Once SCR is turned-on and it operates in the on-state. These devices can also be turned-off by natural commutation and forced commutation. SCRs can be used as switching device up-to about 1 kHz. In off-state, SCRs can able to block high voltages about several kilo volts (kV) and during on-state, a large current about several kA flows from anode to cathode with a small on-state voltage. To incorporate the turn-off capability in SCRs, the structure of SCR should be modified. GTO is just like a SCR but the turn-off feature is incorporated within this device. When a positive gate current is applied in between gate and cathode, GTO will be turned on and operates in conduction state. When a negative gate current of required amplitude is applied to gate-cathode terminals, GTO will be turned off. As GTO has self-turned off capability, this device is suitable device for inverters and choppers.

2.12.1 Construction of GTO

GTO is a four layer *p n p n* semiconductor device and it has three terminals namely gate (G), cathode (K) and anode (A). Fig.2.27 shows the basic structure of GTO. The doping density of different layers of GTO is just like the doping density of different layers of SCR. In GTO, the thickness of P_2 layer is smaller compared to conventional SCR.

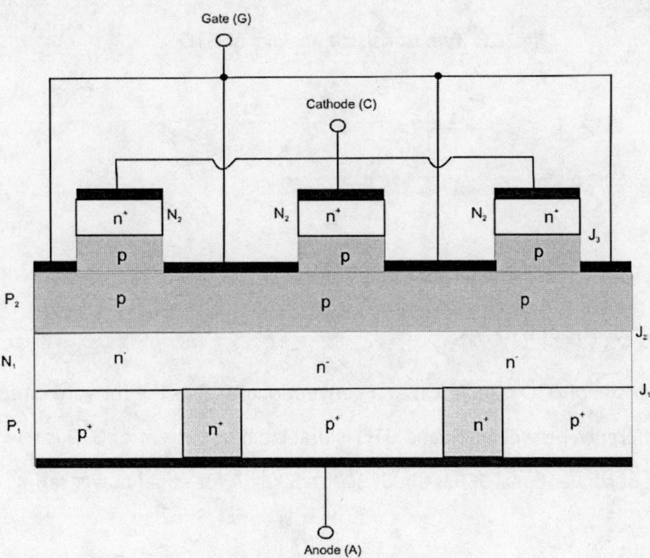

Fig.2.27 Construction of GTO

The n^+ region is penetrated in the p-type anode (P_1 layer) at regular interval so that the n^+ region makes contact with the n^- region (N_1 layer). Since n^+ regions are over laid with the same metallization contacts, the p-type anode is called anode short. Due to anode short structure, the turn-off process of GTO will be very fast.

Just like SCR, GTO can be represented by two transistor analogy (Fig.2.28). Transistor T_1 is p^+np^+ type transistor but transistor T_2 is np^+n^+ type transistor. The emitter of transistor T_1 (p^+) is anode of GTO and emitter of transistor T_2 (n^+) is cathode of GTO. The symbol of GTO is shown in Fig.2.29. The two-way arrow convention on the gate terminal will be used to differentiate GTO from a conventional SCR.

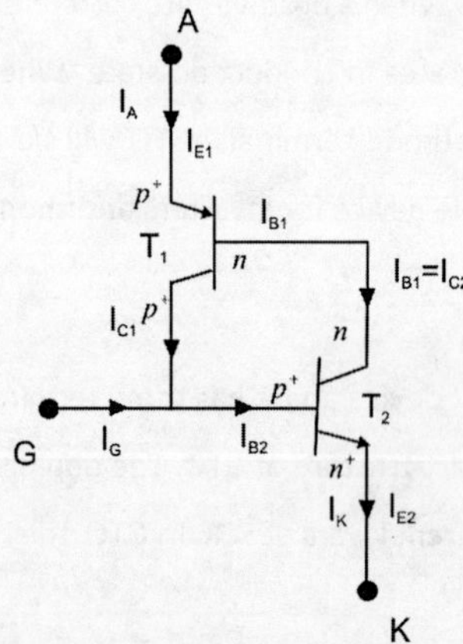

Fig.2.28 Two transistor analogy of GTO

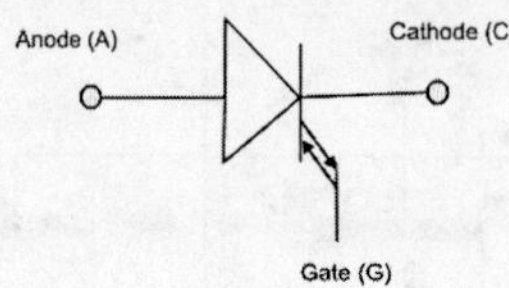

Fig.2.29 Symbol of GTO

2.12.2 V-I Characteristic of GTO

The V-I characteristic of GTO is identical to a conventional SCR in the forward direction as shown in Fig.2.30. The difference between SCR and GTO is that latching current of GTO is few amperes but the latching current of conventional SCR is about 100 to 500 mA for equal power rating.

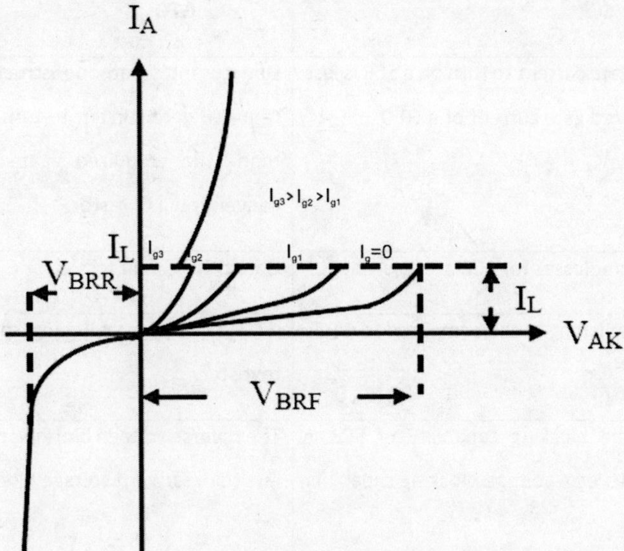

Fig.2.30 V-I characteristic of GTO

If the gate current is not able to turn on the GTO, this device behaves just like a high voltage low gain transistor with sufficient anode current. In this case, there will be certain power loss.

During reverse direction, virtually the GTO has no blocking capability for anode-short structure. Since junction J_3 is in reverse direction block, GTO has low break down voltage about 20-30V as large doping densities exist on both sides of the junction. The difference between SCR and GTO is given in Table 2.1

Table 2.1 Differences between SCR and GTO

SCR	GTO
The magnitude of latching and holding current of SCR is less than GTO.	The magnitude of latching and holding current of GTO is more than thyristor.
The on-state voltage drop of SCR is less than GTO.	The on-state voltage drop of GTO is more than thyristor.
The gate and cathode structure of SCR are not highly interdigitated	The gate and cathode structure of GTO are highly interdigitated with small cathode and gate widths, the use of anode shorts of GTO.

SCR	GTO
The required gate current to turn on a SCR is less than the required gate current of a GTO.	Due to multi-cathode structure of GTO, the required gate current to turn on a GTO is more than the required gate current for a conventional thyristor.
Gate drive circuit losses for SCR are less.	Gate drive circuit losses for GTO are more.
SCR has less switching speed compared to GTO.	GTO has faster switching speed compared to thyristor.
Reverse voltage blocking capability of SCR is greater than Reverse voltage blocking capability of GTO.	The reverse voltage blocking capability of GTO is less than forward voltage blocking capability.
SCR has less di/dt rating at turn-on compared to GTO.	GTO has more di/dt rating at turn-on compared to thyristor.

2.12.3 Operating Principle of GTO

GTO during turn-on: GTO is a monolithic *p n p n* semiconductor structure and its operating principle can be similar to that of a SCR. Fig.2.31 shows that the *p n p n* structure of a GTO which consist of two transistors where one *pnp* and one *npn* transistor are connected in the regenerative configuration.

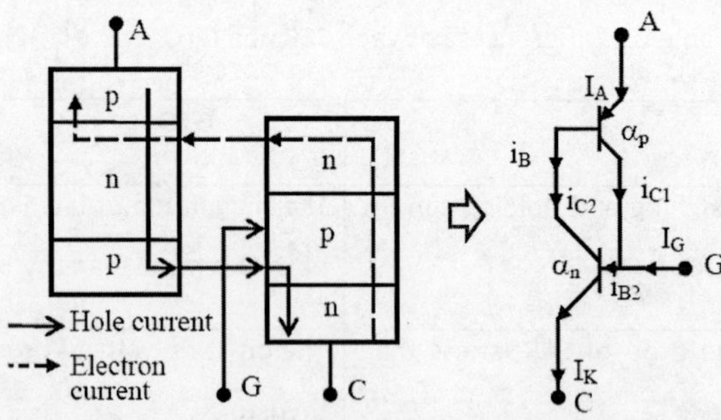

Fig. 2.31 Current distribution in GTO during turn-on

Collector current of T_1 is $i_{C1} = \alpha_1 I_A + I_{CB01}$ (2.13)

Base current of T_1 is $i_{B1} = i_{C2} = \alpha_2 I_K + I_{CB02}$ (2.14) as $\alpha_1 = \alpha_p$ and $\alpha_2 = \alpha_n$

From Fig. 2.31, we get $I_K = I_A + I_G$ and $I_A = i_{B1} + i_{C1}$ (2.15)

After combining the above equation, we obtain

$$I_A = \frac{\alpha_2 I_G + (i_{CBO1} + i_{CBO2})}{1 - (\alpha_1 + \alpha_2)} \quad (2.16)$$

GTO is forward biased, and the applied forward voltage V_{AK} is less than the forward break over voltage V_{BRF}, both the currents I_{CBO1} and I_{CBO2} are small. When gate current I_G is zero, anode current I_A is only slightly higher than ($I_{CBO1} + I_{CBO2}$). At this instant current gain $\alpha_1$and α_2 are small and $\alpha_1 + \alpha_2 << 1$. Subsequently GTO operates in off condition and the device is in the forward blocking state.

To turn-on GTO, the gate current is injected through gate terminal. When GTO is forward biased and a positive gate current I_g input to gate and cathode, the current gain $\alpha_1$and α_2 start to increase rapidly as the emitter current increases. If $\alpha_1 + \alpha_2$ move towards to 1 or $\alpha_1 + \alpha_2 \cong 1$, anode current I_A tends to infinity. When $\alpha_1 + \alpha_2 = 1$, GTO starts to regenerate and transistor T_1 and T_2 operate in saturation. Subsequently all junctions will be forward biased and total potential drop across GTO is equal to that of a single p-n junction diode. Since each transistor is in saturation level, GTO is turned on and anode current starts to flow. When GTO is turned on, anode current is limited by load impedance.

GTO during turn-off: To turn-off the GTO, the gate terminal voltage should be negative with respect to cathode so that gate–cathode is reverse biased. In fact, the holes are injected from anode and these holes are taken out from *p* base through gate metallization into gate terminal. Thus, the voltage drop across *p* base and *n* emitter of transistor T_2 starts reverse biasing the junction J_3 and consequently electron injection stops.

As soon as the electron injection stops completely, the depletion layer starts to grow on both junctions J_2 and J_3. Then once again the device starts blocking forward voltage. The flow of cathode current has stopped and the anode to gate current continues to flow due to the *n* base excess carriers diffuse towards junction J_1. This current is known as "*tail current*". The amplitude of tail current decays exponentially because the *n* base excess carriers will be reduced by recombination. If the tail current becomes zero, GTO device regain its steady state blocking characteristics and it operates in off-state. The current distribution in a GTO during turn-off is shown in Fig.2.32.

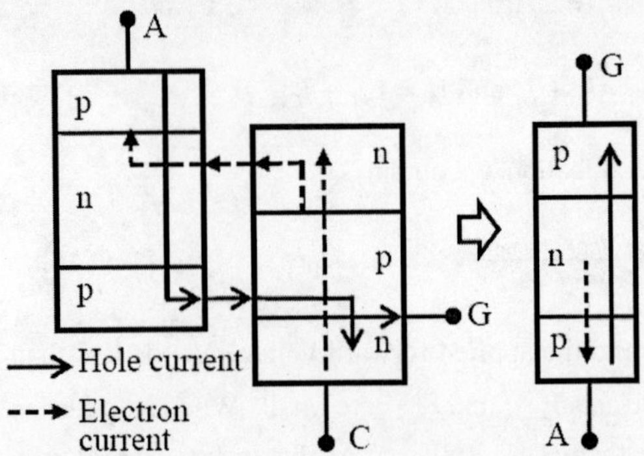

Fig.2.32 Current distribution in GTO during turn-off

2.12.4 Advantages of GTO over BJT

(i) GTO has higher blocking voltage capability

(ii) The ratio of peak current rating to average current is high

(iii) The ratio of peak surge current rating to average current is high and it is about 10:1

(iv) High on-state current gain and it value is about 600

(v) The pulse width of gate signal is short.

2.12.5 Advantages of GTO over SCR

(i) GTO has faster switching speed compared to SCR.

(ii) GTO has more di/dt rating at turn-on compared to SCR.

(iii) The surge current capability is analogous with an SCR

(iv) GTO circuit has lower size and weight as compared to SCR circuit.

(v) Elimination of commutating circuit components in forced commutation as it is not required.

(vi) Improved efficiency of converter circuits.

2.12.6 Applications of GTO

(i) High performance converter fed DC drives

(ii) High performance AC drives such as vector control are used in rolling mills, process control industry, machine tools control, and robotics.

(iii) Electric traction drives

(iv) Variable voltage variable frequency inverter fed ac drives.

2.13 UJT (UNIJUNCTION TRANSISTOR)

2.13.1 Construction of UJT

Fig.2.33 shows the construction of UJT which is a three terminal semiconductor device namely Emitter E, Base B_1 and Base B_2. This device is formed from a lightly doped slab of n-type material which has high resistance. The two base contacts are made at each end of *n* type slab is used as base terminal such as B_1 and B_2 and aluminum rod is inserted on the other side to form a single *p-n* junction. Consequently UJT is known as *uni-junction transistor*. Usually aluminum rod is placed near to base terminal B_2. As shown in Fig.2.33 that B_2 is positive with respect to B_1 by voltage V_{BB} and the symbol of a UJT is shown in Fig.2.34.

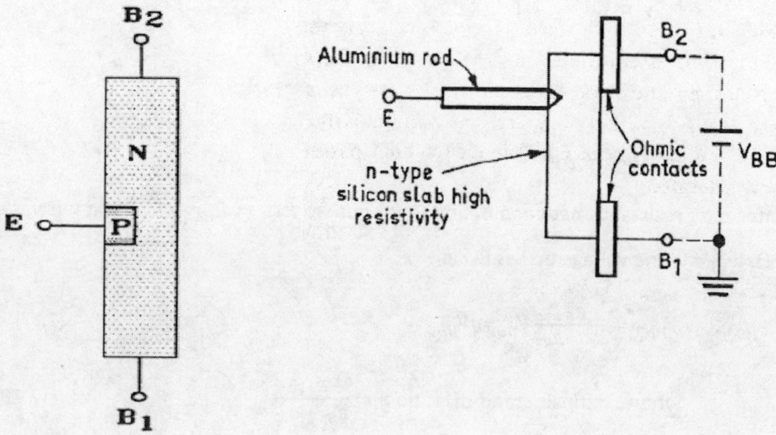

Fig.2.33 Construction of an UJT

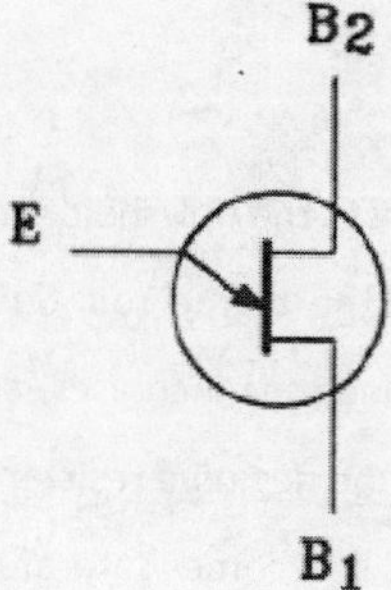

Fig.2.34 Symbol of UJT

The equivalent circuit of UJT is depicted in Fig.2.35. The p-n junction is represented by diode, R_{B1} is a variable resistance, and R_{B2} is a fixed resistance. The value of R_{B1} decreases while emitter current increases. Generally, R_{B1} varies from 50 kΩ to 50 Ω when emitter current changes from 0 to 50 mA.

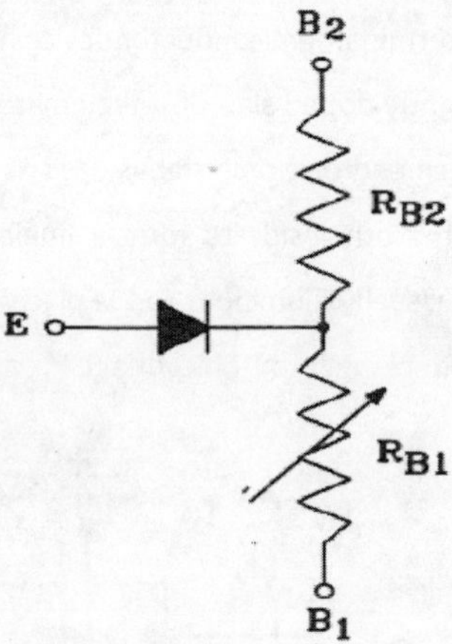

Fig.2.35 Equivalent circuit of UJT

Inter-base resistance between B_1 and B_2 is equal to $R_{BB} = R_{B1} + R_{B2}$ and it is varies from 4 kΩ to 10 kΩ. If $I_E = 0$, the voltage across R_{B1} is

$$V_{R_{B1}} = \frac{R_{B1}}{R_{B1}+R_{B2}} V_{BB} = \eta V_{BB} \qquad (2.17)$$

where, intrinsic stand-off ratio $\eta = \frac{R_{B1}}{R_{B1}+R_{B2}}$. (2.18)

The intrinsic stand-off ratio is controlled by the location of the aluminum rod. The emitter threshold potential is equal to

$$V_P = \eta V_{BB} + V_D \qquad (2.19)$$

2.13.2 Working Principle and V-I characteristics of UJT

Fig.2.36 shows the V-I characteristics of a UJT. When V_E crosses the threshold potential V_P, the emitter fires and holes are injected into the *n*-type slab through the *p*-type rod. Subsequently, the holes content in *n*-type slab increases. As the number of free electrons increases, conductivity increases. Hence V_E drops off when I_E increasing. The UJT operates in the negative resistance region as shown in Fig.2.36. When UJT passes through the valley point (I_V, V_V), it becomes saturated.

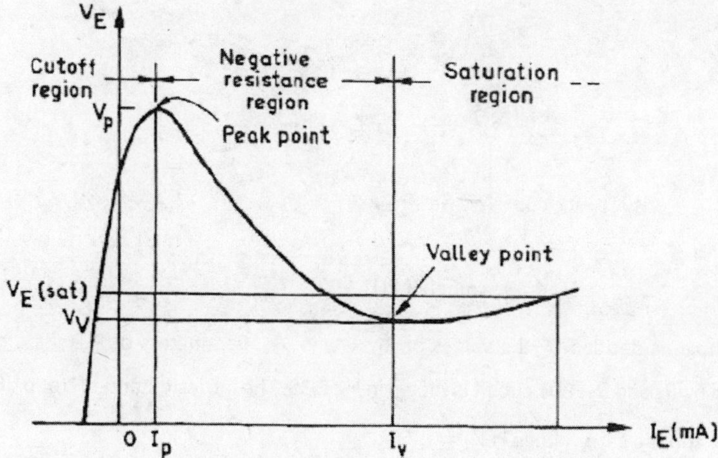

Fig.2.36 V-I characteristics of a UJT

2.14 PUT (Programmable Unijunction Transistor)

Programmable unijunction transistor (PUT) is a four-layer PNPN device. It is an improved version of UJT, and it also called as small version of thyristors. The operating principle is just like to UJT. Therefore, it is permanently considered with UJT. The trigger voltage V_P of PUT can be programmed with the help of two external resistors but V_P is fixed in UJT. As a result, it is known as programmable unijunction transistor (PUT). The construction of PUT is illustrated in Fig.2.37

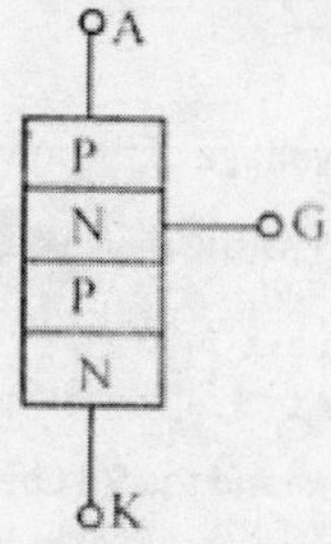

Fig. 2.37 Construction of PUT

The symbol of PUT is shown in Fig 2.38. It has three junctions J_1, J_2 and J_3 and it has three terminals anode, cathode and gate. The gate terminal is connected to the N region, though in thyristor gate is connected with P region. The anode and cathode constitute the PN junction which controls the ON and OFF states of PUT.

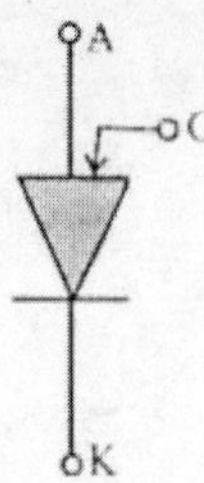

Fig.2.38 Symbol of PUT

The gate voltage is decided by the values of R_1, R_2 and V_{BB}. Parameters of PUT R_{BB}, η and V_P are programmed by R_1 and R_2. PUT circuit is shown in Fig.2.39. The intrinsic stand-off ratio of PUT is equal to $\eta = \frac{R_1}{R_1+R_2}$ and the gate voltage is $V_G = \eta V_{BB}$.

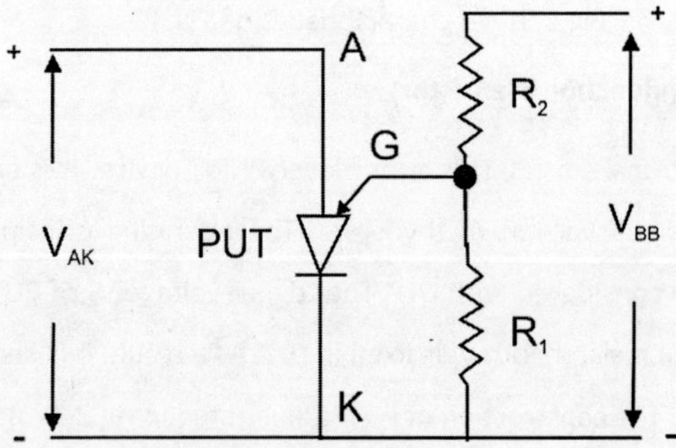

Fig.2.39 PUT circuit

For PUT, the peak point of voltage V_P depends on the voltage of the inner base resistance V_{BB}, intrinsic stand-off ratio η and the anode-gate junction voltage V_D. The peak voltage is expressed as

$$V_P = \eta V_{BB} + V_D \tag{2.20}$$

Fig.2.40 shows the V-I Characteristics of PUT. To understand the V-I Characteristics of PUT, we assume that PUT is forward biased when anode is positive with respect to cathode and the gate is forward biased with respect to cathode. When the anode voltage less than the gate voltage ($V_A < V_G$), the PUT is reverse biased, and it is in *off-state*. During off-state, it has high resistance. If the anode voltage increases and it exceeds the gate voltage and voltage V_D, the PUT is said to be in the ON-state and it offers a low resistance path.

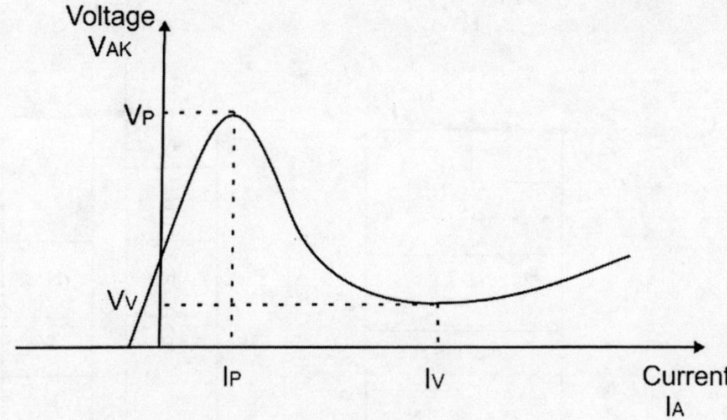

Fig.2.40 V-I Characteristics of PUT

Generally, in PUT positive potential is applied to gate with respect to cathode. When the anode voltage is less than gate voltage V_G, the anode gate junction becomes reverse biased and PUT operates in off-state. When the anode voltage exceeds the gate voltage, anode cathode junction becomes forward biased and PUT is turned-on. In the on-state, PUT behaves just like a four-layer PNPN SCR. Therefore, PUT is also known as *complementary* SCR.

The rating of available PUT is about 100V, 2A or 200V, 1A. It is mainly used in SCR trigger circuits and delay logic relaxation oscillators, high gain phase control circuits. A PUT can be used as *relaxation oscillator* which is discussed in Unit-III. The advantages of PUT over UJT are (i) the switching voltage is easily earned by changing Vg through the potential divider, (ii) PUT can operate at lower voltages than IC's and (iii) Peak current of PUT is lower than UJT.

2.15 DIAC

2.15.1 Construction Of DIAC:

Fig.2.41 shows the cross-sectional view of all existing semiconductor layers and junction of a DIAC. DIAC is a two-terminal semiconductor device namely MT-1(main terminal-1) and MT-2 (main terminal-2). These terminals are also designated as anode-I and anode-II. It is a bidirectional avalanche diode which can be switched from the off-state to the on-state in both positive and negative direction. Since the device is bidirectional and anode current flows in either direction.

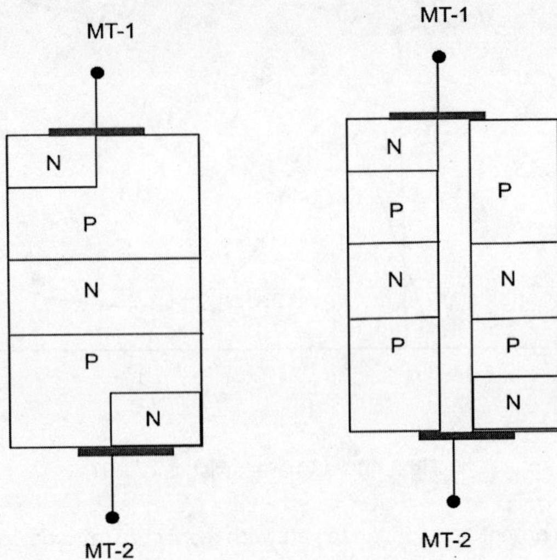

Fig.2.41 Construction of DIAC

2.15.2 Operating Principle and V-I Characteristics

Fig. 2.42 shows the V-I characteristic of a DIAC. This device can work for both positive half cycle and negative half cycle of ac supply. It operates in I and III quadrants, and each direction has a different value of break-over voltage (V_{BO}). As there is no gate in DIAC and the only way to fire a DIAC is to apply its break-over voltage V_{BO}. The symbol of DIAC is shown in Fig.2.43.

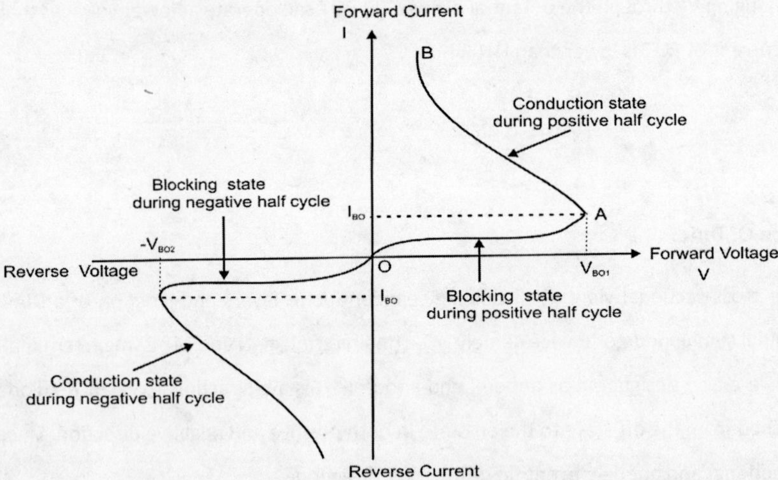

Fig. 2.42 V-I characteristic of DIAC.

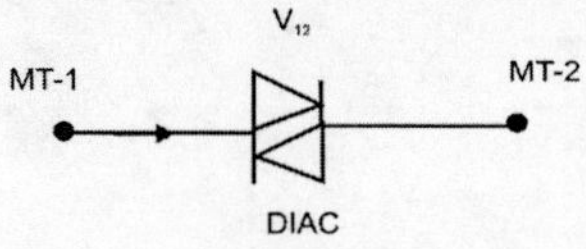

V_{12}

MT-1 MT-2

DIAC

Fig.2.43 The symbol of DIAC

When the applied voltage across two terminals crosses the break over voltage in either direction, DIAC conducts. While MT-1 is positive with respect to MT-2 and V_{12} voltage across two terminals is greater than break-over voltage V_{BO1}, the PNPN structure conducts. Similarly, when MT-2 is positive with respect to MT-1 and V_{12} voltage across two terminals is more than break-over voltage V_{BO2}, again the PNPN structure conducts. The break over voltage of a DIAC is about 30V. In conducting state of DIAC, it acts as a low resistance and on-state voltage drop across it is about 3V.

If voltage across DIAC is less than the break over voltage, a leakage current flows through the device. Then the device operates in non-conducting state i.e., OA region. At operating point, A, voltage just reaches the break over voltage and the device starts to conduct. The conduction state of DIAC express negative resistance characteristics and the current flow through it starts to increase and voltage across it starts to decrease. AB region of V-I characteristic curve is the conduction state. The amplitude of voltage at A is called the *break over voltage* V_{BO}. In the same way, the V-I characteristic curve can be obtained for negative half-cycle of ac supply voltage, and it operates in the third quadrant. When the device starts conducting, the current flow through it is high and its amplitude can be limited by external resistance. But it stops conducting only when the current is reduced below its holding value.

The circuit for DIAC operation is shown in Fig.2.44. The device will operate in on-state when the input voltage is greater than its break-over voltage V_{BO}. In positive half cycle of ac input voltage, if applied voltage is more than the break over voltage V_{BO} at t_1, the diac starts to conduct and it is in on-state from $t = t_1$ to $t = \pi$. During negative half cycle, if the supply voltage is greater the break over voltage $-V_{BO}$ at t_2, the diac is again closed and starts to conduct from $t = t_2$ to $t = 2\pi$. This device can be turned off if the supply voltage is reduced and the current flow through the device falls below its holding value. Usually, DIAC has the same value of break over voltage and the firing angle of DIAC is fixed. In this way diac is used as a two terminal ac switch. Generally, DIAC is used as triggering device for a TRIAC. It can be used as a lamp dimmer, fan regulator, heater control, ac motor control, etc. All these circuits are almost similar but only the load i.e., lamp, heater, fan, motor, etc. is changed.

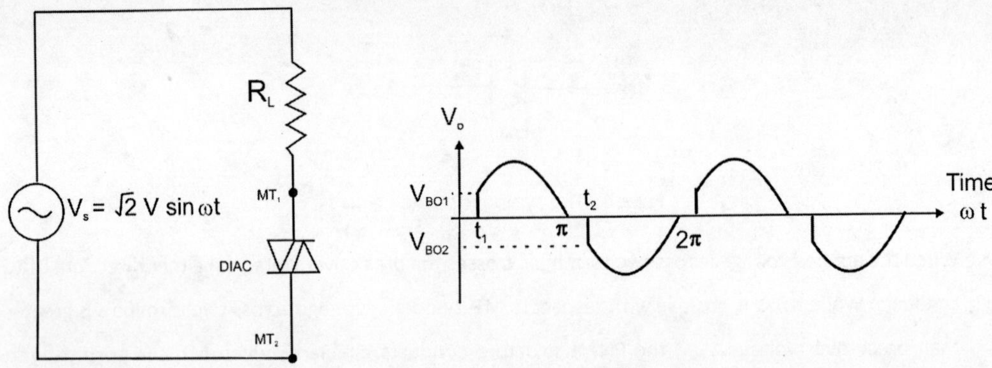

Fig.2.44 The circuit for DIAC operation along with output voltage across R_L

2.16 TRIAC

2.16.1 Construction Of TRIAC

SCR is a unidirectional device since it conducts during forward biased condition when trigger pulse is applied between gate to cathode. SCR has reverse blocking characteristics and current cannot flow from cathode to anode direction. In ac circuit applications, the bidirectional current flow is required. Then two SCRs are connected *back-to-back* or two *anti-parallel* SCR can be integrated into a single chip called as TRIAC (Fig.2.45). TRIAC can conduct in both positive half cycle and negative half cycle of supply voltage. Therefore, TRIAC is a bidirectional thyristor, and it is used for ac voltage controller.

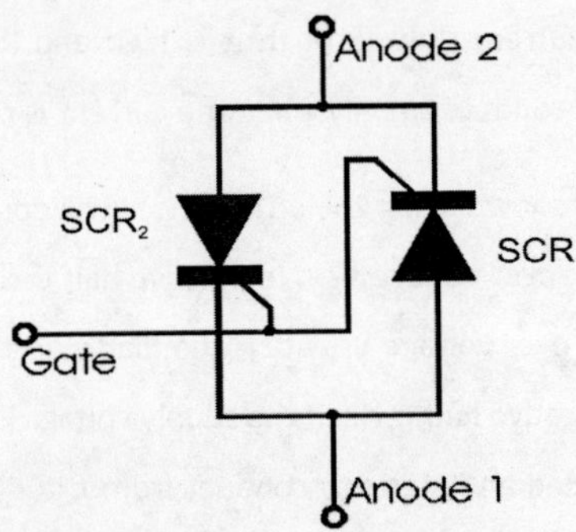

Fig.2.45 Equivalent circuit of TRIAC

TRIAC is equivalent to two SCRs connected in anti-parallel and it conducts in bidirectional. The anode and cathode terminals cannot be a symbol of a TRIAC. TRIAC has three terminals namely MT_1, MT_2 and gate G. MT_1 is used as reference point to measure voltages and currents at gate terminal and MT_2. Always gate G is present near MT_1. The construction of TRIAC is shown in Fig.2.46 and the circuit symbol of TRIAC is shown in Fig.2.47.

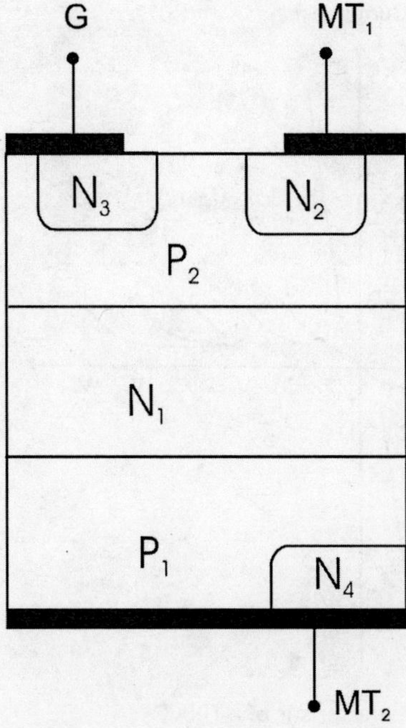

Fig.2.46 Basic construction of TRIAC

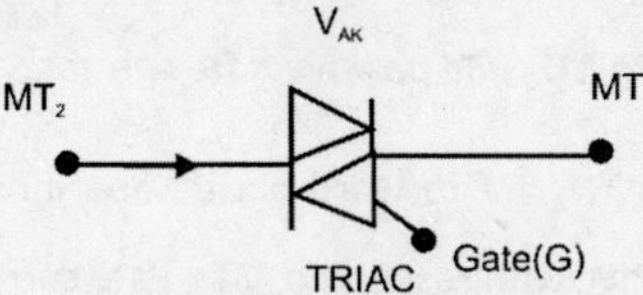

Fig.2.47 Symbol of TRIAC

2.16.2 Operating Principle and V-I characteristic:

Fig.2.48 shows the V-I characteristic of a TRIAC and it operates in first quadrant and third quadrant. During first quadrant operation, MT_2 is positive with respect to MT_1 and in the third quadrant operation, MT_1 is positive with respect to MT_2. If gate pulse is not applied, the TRIAC can block both positive and negative half cycles of ac applied voltage and the peak voltage across MT_1 and MT_2 must be less than break over voltage.

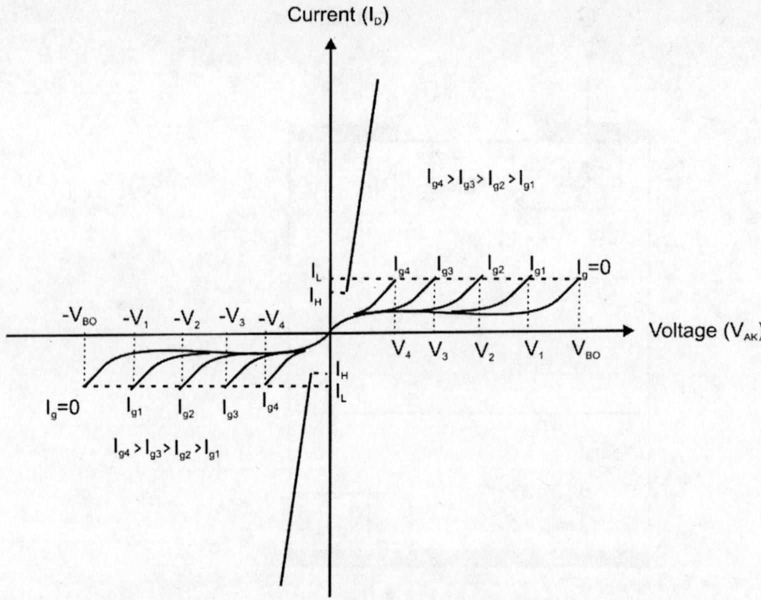

Fig.2.48 V-I characteristic of a TRIAC

Consider the device is in blocking condition. If a gate pulse is applied, a gate current flows to trigger the TRIAC in first or third quadrant. Based on the biasing condition and gate pulse, TRIAC can operate in four different operating modes.

Mode 1: MT_2 positive with respect to MT_1 *and positive gate current:*

In this condition, junctions P_1-N_1, P_2-N_2 are forward biased and junction N_1-P_2 is reverse biased in Fig.2.49. If the gate terminal is positive with respect to MT_1, gate current is positive, and it flows from gate terminal to MT_1 through P_2-N_2 junction J_3 just like SCR. TRIAC will be turn on just like a SCR. To turn on TRIAC, the gate current must be greater than SCR. As ohmic contacts of gate and MT_1 on P_2 layer are present, a small amount of gate current flows from gate to MT_1 through P_2. Since P_2 layer is flooded by electrons when gate current flows through P_2-N_2 junction. Then electrons diffuse the junction J_2 and collected at N_1 and subsequently the reverse bias junction N_1-P_2 break downs just like SCR. So, the $P_1N_1P_2N_2$ structure operates just like SCR in on-state and TRIAC operates in first quadrant.

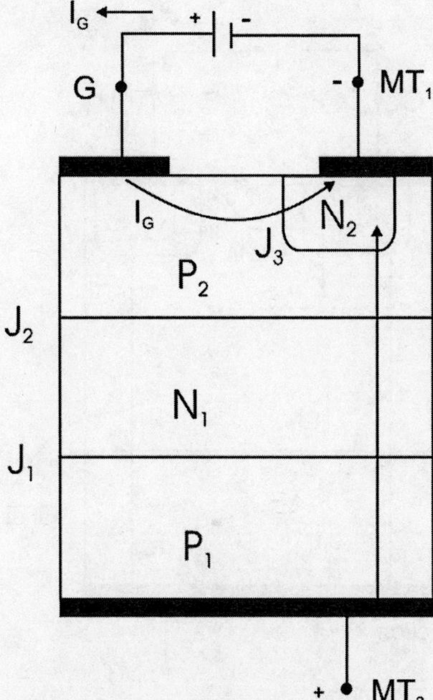

Fig.2.49 *Mode* 1- MT_2 positive and positive gate current

Mode 2: MT_2 positive with respect to MT_1 *and gate current negative:*

If MT_2 is positive and gate terminal is negative with respect to MT_1, gate current starts to flow through P_2-N_3 and TRIAC starts to conduct through P_1 N_1 P_2 N_3 structure as shown in Fig.2.50. Due to the conduction, the voltage drop across the path P_1 N_1 P_2 N_3 falls, and the potential of P_2 N_3 junction rises towards anode potential of MT_2. Since the right-hand portion of P_2 is clamped at potential of MT_1, a potential gradient developed across P_2 and its left-hand region being at high potential than its right hand region. A current flow in P_2 layer from left to right which forward biased P_2N_2 junction and lastly P_1 N_1 P_2 N_2 structure starts to conduct. This turn on method of TRIAC is less sensitive compared to the turn on process of TRIAC when MT_2 is positive, and gate is positive.

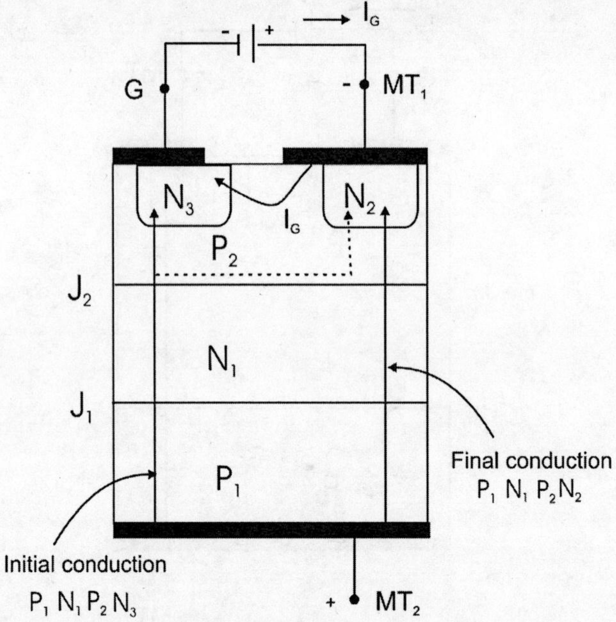

Fig.2.50 Mode 2- MT_2 positive and negative gate current

Mode 3: MT_2 negative with respect to MT_1 and positive gate current:

In this case, the device will be turn on by applying a positive gate voltage across gate and MT_1 as shown in Fig.2.51. TRIAC operates in third quadrant when it is turned on. The $P_2 N_1 P_1 N_4$ structure with N_2 acts as a remote gate. The gate current establishes forward biases P_2N_2 junction. Electrons from N_2 layer injected into P_2 layer and electrons from N_2 are collected by P_2N_1 junction due to increase of current through the junction P_2N_1. Subsequently holes are injected from P_2, diffuse through N_1 and reach at P_1. Therefore, a positive space charge builds up in P_1 region and more electrons from N_4 diffuse into P_1 to neutralize the positive space charge region. As these electrons reach at junction J_2, electrons generate a negative space charge in N_1 region. After that more holes are injected from P_2 into N_1 and the regenerative process continues until $P_2N_1P_1N_4$ completely turned on. When TRIAC is turned on, the current flow through TRIAC is controlled by external load. As TRIAC is turned on by the remote gate N_2, it is less sensitive.

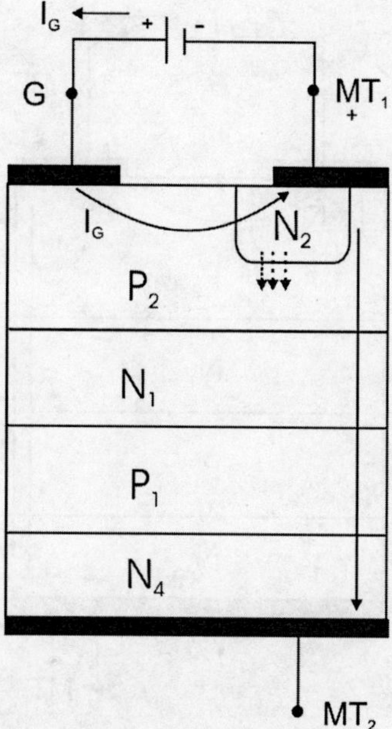

Fig.2.51 Mode 3 - MT_2 negative and positive gate current

Mode 4: MT_2 negative with respect to MT_1 and negative gate current:

In this mode, N_3 acts as a remote gate. The gate current I_G flows through forward biased P_2N_3 junction and electrons injected from N_3 to P_2 as shown in Fig.2.52. Then electrons from N_3 are collected by P_2N_1 and there is an increase of current across P_1N_1. Subsequently the $P_2N_1P_1N_4$ structure turns on by regenerative action. Then TRIAC is turns on due to the increased current in N_1 layer. TRIAC is more sensitive in this mode operation compared to MT_2 negative and positive gate current.

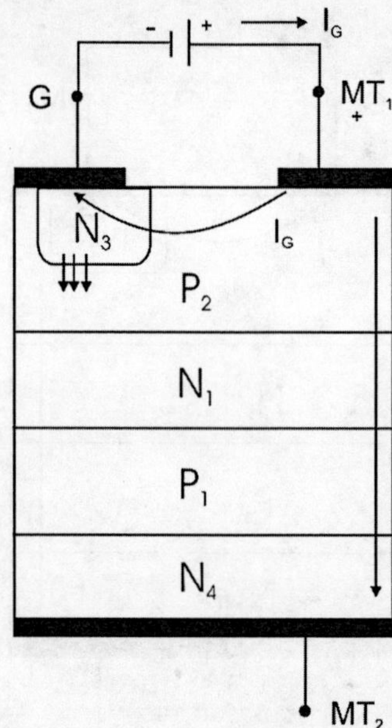

Fig.2.52 Mode 4 - MT_2 negative and negative gate current

It is clear from the above four operating modes, TRIAC can be switched into on-state by either positive or negative gate current, though the device is more sensitive if positive gate current is injected when MT_2 is positive and negative gate current is injected when MT_2 is negative. At present TRIACs are available with voltage rating about 1200V and current rating about 300A. TRIACs are widely used in lamp dimmers, heat control and speed control of single-phase ac motors. The comparison between DIAC and TRIAC is given in Table 2.2.

Table 2.2 Comparison between DIAC and TRIAC

Sr. No.	DIAC	TRIAC
1.	DIAC is a two-terminal device	TRIAC is a three-terminal device
2.	Break over voltage of DIAC cannot be controlled.	Break over voltage of TRIAC can be controlled by varying the gate current.
3.	DIAC is a low power device	TRIAC is a high-power device
4.	This device is used as a triggering device for TRIAC	This device is used in fan speed control and illumination control etc.

2.17 Protection Circuit of Thyristor

Thyristor should always be operating within the specified voltage and current ratings for better and reliable operation. Although in some special applications, thyristors must be subjected to over voltage and over current. When the device is under over voltage and over current, the junction temperature can exceed the maximum allowable temperature $150°C$ and the device may get damage permanently. Consequently efficient cooling methods should be used for cooling thyristor to dissipate the excess heat into atmosphere. In turn-on and turn-off process, there is lot of power loss within the device and device temperature increases. Power loss in the device is extensively high in high frequency applications and the device may get damage due to temperature rise.

During the turn on process of thyristor, $\frac{di}{dt}$ is reasonably large and local heating is possible within the device. Subsequently, thyristor may get destroy. The false triggering of thyristor is possible due to high $\frac{dv}{dt}$ and noise signal across gate to cathode terminals. Always thyristor must be protected from false triggering. For standard operation of thyristor in power electronics converter circuits, the following protections circuits are used such as over voltage protection, over current protection, snubber circuit and crowbar circuit.

2.17.1 Over Voltage Protection

Thyristors are very sensitive to over voltage which is the main cause of failure. The transient over voltage across thyristor can turn it on without any gate signal and the converter circuit perform malfunction. Usually thyristor is able to withstand in external over voltage and internal over voltage.

External over voltage: If the current flow in an inductive circuit is interrupted, external over-voltage appears across thyristor. When the converter circuit is supplied through transformer, transient over-voltage across thyristor occurs while transformer is energized or de-energized. Due to transient over-voltage, thyristor may be turn on abruptly and over voltage appear across load. Accordingly, a large fault current flows though the converter circuit and thyristors may get damage partly or completely.

Internal over voltage: Generally internal over voltage is generated during turn-off process of thyristor. When anode current becomes zero, current starts to flow in reverse direction due to stored charges and reaches the peak reverse recovery current. Then the reverse recovery current starts to fall abruptly with high $\frac{di}{dt}$. Since series inductance L is present in the converter circuit, a large transient voltage $L.\frac{di}{dt}$ is created and thyristor will be damaged due to large transient over-voltage.

Usually voltage clamping devices namely varistor can be used to protect thyristors from over-voltage. In fact varistor is a non-linear resistance when the current increases its resistance value decreases. The V-I characteristics of varistor is shown in Fig.2.53. and action of current limiting fuse is illustrated in Fig.2.54.

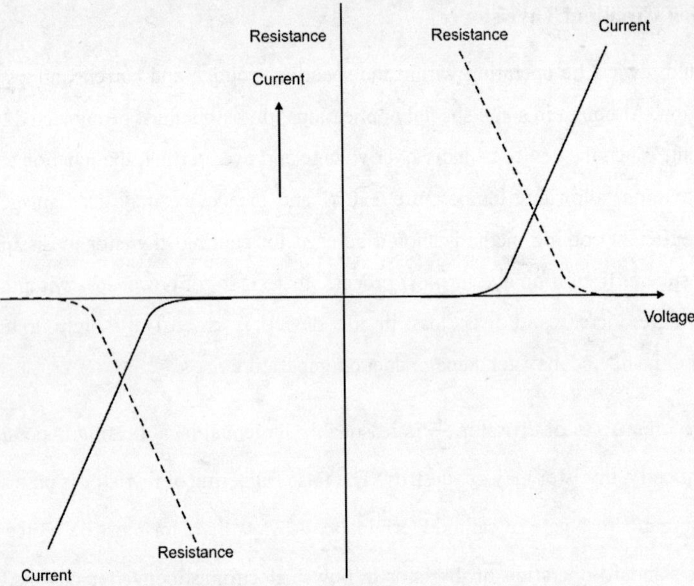

Fig.2.53 V-I characteristics of varistor

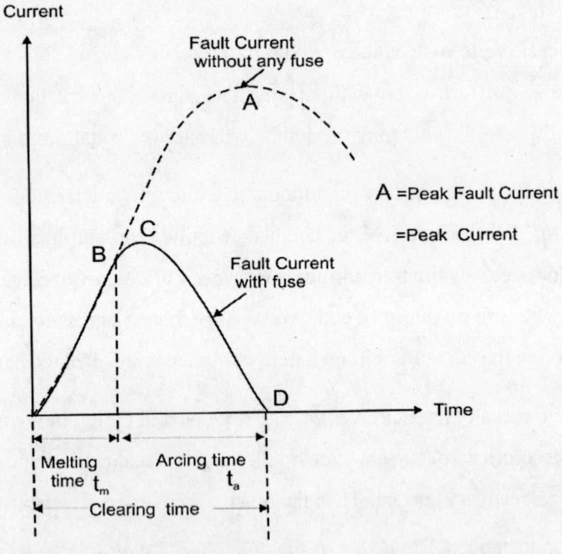

Fig.2.54 Function of current limiting switch

2.17.2 Over Current Protection:

Over current may damage thyristor due to the junction temperature exceeding rated value. Therefore there is a need of over current protection. In any electrical circuit, generally current is limited by fuses or circuit breakers. Each thryristor has I^2Rt rating for protection from over current. In fact a fuse is placed in series with each device to protect it. Fig.2.55 shows the over current protection.

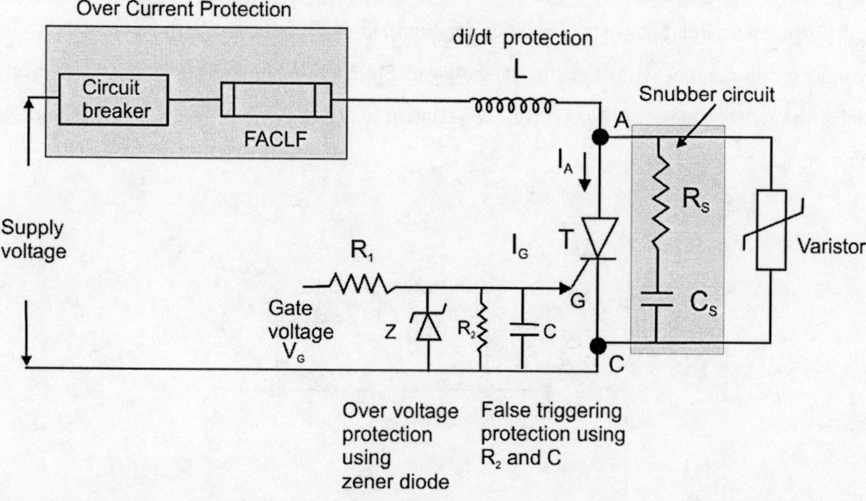

Fig.2.55 Thyristor protection

2.17.3 Snubber Circuit

Once a voltage is applied across anode to cathode of a thyristor and anode is positive with respect to cathode, J_1 and J_3 junctions are forward biased and J_2 junction is reverse biased. In fact the reverse biased junction J_2 behaves as capacitor. Subsequently holes from p-region of junction J_1 are accumulated at junction J_2 and electrons from n-layer of junction J_3 are also accumulated at the other side of junction J_2. Consequently, space charge carriers are present across the junction J_2. If a large voltage is applied abruptly, charging current i_C flows through thyristor and the device will be turned on without gate signal. This is the unnecessary triggering of thyristor.

The charging current is $i_C = \frac{dq}{dt}$ (2.21)

The equation (2.21) can be written as $i_C = \frac{d(C_jV)}{dt} = V\frac{dC_j}{dt} + C_j\frac{dV}{dt}$ (2.22)

(as $q = C_jV$)

Since $\frac{dC_j}{dt} = 0$, $i_c = C_j\frac{dV}{dt}$ (2.23)

Generally, $\frac{dV}{dt}$ rating of thyristor is about 20 to 500 $\frac{V}{\mu s}$. When $\frac{dV}{dt}$ is more than it's rated value, thyristor will be turned on and this is the abnormal triggering of thyristor. Then thyristor based converter circuit starts to malfunction. So a snubber circuit should be connected in parallel with thyristor to protect from $\frac{dV}{dt}$.

Basically a snubber circuit consists of a series combination of a resistance R_s and a capacitance C_s and it is connected in parallel with thyristor as shown in Fig.2.56. When a capacitor C_s is connected in parallel with thyristor, the capacitor C_s itself is sufficient to protect from unwanted abnormal triggering of thyristor.

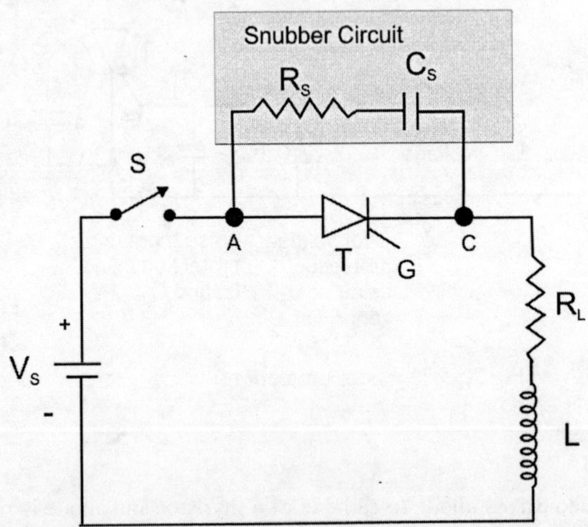

Fig.2.56 Snubber circuit connected across thyristor (T)

If the switch S is closed at $t = 0$, abruptly the input voltage V_s appears across the circuit and the capacitor behaves as short circuit. So the voltage across thyristor is zero. There after the voltage across capacitor builds up slowly with a $\frac{dV}{dt}$, which is much less than the specified maximum $\frac{dV}{dt}$. Accordingly C_s is sufficient to prevent the thyristor from $\frac{dV}{dt}$ triggering.

While the gate pulse is not applied and $\frac{dV}{dt}$ is less than its maximum limit, thyristor will be in off-state and capacitor is charged to supply voltage V_S. If thyristor is turned on by gate pulse, the capacitor starts to discharge through SCR and the current flows through the path formed by capacitor C_S and thyristor is $I_S = \frac{V_S}{R_T}$ (R_T is the resistance of thyristor during forward conduction state).

As R_T is quite low, the $\frac{di}{dt}$ during turn-on will be very high and thyristor may be permanently damaged. Consequently current I_S must be a safe value while a series resistance R_S is connected with the capacitor C_S.

Whenever the switch S is closed, capacitor is short circuit and SCR operates in forward blocking state, the equivalent circuit of Fig.2.56 is shown in Fig.2.57.

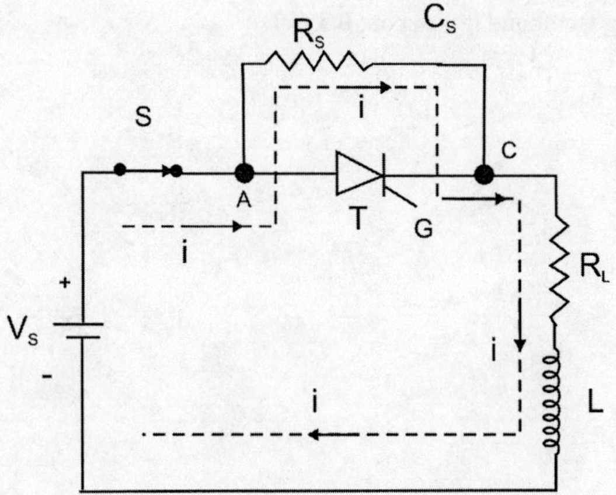

Fig.2.57 Equivalent circuit of snubber circuit

Using KVL in the circuit, we get $V_s = i_s(R_s + R_L) + L.\frac{di_s}{dt}$ (2.24)

The solution of equation (2.22) is $i_s = \frac{V_s}{R_s+R_L}\left(1 - e^{-\frac{t}{\tau}}\right)$ (2.25)

Since $I = \frac{V_s}{R_s+R_L}$ and $\tau = \frac{L}{R_s+R_L}$, $i_s = I\left(1 - e^{-\frac{t}{\tau}}\right)$ (2.26)

After differentiating the equation (2.26), we obtain

$$\frac{di}{dt} = Ie^{-\frac{t}{\tau}}\frac{1}{\tau} = \frac{V_s}{R_s+R_L}\frac{R_s+R_L}{L}e^{-\frac{t}{\tau}} = \frac{V_s}{L}e^{-\frac{t}{\tau}} \quad (2.27)$$

At t=0, $\frac{di}{dt}$ is maximum and $\frac{di}{dt} = \frac{V_s}{L}_{max}$ (2.28)

The voltage across SCR is $v = iR_s$ (2.29)

After differentiating the equation (2.29), we find $\frac{dv}{dt} = R_s\frac{di}{dt}$ (2.30)

Then $\frac{dv}{dt_{max}} = \frac{R_sV_s}{L}$ $\frac{dv}{dt}|_{max} = \frac{R_sV_s}{L}$

So $R_s = \frac{L}{V_s}\frac{dv}{dt}|_{max}$ (2.31)

The R-L-C circuit should be analyzed to find the optimum values of R_s and C_s of snubber circuit. After analyzed, we find the values of R_s and C_s are

$$R_s = 2\xi\sqrt{\frac{L}{C_s}} \text{ and } C_s = \left(\frac{2\xi}{R_s}\right)^2 L \quad (2.32)$$

where, ξ is the damping factor and it is about 0.5 to 1.

For more on snubber circuit

Example 2.5 Consider a SCR is operating with a rms supply voltage of 220V and it has the

repetitive peak current $I_p = 80A, \frac{dv}{dt|_{max}}$, and $\frac{di}{dt|_{max}}$.

Design a snubber circuit for SCR protection. Assume that factor of safety is 2 and minimum value of resistance is 25Ω .

Solution:

Since factor of safety is 2, the permissible value of $I_p = \frac{80}{2A} = 40A, \quad \frac{dv}{dt|_{max}} = \frac{250V}{\mu s}, \frac{di}{dt|_{max}} = \frac{50A}{\mu s}$

To limit the value of $\frac{di}{dt|_{max}}$ within $\frac{50A}{\mu s}$, a inductance L should be connected in series with thyristor

and its value is $L = \frac{V_s}{\frac{di}{dt|max}} = \frac{220\sqrt{2}\times10^{-6}}{50} = 6.22\mu H$

The value of $R_s = \frac{L}{V_s}\frac{dv}{dt}|_{max} = \frac{6.22\times10^{-6}}{220\sqrt{2}} \times \frac{250}{10^{-6}} = 5\Omega$

If the SCR is in the blocking state, capacitor will be charged to maximum voltage $220\sqrt{2}$V. When the SCR is turned on, the peak current flows through the SCR is

$$\frac{220\sqrt{2}}{25} + \frac{220\sqrt{2}}{5} = 77.77A$$

The maximum permissible peak current is 40A. Therefore, the value of R_S should be such that, the peak current is less than 40A.

Assume $R_s = 15\Omega$, then the peak current flows through the SCR is

$$\frac{220\sqrt{2}}{25} + \frac{220\sqrt{2}}{15} = 33.18A$$

If the damping factor $\xi = 0.7$, the value of capacitance is

$$C_s = \left(\frac{2\xi}{R_s}\right)^2 L = \left(\frac{2\times0.7}{15}\right)^2 \times 6.22\times10^{-6}=0.054\mu F$$

The parameter of snubber circuit is $R_s = 15\Omega$ and $C_s = 0.054\mu F$

2.17.4 Crowbar Circuit

Fig.2.58 shows the crowbar protection circuit. A crowbar thyristor is connected across dc supply. If over current flows through converter circuit and current sensing resistance generates voltage which is greater than preset value, triggering circuit generates triggering pulse which is applied to the crowbar thyristor. Then crowbar thyristor will be turn-on. Since thyristor becomes turn-on and dc supply will be short circuited, fuse will be blown due to high current and interrupt the fault current to protect thyristor. Sometimes fuse may be replaced by circuit breaker for thyristor protection.

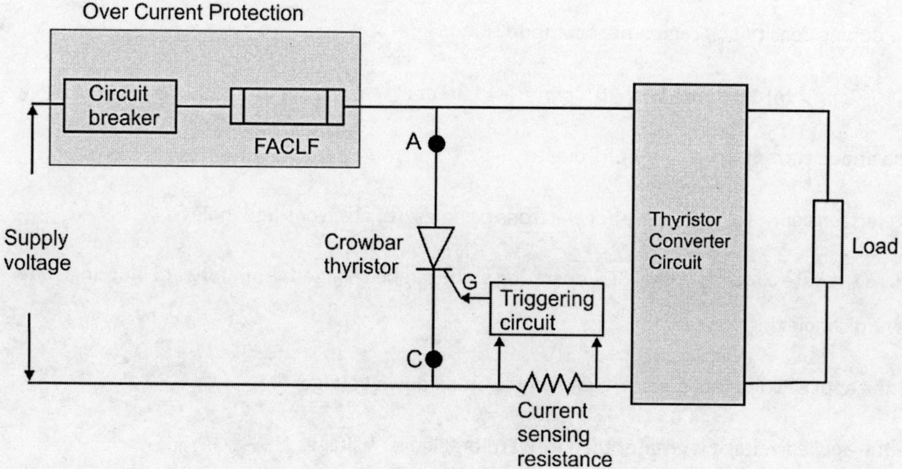

Fig.2.58 Crowbar protection circuit

EXERCISE

MULTIPLE CHOICE QUESTIONS

2.1 SCR is a _______ layer device

(a) two (b) three (c) four (d) five

2.2 The number of *pn* junctions in a SCR is

(a) 4 (b) 3 (c) 2 (d) 1

2.3 A SCR is a _______ layer device

(a) $pnpn$ (b) $p^+n^+n^-n^+$ (a) $n^-n^+p^-p^+$ (a) $n^+p^-p^+n^-$

2.4 When anode terminal of SCR is positive with respect to cathode, the number of forward biased junctions are

(a) 3 (b) 2 (c) 1 (d) 4

2.5 If SCR is forward biased, the number of reverse biased junctions are

(a) 1 (b) 2 (c) 3 (d) 4

2.6 If anode terminal of SCR is negative with respect to cathode, the number of blocked *p-n* junctions are

(a) 4 (b) 2 (c) 3 (d) 1

2.7 The doping density of *n* region near cathode is

(a) $10^{15} cm^{-3}$ (b) $10^{14} cm^{-3}$ (c) $10^{16} cm^{-3}$ (d) $10^{19} cm^{-3}$

2.8 The anode current flows in a SCR due to

(a) electrons only (b) holes only (c) electrons or holes (d) electrons and holes

2.9 When the gate circuit is open, SCR operating state can be changed from forward blocking state to forward conducting state if

(a) the applied voltage is greater than reverse break down voltage

(b) the applied voltage is greater than forward break over voltage

(c) the applied voltage is 1.0 V

(d) the applied voltage is greater than peak repetitive reverse voltage

2.10 In a SCR, the latching is

(a) equal to holding current (b) greater than holding current

(c) less than holding current (d) None of these

2.11 The turn on time of SCR is equal to

(a) delay time (b) rise time (c) spread time

(d) sum of delay time, rise time and spread time

2.12 In a SCR, turn on time is

(a) equal to turn-off time (b) greater than turn-off time (c) less than turn-off time

2.13 The ratio of latching current to holding current of thyristor is

(a) 1 (b) 1.5 (c) 2 (d) 2.5

2.14 The on-state voltage drop of a SCR is

(a) 0.7 V (b) 1V to 1.5V (c) 5V (d)10V

2.15 The typical $\frac{di}{dt}$ rating of thyristor is

(a) 0.1 μA to $2\mu A$ (b) $0.5\mu A$to $10\mu A$ (c) $20\mu A$to $500\mu A$ (d) $1500\mu A$

2.16 The typical $\frac{dv}{dt}$ rating of thyristor is

(a) 0.1 μV to 10 μV (b) $0.5\mu V$to $5\mu V$ (c) $20\mu V$to $500\mu V$ (d) $1500\mu V$

2.17 Snubber circuit is used for ________ protection of thyristor

(a) $\frac{di}{dt}$ (b) $\frac{dv}{dt}$ (c) over current (d) over voltage

2.18 Thyristor can be used as

(a) bidirectional switch (b) unidirectional switch (c) either bidirectional or unidirectional switch

(d) none of these

2.19 The parameters of snubber circuit are

(a) $R_s = 2\xi\sqrt{\frac{C_s}{L}}$ and $C_s = \left(\frac{2\xi}{R_s}\right)^2 L$ (b) $R_s = 2\xi\sqrt{\frac{L}{C_s}}$ and $C_s = \left(\frac{2\xi}{L}\right)^2 R_s$

(c) $R_s = 2L\sqrt{\frac{\xi}{C_s}}$ and $C_s = \left(\frac{2\xi}{R_s}\right)^2 L$ (d) $R_s = 2\xi\sqrt{\frac{L}{C_s}}$ and $C_s = \left(\frac{2\xi}{R_s}\right)^2 L$

2.20 In a UJT with V_{BB} voltage across two base terminals, the emitter potential at the peak point is given by

(a) $V_P = V_{BB} + \eta V_D$ (b) $V_P = \eta V_{BB}$ (c) $V_P = \eta V_{BB} - V_D$ (d) $V_P = \eta V_{BB} + V_D$

2.21 A DIAC is used

(a) in triggering circuit of a TRIAC (b) for protection of a TRIAC

(c) to increase efficiency of a TRIAC (d) to decrease efficiency of a TRIAC

2.22 The snubber circuit is used in a thyristor

(a) to limit the value of $\frac{di}{dt}$ (b) to limit the value of $\frac{dv}{dt}$

(c) to limit both the value of $\frac{di}{dt}$ and $\frac{dv}{dt}$ (d) to turn on and turn off

2.23 The three terminals of TRIAC are

(a) Base, emitter, collector

(b) Base 1, Base 2 and emitter (c) Anode, cathode and gate

(d) MT_1, MT_2 and gate

2.24 Thyristor is equivalent to

(a) two transistors (b) two diodes (c) two DIAC (d) two MOSFET

2.25 TRIAC is equivalent to

(a) two transistors (b) two Thyristors (c) two DIAC (d) two MOSFET

Answers of multiple choice questions

2.1 (c) 2.2 (b) 2.3 (a) 2.4 (b) 2.5(a) 2.6 (b) 2.7 (d) 2.8 (d) 2.9 (b) 2.10 (b)
2.11 (d) 2.12 (c) 2.13 (d) 2.14 (b) 2.15(c) 2.16 (c) 2.17 (b) 2.18 (b) 2.19 (d) 2.20 (d)
2.21 (a) 2.22 (b) 2.23 (d) 2.24 (a) 2.25(b)

REVIEW QUESTIONS

Short Answer Type Questions

2.26 What is SCR?

2.27 What is thyristor?

2.28 What are the different names of thyristors ?

2.29 What is the symbol of thyristor?

2.30 What are the necessary conditions for turning on of a thyristor?

2.31 What are the types of switching characteristics of SCR?

2.32 What are the different modes of operation of a thyristor?

2.33 Define (a) delay time (b) rise time (c) spread time (d) reverse recovery time

2.34 Define the following ratings of thyristor

(a) on state voltage drop (b) voltage safety factor

(c) latching current (d) holding current (e) I^2t rating

2.35 Why turn-on time is less than turn-off time in a thyristor?

2.36 Why protection circuit is required during operation of thyristor?

2.37 Draw the V-I characteristics of LASCR and GTO

2.38 What are the advantages of GTO over thyristors?

2.39 What is a summer circuit?

2.40 State the applications of UJT and PUT

Long Answer Type Questions

2.41 Draw the structure of a SCR and explain its operating principle briefly.

2.42 Describe the behaviour of thyristor using a two-transistor model.

2.43 In two transistor model of SCR, prove that anode current of SCR is equal to

$$I_A = \frac{\alpha_2 I_G + I_{CB01} + I_{CB02}}{1-(\alpha_1+\alpha_2)}$$

2.44 Draw the V-I characteristics of a SCR and explain different operating regions.

2.45 What is the effect of gate current on the V-I characteristics of a SCR?

2.46 Draw the turn-on characteristics of SCR and explain briefly?

2.47 Draw the turn-off characteristics of SCR and explain in brief?

2.48 State the turn-on and turn-off time of SCR.

2.49 Draw the gate characteristics of a thyristor and explain its importance.

2.50 What are the different protection schemes of thyristor?

2.51 What is purpose of $\frac{di}{dt}$ protection and $\frac{dv}{dt}$ protection?

2.52 Write short notes on

(a) over voltage protection (b) over current protection

2.53 Explain the design method of a snubber circuit.

2.54 Describe the crowbar protection circuit for the over current protection of thyristors.

2.55 Define (a) peak voltage (b) valley voltage (c) stand-off ratio of UJT

2.56 Explain the operating principle of DIAC with proper diagram.

2.57 Draw the V-I characteristics of a DIAC. What are the applications of DIAC?

2.58 What is TRIAC? Explain the operating principle of TRIAC with proper diagram.

2.59 Draw the V-I characteristics of a TRIAC and explain different operating regions.

2.60 Give a list of applications of TRIAC

Problems

2.61 The turn-on and turn-off time of SCR are $3.7\mu s$and $6.3\mu s$ respectively, compute the maximum switching frequency of SCR in a converter circuit.

2.62 The average gate power dissipation of a SCR is $P_{gav} = 0.5\ Watt$. When the gate voltage varies from 2.5 V to 9.5V, plot the curve where gate voltage is a function of gate current. Assume average gate power dissipation is constant.

2.63 The gate characteristics of a SCR is $V_g = 1 + 4.5I_g$. If a rectangular pulse of 10V with $25\mu s$is applied to gate, determine (a) the series connected resistance in gate, (b) triggering frequency and (c) duty cycle. Assume average power dissipation is 0. 45Watt and peak gate drive power is 4Watt

2.64 If a SCR is operating with a peak supply voltage of $210\sqrt{2}$V and it has the following parameters:

repetitive peak current $I_p = 100A, \frac{dv}{dt|_{max}}$, and $\frac{di}{dt|_{max}}$.

Design a snubber circuit for thyristor protection. Assume the factor of safety is 2 and the minimum value of resistance is 20Ω .

2.65 In a thyristor the capacitance value of reverse-biased junction J_2 is $C_{J_2} = 50pF$ and it is independent of the off-state voltage. The limit value of the charging current to turn the thyristor is about 20mA. Determine the critical value of $\frac{dv}{dt}$.

PRACTICAL

1. Experiment on Test the proper functioning of DIAC to determine the break over voltage

Aim:

To study the V-I characteristics of a DIAC in both directions, first and third quadrant operation of DIAC and determine break over voltages.

Apparatus:

Regulated dc power supply 0-50V, Digital voltmeter/multi-meter 0-50V DC, Digital ammeter/multi-meter 0-100mA; DIAC-DB3/DB4, and Resistance R_1=1kΩ, 10W

Circuit Diagram:

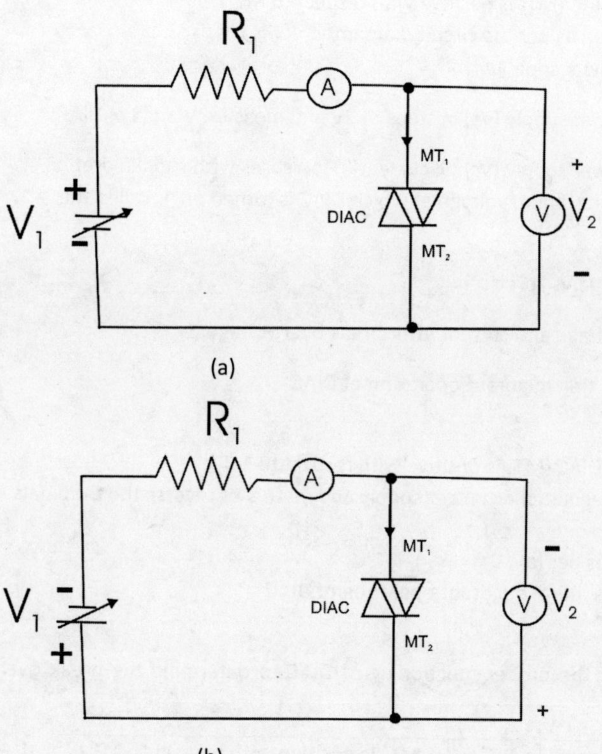

Fig.2.59 Circuit diagram for test the proper functioning of DIAC to determine the break over voltage (a) MT_1 is positive with respect to MT_2 (b) MT_2 is positive with respect to MT_1

Theory:

DIAC is a member of the thyristor family. This device consists of three layers and two terminals MT_1 and MT_2. The construction of DIAC is almost same as that of the transistor, but there are certain deviations from the construction of the transistor. The discriminating points are: (i) there is no base terminal in the DIAC, (ii) the three regions have almost same level of doping, and (iii) DIAC has symmetrical switching characteristics for either polarity of voltages. DIAC can be turned on or off for both the polarity of voltages. This device operates when avalanche breakdown occurs. The detail construction of DIAC is explained in section 2.16.1 and V-I characteristics described in section 2.16.2 in detail. This device can be used in the TRIAC triggering circuit, lamp dimmer circuit, fan regulator, and temperature controller. The advantage of using DIAC is that it can be turned on or off simply by controlling the applied voltage level about break over voltage.

Precautions:

1.Initially all the knobs of the power supplies are at zero value.

2.The applied voltage and current should not be more than the maximum rating of Power BJT.

3.Reading of observation should be error free.

Procedure

(a) First Quadrant Operation of DIAC (MT_2 is positive with respect to MT_1)

(i) Make the circuit connection as per the circuit diagram

(ii) Switch on regulated dc power supply

(iii) Increase voltage of dc power supply (Vs) in steps of 2V and measure V and I of DIAC

(iv) Increase voltage of dc power supply (Vs) till current (I) increases with sudden drop in voltage across DIAC. The maximum voltage at which DIAC is turned on, is called the break over voltage V_{BO}.

(v) Measure break over voltage V_{BO} precisely

(vi) Take some readings of voltage and current after break over voltage

(vii) Plot V-I Characteristics in first quadrant operation of DIAC

(b) Third Quadrant Operation of DIAC (MT_1 is positive with respect to MT_2)

(i) Reverse the polarity of regulated dc power supply and meters or reverse the terminals DIAC

(ii) Repeat steps (iii) to (vi) as per (a)

(iii) Plot V-I Characteristics in third quadrant operation of DIAC

Observation Table:

Table 2.3 Observation table to test the proper functioning of DIAC to determine the break over voltage

Sr.No.	**MT_2 is positive with respect to MT_1**		**MT_1 is positive with respect to MT_2**	
	V (Volts)	I (mA)	V (Volts)	I (mA)

VIVA Questions Related to DIAC

1. Explain the different working mode of operation of DIAC
1. Determine first quadrant break over voltage and third quadrant break over voltage
2. Write specifications of the DIAC from the manufacturer data sheet

2. Experiment on determine the latching and holding current using V-I characteristics of SCR

Aim:

To study the V-I characteristics of a SCR and determine the latching and holding current using V-I characteristics of SCR.

Apparatus:

Regulated dc power supply 0-50V; Regulated dc power supply 0-30V; Digital voltmeter 0-50V DC, Digital ammeter 0-30mA; Digital ammeter 0-150mA; SCR-SN 104/TYN604, Resistance R_2=R_L=10kΩ,10W, R_3= 1kΩ,10W, R_1=R_G=3.3kΩ,0.5W

Circuit Diagram:

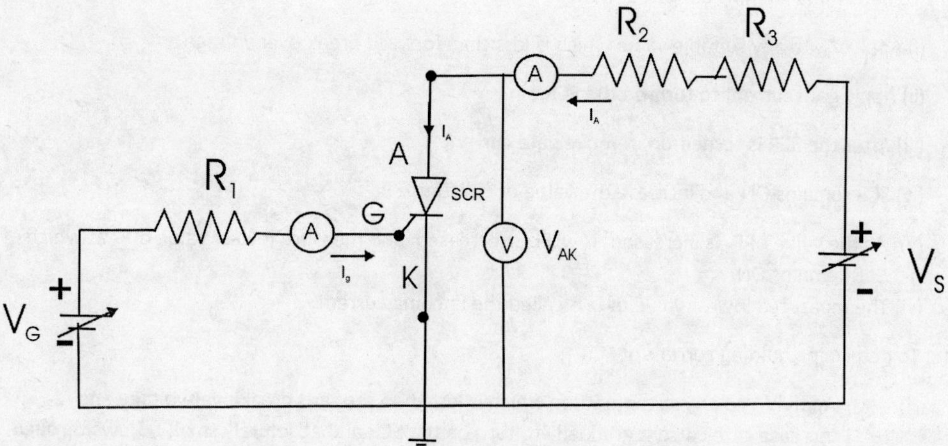

Fig.2.60 Circuit diagram to determine the latching and holding current using V-I characteristics of SCR.

Theory:

SCR is a unidirectional semiconductor device. It has three terminals namely anode (A), cathode (K) and gate (G). SCR is a four-layer as *p-n-p-n* and three junctional devices. The construction of SCR is described in section 2.2 and V-I characteristics of SCR is explained in section 2.3. SCRs are commonly used for home appliance control such as lighting, temperature control, fan speed regulator etc. and for industrial applications, SCRs are used to control motor speed, battery charging and power conversions.

Precautions:

1.Initially all the knobs of the power supplies are at zero value.

2.The applied voltage and current should not be more than the maximum rating of SCR.

3.Reading of observation should be error free.

Procedure

(a) Forward characteristics of SCR

(i) Make the circuit connection as per the circuit diagram

(ii) Switch on regulated dc power supply

(iii) SCR must be in off state with open gate and I_G=0 mA

(iv) Increase power supply voltage across anode to cathode (V_{AK}) gradually and when I_A flows then V_{AK} is called as forward break over voltage V_{BO}

(v) Set I_G=2mA, increase V_{AK} gradually till SCR turns on and measure V_{AK} and I_A

(vi) Set I_G at different value and repeat step (v)

(vii) Plot I_A versus V_{AK} for various values of I_G on graph paper

(b) To determine latching current of SCR

(i) Apply V_{AK} to any suitable value which is less than forward break over voltage

(ii) Apply gate current to turn-on the SCR

(iii) After the SCR is turned-on, remove gate current

(iv) SCR remains ON and increase the value of resistance R_L

(v) As the value of R_L is increased, I_A will be decreased and measure lowest value of I_A at which SCR remains ON

(vi) The measured lowest value of I_A is called the latching current.

(c) To determine holding current of SCR

i) Apply V_{AK} to any suitable value which is less than forward break over voltage and increase gate current gradually to turn on the SCR and absolutely small I_A flows through it.

ii) Remove gate current if SCR turns off, then as per step-(i) a minimum I_A flows through the SCR., and it is called holding current I_H.

iii) Gradually increase R_L and I_A will decrease

iv) Measure the lowest value of I_A at which SCR will be turns off. This is called holding current I_H.

Observation Table:

Table 2.4 Observation table for determine the latching and holding current using V-I characteristics of SCR.

Sr. No.	I_{G1}= in mA		I_{G2}= in mA		I_{G3}= in mA	
	V_{AK} in Volt	I_A in mA	V_{AK} in Volt	I_A in mA	V_{AK} in Volt	I_A in mA

Related Questions

1. Write specifications of SCR from manufacturer data sheet
2. Define forward break down voltage, holding current and latching current
3. Mention some applications of SCR

KNOW MORE

Thyristors can take several forms, but they have certain things in common. All of thyristor family devices are solid state switches which act as open circuits with forward rated voltage blocking capability until triggering pulse is applied. Whenever thyristors are triggered, they provides low-impedance current paths and remain in conduction state until the current stops or falls below the holding level. Most commonly used thyristor family devices such as SCR, LASCR, SCS, GTO, UJT, PUT, DIAC and TRIAC are already described in this unit. Field-controlled thyristors (FCTs) is a new family of power semiconductor device. To study the operating principle and application circuits of FCTs, please scan the QR code.

For more on Field Controlled

Thyristor (FCT)

References:

1. Ramamoorty M., An Introduction to Thyristor and their applications, East-West Press Pvt. Ltd, New Delhi
2. Sugandhi, Rajendra Kumar and Sugandhi, Krishna Kumar, Thyristors: Theory and Applications, New Age International (P) ltd. publishers, New Delhi
3. Bhattacharya, S.K., Fundamental of Power Electronics, Vikas Publishing House Pvt. Ltd. Noida
4. Jain & Alok, Power Electronics and its Applications, Penram International Publishing(India) Pvt. Ltd, Mumbai
5. Rashid, Muhammad, Power Electronics Circuits Devices and Applications, Pearson Education India, Noida
6. Sing, M.D. and Khanchandani, K.B., Power Electronics, Tata McGraw Hill Publishing Co. Ltd, New Delhi
7. Zbar, Paul B., Industrial Electronics: A Text-Lab manual, McGraw Hill Publishing Co. Ltd, New Delhi
8. Grafham D.R., SCR Manual, General Electric Co.

Dynamic QR Code for Further Reading

-QR codes embedded in the unit.

3 Turn-on and Turn-off Methods of Thyristors

UNIT SPECIFIES

Through this unit we have discussed the following aspects:

- SCR turn-on methods
- High voltage triggering, thermal triggering, illumination triggering, dv/dt triggering and gate triggering
- Gate triggering circuits- R and RC circuits, SCR triggering using UJT
- PUT as Relaxation oscillator and synchronized UJT circuit
- Pulse transformer and opto-coupler based triggering
- SCR turn-off methods

The practical applications of the topics are discussed for generating further curiosity and creativity as well as improving problem solving capacity of learners.

Not only a large number of multiple choice questions are incorporated in this chapter but also questions of short and long answer types marked in two categories following lower and higher order of Bloom's taxonomy, assignments through a number of numerical problems. A list of references and suggested readings are illustrated in this unit so that learners can go through them for practice and get detailed knowledge. It is also be noted that some QR codes have been incorporated in different sections of this unit for getting more information on various topics of interest.

After discussion the content related to theory, there is a "Know More" section at end of this unit which is related to laboratory experiments. This section has been carefully designed so that the supplementary information provided in this section becomes beneficial for the earners. Usually, this section highlights the initial activity, examples of some interesting facts, analogy, and history of the development of the power electrics devices focusing the salient observations and finding, timelines starting from the development of the concerned topics up to the recent time, applications of the subject matter for our day-to-day life, and industrial applications of power electronics devices.

RATIONALE

This unit on "turn-on and turn-off methods of thyristors" can be able to provide detail information regarding different turn-on methods of SCR such as high voltage triggering, thermal triggering, illumination triggering, dv/dt triggering and gate triggering. Gate triggering is most commonly used in modern industry for the control and conversion of electrical power. This unit can help students to get a primary idea about gate trigger circuits, R and RC circuits, SCR triggering using UJT, PUT as Relaxation oscillator and synchronized UJT circuit and Pulse transformer and opto-coupler based triggering. The different SCR turn–off methods such as class A- resonant commutation circuit, class B-shunt resonant commutation circuit, class C-complimentary symmetry commutation circuit, class D-auxiliary commutation, class E-external pulse commutation and class F-line or natural commutation are described in detail to help students to know about the recent development of turn-off methods of thyristors.

PRE-REQUISITES

Semiconductor physics

Analog and digital electronics

Electrical circuit theory

Electrical Machines and power system

UNIT OUTCOMES

List of outcomes of this unit is as follows:

U3-O1: To explain turn-on methods of SCR

U3-O2: To understand the concept of high voltage triggering, thermal triggering, illumination triggering, dv/dt triggering and gate triggering

U3-O3: To design gate triggering circuits, R and RC circuits and SCR triggering using UJT

U3-O4: To discuss PUT as Relaxation oscillator and synchronized UJT circuit

U3-O5: To explain pulse transformer and opto-coupler based triggering

U3-O6: To provide a list of SCR turn-off methods

U3-O5: To discuss about construction, working principle and V-I characteristics of IGBT

U3-O6: To provide a list of SCR turn-off methods

U3-O7: To understand the concept of class A- resonant commutation, class B-shunt resonant commutation and class C-complimentary symmetry commutation

U3-O8: To explain class D-auxiliary commutation, class E-external pulse commutation and class F-line or natural commutation

Unit-3 Outcomes	EXPECTED MAPPING WITH COURSE OUTCOMES (*1-Weak Correlation; 2- Medium Correlation; 3-Strong Correlation*)				
	CO-1	CO-2	CO-3	CO-4	CO-5
U3-O1	-	-	3	-	-
U3-O2	-	2	3	-	-
U3-O3	-	2	3	-	-
U3-O4	-	2	3	-	-
U3-O5	-	2	3	2	2
U3-O6	-	-	3	-	-
U3-O7	-	2	3	-	-
U3-O8	-	2	3	2	2

3.1 Introduction

Thyristor is a unidirectional switching device. It can be switched from off-state to on-state by forward voltage triggering, dv/dt triggering, temperature triggering, high voltage thermal triggering and gate triggering. Thryristor can be turned-off by natural commutation and forced commutation techniques. In this chapter, different turn-on and turn-off methods of thyristors are discussed elaborately.

3.2 TURN-ON METHODS of Thyristors

Thyristors can be turned on by different methods such as (i) High forward voltage triggering (ii) Thermal triggering, (iii) illumination triggering (iv) dv/dt triggering and (v) gate triggering. In this section, the above turn-on methods of thyristors are explained in detail.

3.2.1 High Forward Voltage Triggering

Whenever the applied voltage across anode to cathode is greater than the forward break-down voltage V_{BD}, the leakage current flow through anode to cathode and it can lead to a regenerative action. Subsequently, thyristor will be turned on. In reality, high forward voltage triggering type of turn on method is destructive and must be avoided.

3.2.2 Thermal Triggering

If the temperature of thyristor increases, the large numbers of electron-hole pairs are generated. Since the minority carriers moves freely and cross the junction J_2, the reverse leakage current increases. After crossing the junction, these minority carriers are accelerated and strike the covalent bond and generate more electron hole pairs. This increases the leakage currents and α_1 and α_2 increases. Due to regenerative action, $\alpha_1 + \alpha_2$ increases towards unity and the thyristor will be turned on. In fact, N_1 and P_2 layers have large number of thermally generated carriers and the device is turned on. Generally this method of triggering is not used for turn-on thyristors as it increases junction temperature and the voltage withstand capability is reduced. In this method of triggering, thermal runway is developed with in the device and therefore this method of triggering is normally avoided.

3.2.3 Illumination or Optical Triggering

When a beam of light falls on the junction J_2 of the thyristor, it generates the electron-hole pairs. The generated minority carriers cross the blocking junction J_2 and reach P_2. Due to high electric field, minority carriers are accelerated and strike the covalent bond of silicon wafer and generate more electron-hole pairs. Therefore the regenerative action takes place and the device is turned on. In fact this method of triggering is used in LASCR. This method of triggering is used for HVDC transmission system, static VAR compensation system where current as well as voltage rating are high.

3.2.4 dV / dt Triggering

When thyristor is forward biased, the junction J_2 is reverse biased and Junction J_1 and J_3 are forward biased. The reverse biased voltage across J_2 blocks the conduction of device. The depletion of junction J_2 provides large capacitance. If the rate of change of voltage across anode to cathode is high, the charging current $I_{CJ2} = C_{J2}(dV_{AK} / dt)$ increases and it flows though P_2 just like the gate current is supplied from gate terminal. Then the device is turned on. A high value of charging current may destroy the device. So, thyristors must be protected from high dV / dt. In fact, the snubber circuit is used for dV / dt protection.

3.2.5 Gate Triggering

The gate triggering is most commonly used to turn on thyristors. When the thyristor is forward biased and a positive gate voltage is applied between gate and cathode, gate current flows through gate and the thyristor will be turn on. For successful triggering of thyristor, the applied gate voltage V_g and gate current I_g must be close to their maximum value but always less than maximum value. If the gate current increases, the forward blocking voltage capability of thyristor decreases.

For more on Turn-on methods of SCR

Example 3.1 What are the different conditions to turn on a thyristor?

Solution: The following conditions must be satisfied to turn-on a thyristor

(i) Thyristor must be forward biased and anode potential must be positive with respect to cathode.

(ii) The width of gate pulse must be greater than the turn-on time of thyristor.

(iii) Anode current must be greater than latching current when the gate signal is removed.

(iv) Anode to cathode voltage must be greater than finger voltage.

(v) Amplitude of gate current must be more than the minimum gate current which is required to turn on SCR.

(vi) Amplitude of gate current must be less than the maximum permissible gate current so that gate circuit may not be damaged.

(vii) The gate triggering must be synchronized with ac supply in case of applications like controlled rectifier.

3.3 GATE TRIGGER CIRCUITS FOR THYRISTORS

Thyristor can be turned-on by forward-voltage triggering, dv/dt triggering, temperature triggering, light triggering and gate triggering. From these triggering methods, forward-voltage triggering, dv/dt triggering, and temperature triggering are not used to control output of any controlled rectifier circuit as these methods are not normal triggering. Illumination triggering is used in some special applications such as series-parallel connected thyristors. The most commonly used triggering method is gate triggering as by using this method, we can properly control the turn-on of thyristors as well as output voltage of converter. The gate triggering method is most efficient and reliable method to turn-on thyristors. To turn on thyristor from its forward blocking state, a gate signal of appropriate wave shape and frequency should be applied between gate and cathode.

During gate triggering, thyristor is forward biased and anode is positive with respect to cathode. When a voltage is applied at the gate terminal, thyristor will be turned-on. The gate voltage may be a slow-rising rectified voltage or a sharp single pulse with constant amplitude dc voltage and high frequency pulse voltage as shown in Fig.3.1. Thyristor will be turned-on as soon as gate voltage V_g is greater than the required critical gate trigger voltage V_{gT}. At $\omega t = \alpha$, voltage V_g is equal to V_{gT} and the thyristor will be triggered at $\omega t = \alpha$. Subsequently the firing angle of thyristor is α.

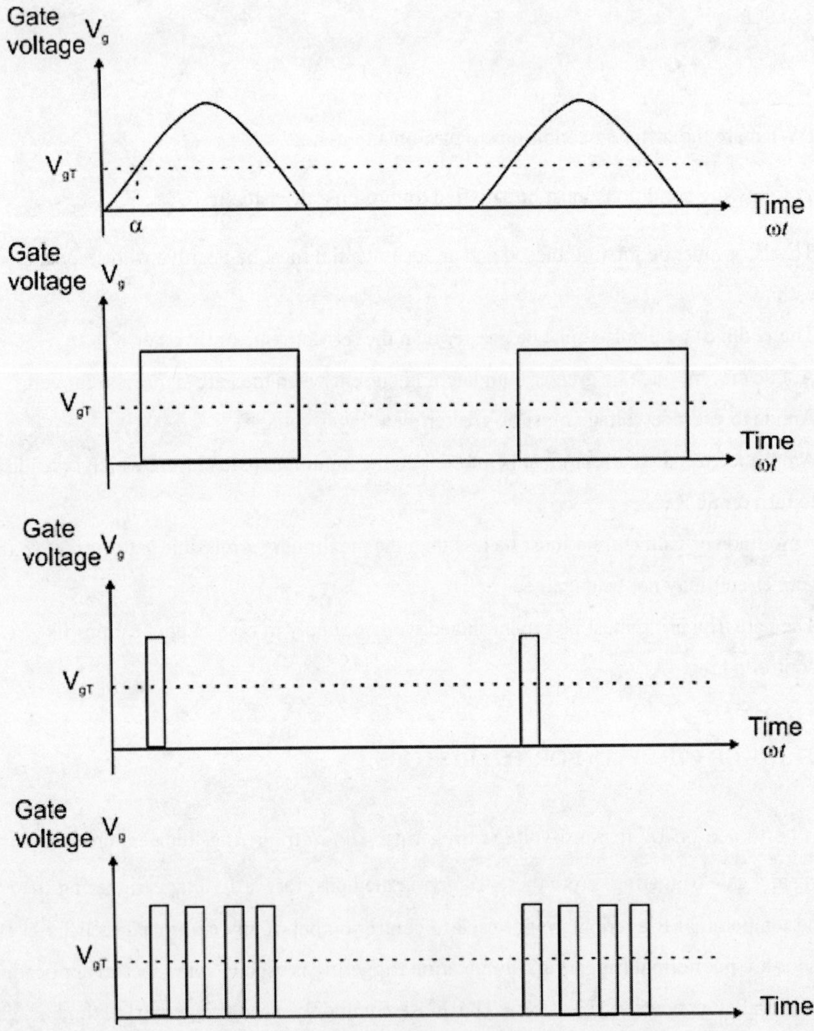

Fig.3.1 Triggering signals of a thyristor

Fig.3.2. shows an ideal gate current waveform. The initial high amplitude and fast rise in gate current can able to turn-on the thyristor. Generally, a continuous gate voltage is required for thyristor. If thyristor is already turned-on, there is no requirement of gate signal after successful triggering of thyristor. The gate voltage signal can be generated by a gate-drive which is called *firing* circuit for thyristors. The different type of triggering circuits are (i) Resistance (R) triggering circuit, (ii) Resistance capacitance (RC) triggering circuit, (iii) Uni-junction transistor triggering circuit, (iv) UJT relaxation oscillator triggering circuit, (v) synchronized UJT circuit, and (vi) pulse transformer and opto-coupler based triggering.

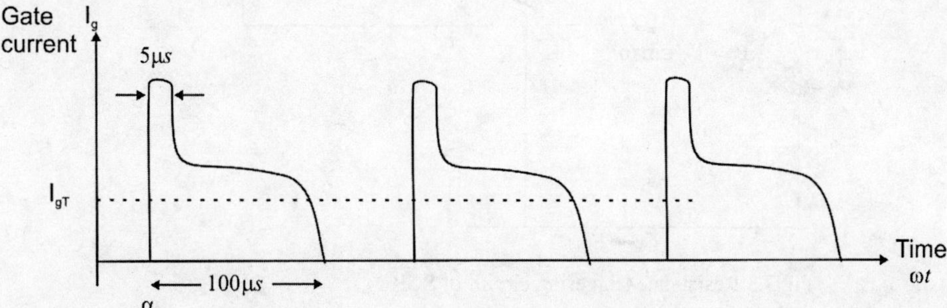

Fig.3.2 Gate current waveform

3.4 Resistance Triggering Circuit

The simplest triggering circuit of thyristor is resistance triggering circuit as shown in Fig.3.3. R_1 is the variable resistance, R_2 is the stabilizing resistance. If R_1 is zero, the gate current flows through R_{min}, D, gate–cathode, load and source. This current should not be greater than maximum permissible gate current I_{gm}

$$\frac{V_m}{R_{min}} \leq I_{gm} \tag{3.1}$$

Therefore $$R_{min} \geq \frac{V_m}{I_{gm}} \tag{3.2}$$

where, V_m is the maximum source voltage.

Fig.3.4 shows the voltage waveforms of resistance triggering circuit of SCR. In positive half cycle of input voltage and at $\omega t = \alpha$, the voltage applied at gate terminal is greater than V_{gT}. Before turn-on of thyristor, the input voltage is applied to thyristor and operates in forward blocking state. The value of R_2 should be such that the maximum voltage across it is not greater than gate voltage V_{gm}.

So, $$V_m \frac{R_2}{R_{\min} + R_1 + R_2} \le V_{gm} \quad (3.3)$$

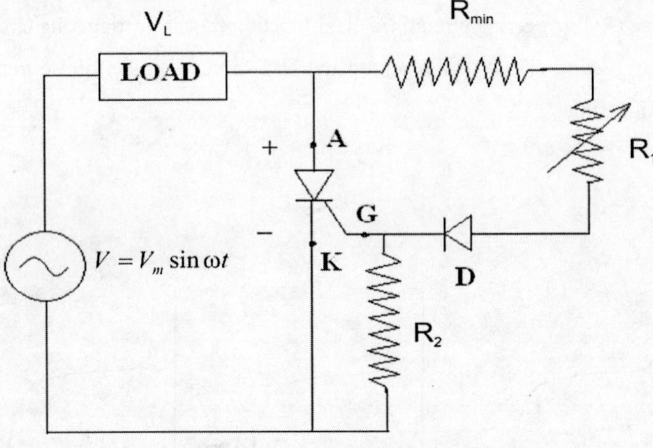

Fig.3.3 Resistance triggering circuit of SCR

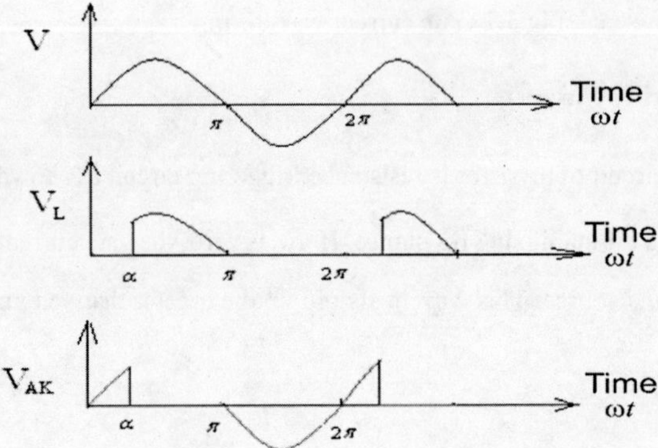

Fig.3.4 Voltage waveforms of resistance triggering circuit of SCR

As R_1 and R_2 are very large, the gate trigger circuit draws small current. Due to presence of diode D, the current flows in positive half cycle only. The amplitude of voltage is controlled by V_{gT}. If R_1 is very large and voltage across R₂ is $V_g = iR_2$. When the peak value of V_g is less than V_{gT}, thyristor will not be turned-on. Consequently, the output voltage is equal to zero.

While R_1 is reduced, i_g current increases, V_g exceeds V_{gT} and thyristor can be turn-on.

At $\omega t = \alpha$, $$V_{gp} \sin \alpha = V_{gT} \quad (3.4)$$

and $\alpha = \sin^{-1}(\frac{V_{gT}}{V_{gp}})$ (3.5)

If $V_{gp} = V_m \frac{R}{R + R_1 + R_2}$ (3.6)

Firing angle will be $\alpha = \sin^{-1}(\frac{V_{gT}}{V_m} \frac{R + R_1 + R_2}{R})$ (3.7)

Since V_{gT}, R_1, R, V_m are constant, $\alpha \propto \sin^{-1}(R_2)$ (3.8)

Therefore, $\alpha \propto R_2$ (3.9)

Thus the firing angle of thyristor is directly proportional to R_2. When R_2 increases, the firing angle of thyristor α increases. The maximum limit of firing angle though resistance triggering circuit is $\alpha = 90°$. Here $\alpha = 0°$ is not possible, the range of firing angle will be represented by $90° \geq \alpha > 0°$. The resistance triggering circuit can be used for Triac. If diode D is removed and the triggering signal will be available at the gate terminal during both positive and negative half cycles.

3.5 Resistance Capacitance Triggering Circuit

Fig.3.5 shows a resistance capacitance triggering circuit and voltage wave forms are given in Fig.3.6. During the negative half cycle of supply voltage, capacitor C charges through diode D_2 to the negative peak value of supply voltage ($-V_m$). At $\omega t = -90°$, $V_c = -V_m$. After $\omega t = -90°$, the supply voltage starts to increase from $-V_m$ to zero. At $\omega t = 0°$, the supply voltage is equal to zero. During $-90° \leq \omega t \leq 0°$, the capacitor voltage decreases and finally falls to OA. At $\omega t = 0°$, $V_c = OA$. Just after $\omega t > 0°$, the supply voltage is positive and capacitor starts to charge though variable resistance R. Whenever the capacitor voltage reaches B and holds the positive voltage during the positive half cycle of supply voltage, the capacitor voltage V_c at point B is greater than V_{gT} and thyristor will be turned on. Thus the firing angle can be controlled by varying resistance R.

In this circuit, diode D_1 is used to flow current in positive direction only. Therefore it prevents the breakdown of cathode to gate junction in negative half-cycle. Using the resistance capacitance triggering circuit, firing angle will never become $0°$ and $180°$. The control range of firing angle is $0° < \alpha < 180°$. In a resistance capacitance triggering circuit firing circuit, the following condition must be satisfied:

$$RC \geq \frac{1.3T}{2} \cong \frac{4}{\omega} \quad (3.10)$$

where, $T = \frac{1}{f}$ = time period of ac line frequency in seconds

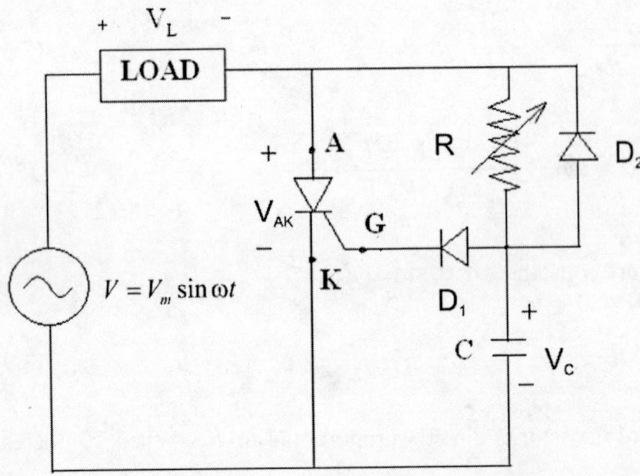

Fig.3.5 Resistance capacitance triggering circuit of SCR

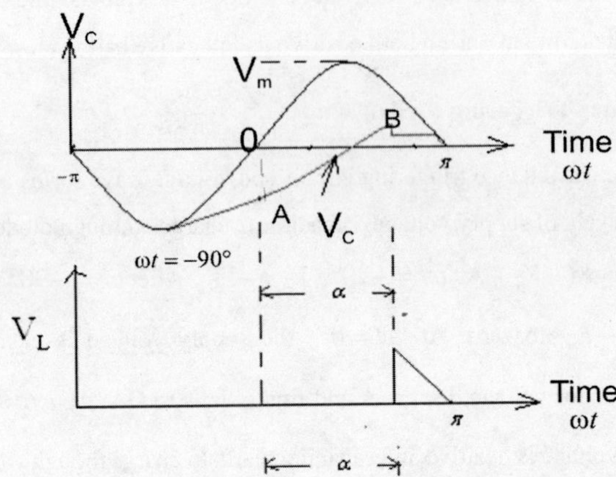

Fig.3.6 Voltage waveforms of resistance capacitance triggering circuit of SCR

Thyristor will be turned on when capacitor voltage is equal to $V_c = V_{gT} + V_d$ (3.11)

where, V_d is the voltage drop across diode D_1.

Since the capacitor voltage drop is constant at the instant of triggering, the gate current I_{gT} must be flow through R, D_1 and gate-cathode of thyristor.

The maximum value of R can be determined by the following equation

$$V \geq I_{gT} R + V_c \qquad (3.12)$$

Therefore, after substituting $V_c = V_{gT} + V_d$ in the equation (3.12), we obtain

$$V \geq I_{gT} R + V_{gT} + V_d \tag{3.13}$$

$$\text{So, } R \leq \frac{V - V_{gT} - V_d}{I_{gT}} \tag{3.14}$$

Where, V is the voltage at which thyristor will be turned on.

If the thyristor is turned on and operates in conduction state, the on-state voltage drop across thyristor is about 1V to 1.5V. Consequently, the voltage drop across R and C will also be reduced to the value of 1V to 1.5V until the negative half cycle voltage appears across C. During the negative half cycle of supply voltage, the capacitor is charged to the maximum voltage ($-V_m$). When the value of R is less, firing angle of thyristor is less and conduction angle is more. If R is increased, firing angle increases and conduction angle decreases.

For more on Triggering Circuits

Example 3.2 In resistance triggering circuit as shown in Fig.3.3, $I_{g(\min)} = 0.12mA$ and $V_{g(\min)} = 0.6V$. If the peak amplitude of input voltage is 210V, find the trigger angle α for $R_1 = 100k\Omega$ and $R_{\min} = 10k\Omega$

Solution:

The KVL equation of gate circuit is

$$V = I_g (R_{\min} + R_1) + V_D + V_g$$

At the point of trigger,

$$V = I_g (R_{\min} + R_1) + V_D + V_g = 0.12 \times 10^{-3} (10 \times 10^3 + 100 \times 10^3) + 0.7 + 0.6 \text{ V}$$

$$= 13.8 \text{ V}$$

If the firing angle of thyristor is α, $13.8 = 210\sin\alpha$

Then $\alpha = \sin^{-1}\left(\frac{13.8}{210}\right) = 3.76°$

Example 3.3 A 220V, 50Hz ac supply is connected to a resistance capacitance triggering circuit. If the resistance R is variable from $2k\Omega$ to $20k\Omega$, $V_{gT} = 2V$ and $C = 0.47\mu F$, what is the minimum and maximum firing angle?

Solution:

The current flow through capacitance C is equal to

$$I = \frac{V}{Z} \quad \text{where } Z = \sqrt{R^2 + (\frac{1}{\omega C})^2}$$

(i) If $R_2 = 2k\Omega$, $Z = \sqrt{R^2 + (\frac{1}{\omega C})^2} = \sqrt{2000^2 + (\frac{1}{2\pi \times 50 \times 0.47 \times 10^{-6}})^2} = 7064.98\,\Omega$

and $\phi = \tan^{-1}\frac{1/\omega C}{R_1} = \tan^{-1}\frac{6775.98}{2000} = 73.55°$

The current I leads V by an angle ϕ, $I = \frac{V}{Z} = \frac{220\angle 0°}{7064.98\angle -73.55°} = 0.03113\angle 73.55°$

The voltage across capacitor is

$$V_C = IX_C = 0.03113\angle 73.55° \times 6775.98\angle -90° = 210.93\angle -16.45°$$

So, $v_C = \sqrt{2} \times 210.93 \sin(\omega t - 16.45)$

At $\omega t = \alpha_1$, $v_C = \sqrt{2} \times 210.58 \sin(\alpha_1 - 16.45) = 2$

Therefore, $\alpha_1 = \sin^{-1}\frac{2}{\sqrt{2} \times 210.93} + 16.45 = 16.75°$

(ii) If $R_2 = 20k\Omega$, $Z = \sqrt{R^2 + (\frac{1}{\omega C})^2} = \sqrt{20000^2 + (\frac{1}{2\pi \times 50 \times 0.47 \times 10^{-6}})^2}$

$= 21116.67\,\Omega$

and $\phi = \tan^{-1}\frac{1/\omega C}{R_1} = \tan^{-1}\frac{6775.98}{20000} = 18.71°$

The current I leads V by an angle ϕ, $I = \frac{V}{Z} = \frac{220\angle 0°}{21116.67\angle -18.71°} = 0.010417\angle 18.71°$

The voltage across capacitor is

$$V_C = IX_C = 0.010417\angle 18.71° \times 6775.98\angle -90° = 70.58\angle -71.29°$$

So, $v_C = \sqrt{2} \times 70.58 \sin(\omega t - 71.29)$

At $\omega t = \alpha_1$, $v_C = \sqrt{2} \times 70.58 \sin(\alpha_1 - 71.29) = 2$

Therefore, $\alpha_1 = \sin^{-1} \frac{2}{\sqrt{2} \times 70.58} + 71.29 = 72.43°$

3.6 SCR Triggering Circuit Using UJT

SCR triggering circuit using UJT is shown in Fig.3.7. In this circuit, the values of external resistances R_1 and R_2 are smaller than the values of internal resistances R_{B1} and R_{B2} of UJT. The charging resistance R should be variable one and value of R should be such that load line must intersect the negative resistance region of UJT characteristics. If a dc voltage V is applied, the capacitor C starts to charge through R. During charging of capacitor C, emitter of UJT behaves as open circuit. The capacitor voltage V_c can be expressed by

$$V_c = V(1 - e^{-\frac{t}{\tau_1}}) \tag{3.15}$$

Therefore, $$V_c = V(1 - e^{-\frac{t}{RC}}) \tag{3.16}$$

Where, $\tau_1 = RC$ is the charging time constant of capacitor.

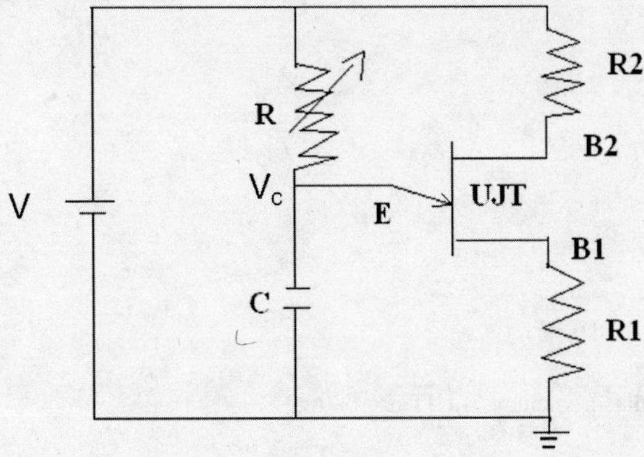

Fig.3.7 SCR triggering circuit using UJT

Voltage waveforms of SCR triggering circuit using UJT are shown in Fig.3.8. Whenever the capacitor voltage Vc reaches the threshold voltage V_p, the junction between E-B1 will get break down. Accordingly, UJT will be turned-on and capacitor C starts to discharge through resistance R1. The discharging time constant is $\tau_2 = R_1 C$. Here τ_2 is smaller than τ_1. As the emitter voltage decreases to valley voltage V_v, emitter current falls below I_v. Subsequently, UJT will be turned off.

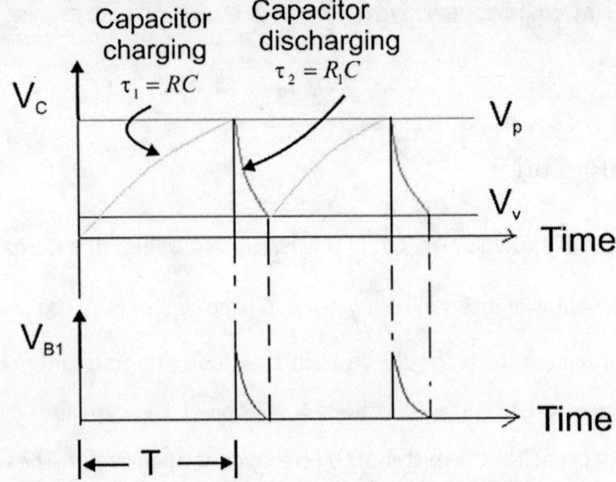

Fig.3.8 Voltage waveforms of SCR triggering circuit using UJT

The threshold voltage is equal to

$$V_p = \eta V + V_D = V_v + V(1 - e^{-\frac{t}{RC}}) \tag{3.17}$$

Where, $V_D = V_v$ and $\eta = (1 - e^{-\frac{t}{RC}})$

At $t = T$, $\quad \eta = (1 - e^{-\frac{T}{RC}})$ (3.18)

So, $\quad T = \frac{1}{f} = RC\ln(\frac{1}{1-\eta})$ (3.19)

The firing angle is equal to

$$\alpha_1 = \omega T = \omega RC\ln(\frac{1}{1-\eta}) \tag{3.20}$$

where, ω is the angular frequency of UJT oscillator.

When the output pulse of UJT oscillator is used to trigger thyristor, R_1 should be small so that the normal leakage current drop must be less than V_p and UJT will not be triggered.

So, $\quad V\frac{R_1}{R_{BB} + R_1 + R_2} <$ SCR trigger voltage V_{gT} (3.21)

Where, $R_{BB} = R_{B1} + R_{B2}$

The value of $R_2 = 10^4 / \eta V$ and the width of triggering pulse is equal to R_1C.

The maximum value of R can be computed from the peak voltage V_P and peak current I_P. When the voltage across capacitor is V_P, the voltage across R is equal to $V - V_p$.

$$\text{So,} \quad R_{max} = (V - V_p)/I_p = [V - (\eta V + V_D)]/I_p \qquad (3.22)$$

The minimum value of R can be determined from valley point voltage and current values V_v and I_v

$$R_{min} = (V - V_v)/I_v \qquad (3.23)$$

3.7 PUT: Relaxation Oscillator

Relaxation oscillator is the most common application of a programmable UJT. PUT relaxation oscillator can be used to generate a wide range of saw tooth wave forms. It is known as a relaxation oscillator as the timing interval is started by the gradual charging of a capacitor and the timing interval is terminated by the sudden discharge of the same capacitor. The circuit diagram of PUT as relaxation oscillator is shown in Fig.3.9.

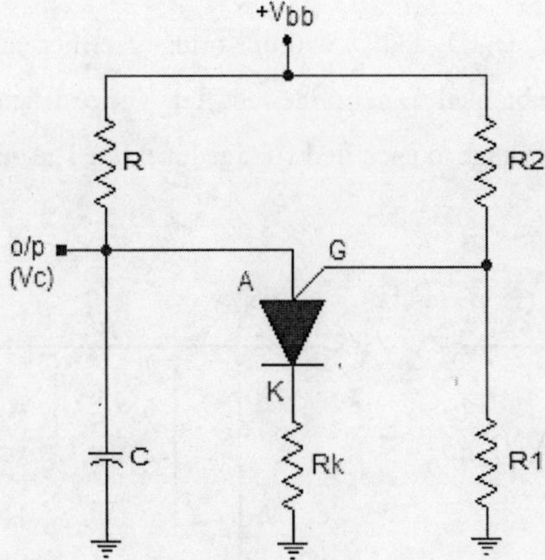

Fig.3.9 PUT as relaxation oscillator

Resistors R1 and R2 are connected in series and set the peak voltage (Vp) and intrinsic standoff ratio (η) of the PUT. Resistor Rk is used to limit cathode current of the PUT. Resistor R and capacitor C are used to set the frequency of the oscillator. Once the supply voltage Vbb is applied, the capacitor C starts to charge through resistor R. When the voltage across the capacitor C exceeds the peak voltage (Vp), the PUT operates into negative resistance region of V-I characteristics of PUT and it creates a low resistance path from anode to cathode. Thereafter the capacitor discharges through the path R-C. Whenever the voltage across the capacitor C is less than the valley point voltage (Vv), the PUT revert back to its initial condition. Then capacitor C starts to charge again and the cycle is repeated accordingly. Fig.3.10 shows the charging and discharging of capacitor as a saw tooth waveform.

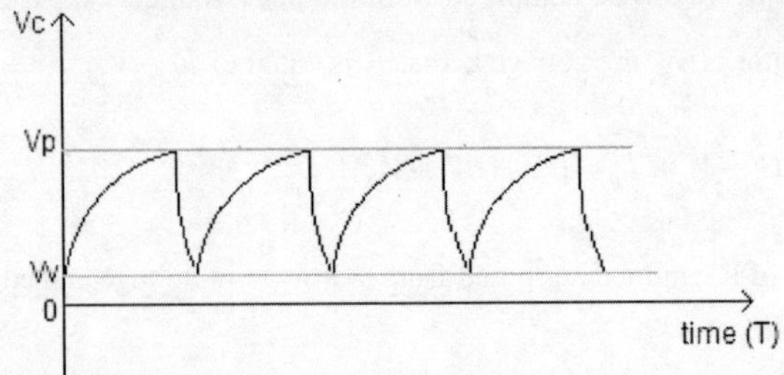

Fig.3.10 Charging and discharging of capacitor as a saw tooth waveform

The frequency of oscillation of a PUT relaxation oscillator is equal to

$$f = 1/(RC\ln(1/(1-\eta))) \qquad (3.24)$$

where, f is the frequency, η is the intrinsic standoff ratio, R is the resistance and C is the capacitance.

3.8 Synchronized UJT Circuit.

A synchronised UJT Circuit for triggering thyristor is shown in Fig.3.11. This circuit consists of a bridge rectifier using four diodes D_1, D_2, D_3 and D_4 and this bridge rectifier converts ac to dc. The full wave rectified dc output voltage is obtained from bridge-rectifier. The resistance R_1 and zener diode Z are used to clip the rectified output voltage to specified voltage level V_Z. The output voltage waveforms are illustrated in Fig.3.12.

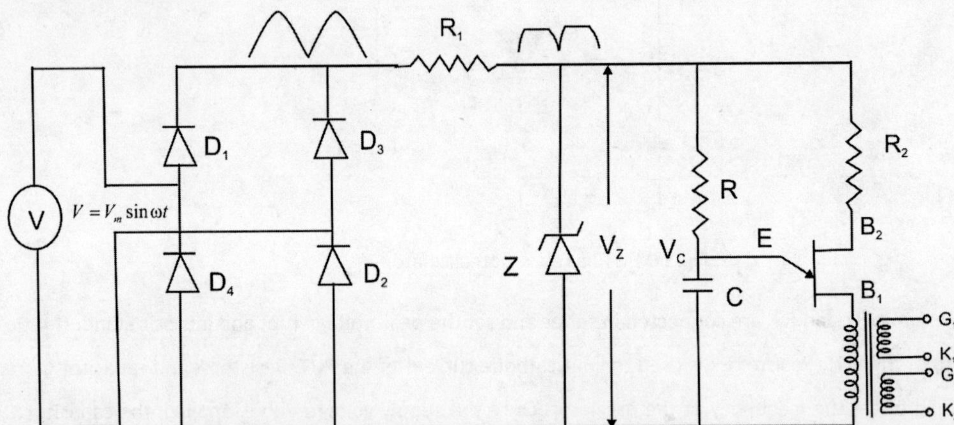

Fig. 3.11 Synchronised UJT Circuit

As output voltage is applied across RC circuit, the capacitor C is charged and its charging rate depends on the value of resistance R. When the capacitor voltage V_C reaches the threshold voltage of uni-junction transistor (ηV_Z), the emitter-base (E-B_1) junction of UJT will be breakdown and the capacitor will be discharged through primary winding of pulse transformer. Then current *i* flows through primary winding of pulse transformer, a voltage will be induced in the secondary winding of pulse transformer and it is used as triggering signal of thyristor.

During discharging of capacitor, once the capacitor voltage is less than valley voltage of UJT, E-B_1 junction becomes open circuit and capacitor starts again to charge through R. The rate of capacitor voltage can be controlled by varying the value of resistance R. By using this triggering circuit, the firing angle is varied within range of $0°$ to $150°$. Fig.3.13 shows the triggering circuit for single phase half wave controlled rectifier and the generation of triggering pulse and output voltage of single phase half wave controlled rectifier are depicted in Fig.3.12.

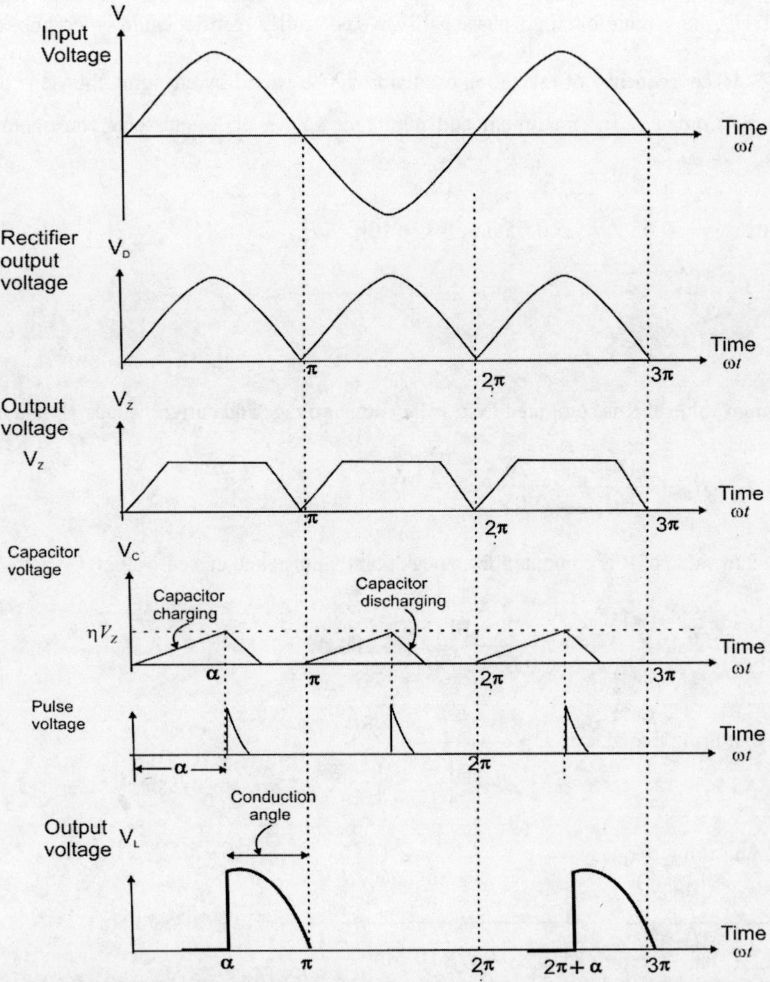

Fig.3.12 Generation of triggering pulse and output voltage of single phase half-wave controlled rectifier

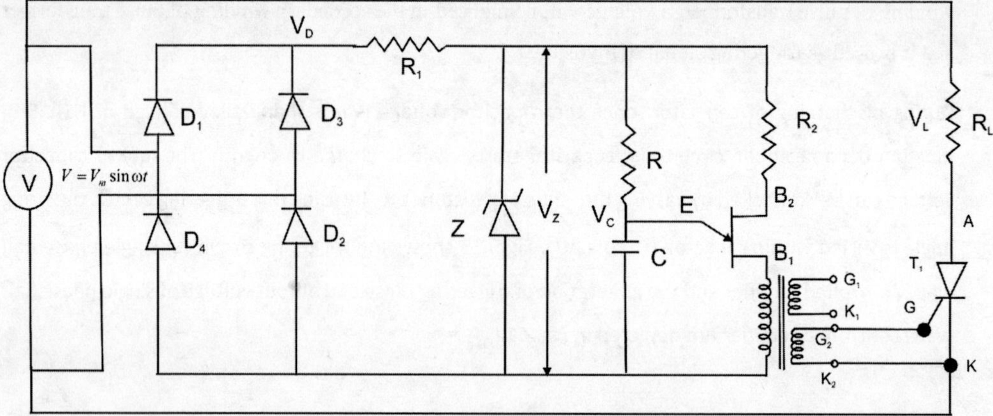

Fig.3.13 Triggering circuit for single phase half-wave controlled rectifier using synchronized UJT

Example 3.4 The frequency of relaxation oscillator can be varied by changing the value of charging resistance R. Compute the maximum and minimum values of R and their corresponding firing frequencies.

Assume that $\eta = 0.65$, $I_p = 0.65mA$, $V_p = 10V$, $I_v = 2.0mA$, $V_v = 2V$, $V_{BB} = 20V$, and $C = 0.047\mu F$

Solution:

The minimum value of R is computed from valley point voltage and current values V_v and I_v

$$R_{\min} = \frac{V - V_v}{I_v} = \frac{20 - 2}{2.0 \times 10^{-3}} = 9k\Omega$$

The maximum value of R is computed from peak point voltage and current values V_p and I_p

$$R_{\max} = \frac{V - V_p}{I_p} = \frac{20 - 10}{0.65 \times 10^{-3}} = 15.384k\Omega$$

We know that firing frequency $f = \dfrac{1}{RC\ln\left(\dfrac{1}{1-\eta}\right)}$

The maximum firing frequency

$$f_{\max} = \frac{1}{R_{\min} C \ln\left(\dfrac{1}{1-\eta}\right)} = \frac{1}{9\times10^{3} \times 0.047\times10^{-6} \ln\left(\dfrac{1}{1-0.65}\right)} \text{Hz} = 2075.73 \text{ Hz}$$

The minimum firing frequency

$$f_{min} = \frac{1}{R_{max} C \ln\left(\frac{1}{1-\eta}\right)} = \frac{1}{15.384\times10^{3}\times0.047\times10^{-6}\ln\left(\frac{1}{1-0.65}\right)} \text{ Hz} = 1317.41 \text{ Hz}$$

3.9 Pulse Transformer and Opto-coupler Based Triggering

The gate drive circuit operates at low voltage with lower power but the load/power circuit is connected with high voltage and power circuit is used to control power. If the gate drive circuit is not isolated from power circuit, the gate drive or control circuit will be damaged. So, the isolation between the gate drive circuit and the power circuit is required for high power applications. Generally, isolation can be provided either by using pulse transformers or by using opto-couplers. In this section, pulse transformer and opto-coupler base triggering circuits are discussed elaborately.

3.9.1 Pulse Transformer Based Triggering

Pulse transformers are used in firing circuits for SCR. Generally pulse transformer has two secondary windings and one primary. The turn ratio from primary to secondary is 1:1:1 or 2:1:1. The pulse transformer has low winding resistance; low leakage reactance and low inter winding capacitance. The pulse transformers which are used in firing circuit of SCR, have the following advantages: (i) the isolation between the low voltage gate circuit and high-voltage anode circuit, and (ii) the triggering pulses from same trigger source can be used to turn-on two or more number SCR.

Fig.3.14 shows the pulse transformer triggering circuit for SCR. While a square pulse is applied at the primary terminals of pulse transformer, it will be transmitted at its secondary terminals as a square wave or the transmitted signal will be the derivative of the input voltage signal.

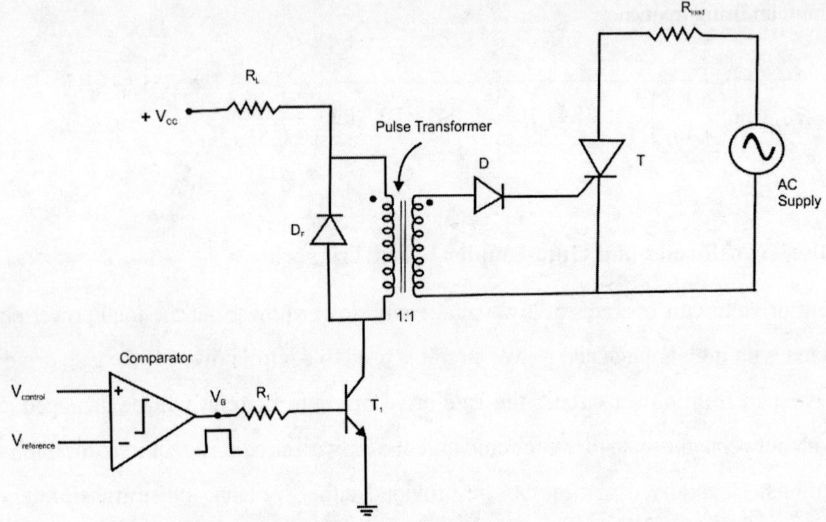

Fig.3.14 Pulse transformer triggering circuit for SCR

In fact a square waveform is applied to base of a transistor T_1 which is act as a switch. When the input pulse is high, transistor T_1 will be turned on and the dc voltage is applied to primary winding of pulse transformer. The advantage of pulse transformer based triggering circuit is that (i) it is not required a strength pulse generator since the amplitude of pulse is same; the strength of generated pulses may be increased by increasing the amplitude of dc bias voltage; (ii) the circuit operation is independent of the pulse characteristics as the pulse is used to turn-on or turn-off the transistor. The pulse distortion has no effect on the circuit operation. The resistance R_L is connected in series with the primary winding of pulse transformer to limit the current flow; and (iii) when the amplitude of applied pulse at base of transistor T_1 is low, T_1 will be turned off and then diode D starts to operate. Subsequently diode D allows to flows the current through pulse transformer.

3.9.2 Opto-coupler Based Triggering

Opto-couplers are made up of a light emitting device and a light sensitive device, all wrapped up in one package but with no electrical connection between the two. An opto-coupler allows two circuits to exchange signals yet remain electrically isolated. The standard opto-coupler circuits design uses a LED shining on a phototransistor. When the signal is applied to the LED, then shines on the transistor in the IC.

It is implemented using a simple circuit. Simply provide a small pulse at the right time to the light emitting diode in the package. The light produced by the LED activities the light sensitive properties of diac and the power is switched on. The isolation between the low power and high power circuits in these optically connected devices is typically about several thousand volts.

The most commonly used standard single channel phototransistor couplers is MCT2E. MCT2E family is an industry standard single channel phototransistor. Each opto-coupler consists of gallium arsenide infrared LED and a silicon NPN phototransistor as shown in Fig.3.15.

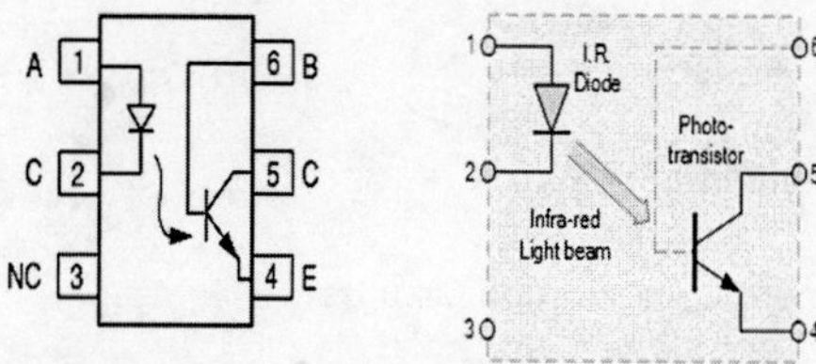

Fig.3.15 Opto-coupler IC- MCT2E

Fig.3.16 shows the opto-coupler based triggering circuit. The externally connected 270kΩ resistor is used to control the sensitivity of the photo-transistors base region. The value of the resistor can be chosen in such a way that it can able to suit the selected photo-coupler device and the amount of switching sensitivity required. The capacitor stops any unwanted spikes or transients from false triggering the opto-transistors base.

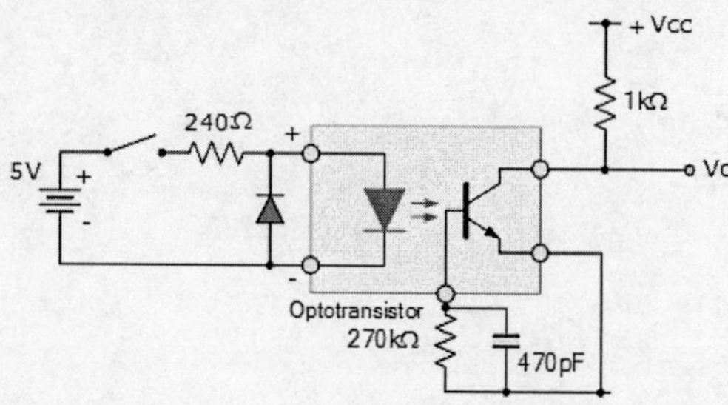

Fig.3.16 shows opto-coupler based triggering circuit

Usually opto-coupler used in lamp ballasts, motor control and incandescent lamp dimmers. The opto-coupler can also be used in switch mode power supply circuit in many electronics equipment. It is connected in between the primary and secondary section of supplies. The applications of opto-coupler in (i) monitor high voltage (ii) output voltage sampling for regulation (iii) system control using microprocessor and microcontroller and (iv) ground isolation.

3.10 SCR Turn-off Methods

SCR can be turned-off when its forward current I_A becomes below the holding current I_H or when a reverse biased voltage is applied across the thyristor for a specified time so that the device operates in the blocking state. After the thyristor turned-off, if the forward voltage is reapplied after certain time, then the excess carriers in the outer and inner layers of the SCR have to decay adequately. The decay and recombination of excess carriers can be accelerated by applying a reverse voltage across the SCR. The process of turning-off of SCR is called *commutation.* The different turn-off methods of SCR are

(i) Class A- Resonant Commutation Circuit

(ii) Class B- Shunt Resonant Commutation Circuit

(iii) Class C- Complimentary Symmetry Commutation Circuit

(iv) Class D- Auxiliary Commutation

(v) Class E- External Pulse Commutation

(vi) Class F- Line or Natural Commutation

In this section, all above turn-off methods of SCR are explained elaborately.

For more on SCR Turn-Off methods

3.10.1 Class A- Resonant Commutation Circuit

Fig.3.17 (a) shows the Class A-resonant commutation circuit where L and C are connected in series with load resistance R_L and L and C are used as commutation components and Fig.3.17(b) shows the Class A-resonant commutation circuit where load resistance R_L is connected in parallel with C.

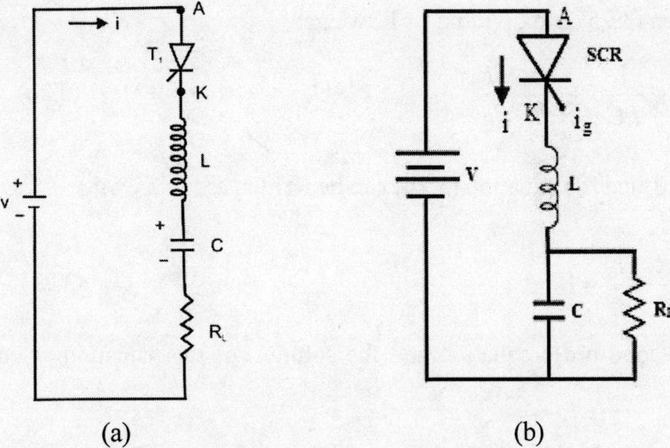

Fig.3.17 Class A resonant commutation circuit

When a dc voltage is applied to the circuit and a gate pulse is also applied between gate and cathode, SCR will be turned on and the charging current of capacitor flows through SCR. After certain time, the charging current decays to a value which is less than holding current when the capacitor is charged to supply voltage V. Fig.3.18 shows the current wave from. At point A, the current becomes zero and the commutation of SCR occurs at this point. The time taken to turn-off the SCR depends on the resonant frequency. The resonant frequency is function of commutating components and load resistance. In high frequency applications above 1 kHz, the LC resonant circuit is used as commutation circuit. Usually this circuit is widely used in series inverter circuit. This commutation is also known as *self-commutation.* Fig.3.19 Voltage and current waveforms of Class A series resonant commutation circuit (when load R_L is parallel with capacitor).

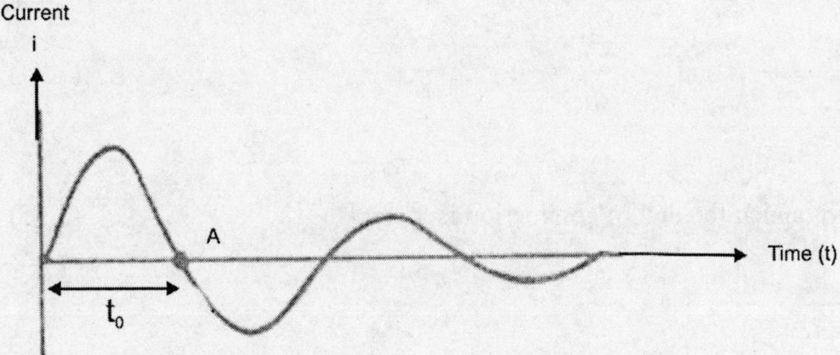

Fig.3.18 Current flows through SCR in Class A-series resonant commutation

3.10.1.1 Design of Class A-Commutation when L and C are connected in series with Load R_L

Assume that SCR T_1 is turned-on at $t = 0$ and initial capacitor voltage is about zero. The KVL equation of circuit is

$$V = L\frac{di}{dt} + \frac{1}{C}\int i.dt + i.R_L \qquad (3.25)$$

After differentiating equation (3.25) and dividing by L, we get

$$\frac{1}{L}\frac{d}{dt}(V)=\frac{d^2i}{dt^2}+\frac{1}{LC}i+.\frac{R_L}{L}\frac{di}{dt} \tag{3.26}$$

As V is constant, $\frac{d}{dt}(V)=0$ and the equation (3.26) can be written as

$$\frac{d^2i}{dt^2}+\frac{1}{LC}i+.\frac{R_L}{L}\frac{di}{dt}=0 \tag{3.27}$$

The equation (3.27) is a second order equation and the solution of this equation at under damped condition is equal to

$$i=e^{-\delta t}\left(A_1\cos\omega t+A_2\sin\omega t\right) \tag{3.28}$$

$$\text{Where,}\quad \delta=\frac{R_L}{2L}, \tag{3.29}$$

$$\text{and } \omega_o=\frac{1}{\sqrt{LC}} \tag{3.30}$$

$$\omega=\omega_o\sqrt{1-\delta^2}=\omega_o\sqrt{\frac{1}{LC}-\frac{R^2}{4L^2}} \tag{3.31}$$

$$\text{If } i(0^+)=i(0^-)=0, A_1=0 \text{ and } A_2=\frac{V}{L} \tag{3.32}$$

$$\text{Therefore,}\qquad \text{current } i(t)=e^{-\frac{R_L}{2L}t}\left[\frac{V}{\omega L}\sin\omega t\right] \tag{3.33}$$

at $\omega t=\pi$, SCR current becomes zero.

$$\text{So } t=\frac{\pi}{\sqrt{\frac{1}{LC}-\frac{R^2}{4L^2}}}\quad\text{and}\quad\frac{di}{dt}=-e^{-\frac{\pi R_L}{2\omega L}}\left(\frac{V}{L}\right) \tag{3.34}$$

The capacitor voltage at the end of conduction is $V_C=V-V_L$ (3.35)

where, $V_L=L\frac{di}{dt}$

$$\text{Therefore, } V_C=V\left[1+e^{-\frac{\pi R_L}{2\omega L}}\right] \tag{3.36}$$

If V_O is the initial state voltage of the capacitor, the equation (3.33) can be written as

$$i(t)=e^{-\frac{R_L}{2L}t}\left[\frac{V-V_o}{\omega L}\sin\omega t\right] \tag{3.37}$$

$$\text{and } V_C=V+e^{-\frac{\pi R_L}{2\omega}t}\left(V-V_o\right) \tag{3.38}$$

Since $\omega > 0$, the condition for under damped is equal to

$$\frac{1}{LC} - \frac{R^2}{4L^2} > 0 \tag{3.39}$$

$$\text{Therefore} \quad R < \sqrt{\frac{4L}{C}} \tag{3.40}$$

The voltage and current waveforms of class A-resonant commutation circuit(when load R_L is parallel with capacitor is illustrated in Fig.3.19.

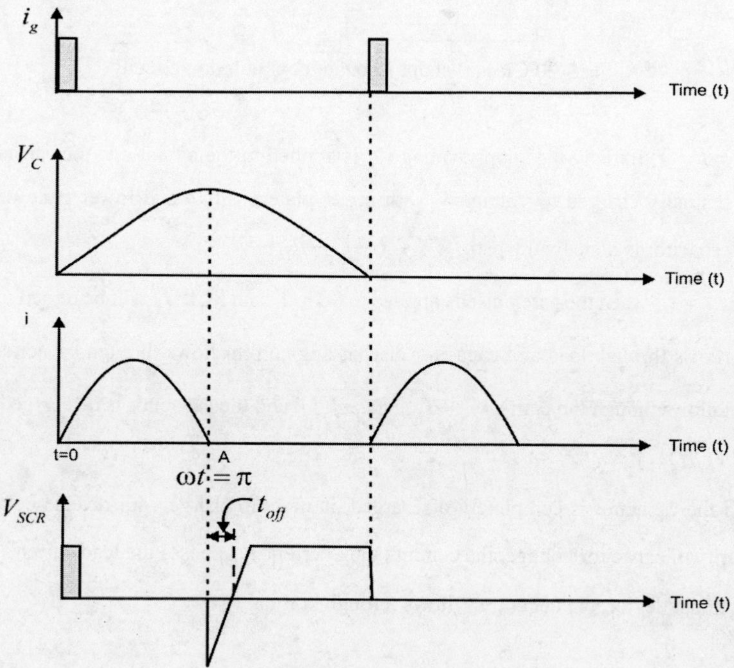

Fig.3.19 Voltage and current waveforms of Class A series resonant commutation circuit (when load R_L is parallel with capacitor)

3.10.2 Class B - Shunt Resonant Commutation Circuit

Fig.3.20 shows the class B-shunt resonant commutation circuit which consist of SCR T_1 and L-C circuit. Class B-shunt resonant commutation is a self-commutation process by an L-C circuit. The L-C resonating circuit is connected across the SCR. This circuit is also called as *resonant-pulse commutation*. This commutation circuit operates four modes.

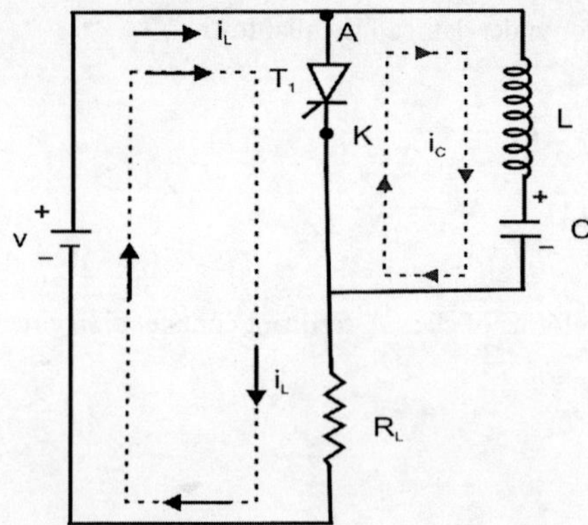

Fig.3.20 Class B shunt resonant commutation circuit

MODE-1: At $t = 0$, initially the supply voltage V is applied to the circuit and the capacitor starts to charge and it finally charged to voltage V with upper plate positive and lower plate negative. The charging of capacitor is done by the path "$V^+ - L - C - R_L - V^-$".

MODE-2: At $t = t_1$, when the gate pulse is applied to SCR T_1 and SCR T_1 will be turned on, a constant current I_L flows through load and capacitor discharging current flows through capacitor. Then load current I_L follows though the path "$V^+ - T_1 - R_L - V^-$" and the capacitor is discharged through the path "$C^+ - L - C - T_1 - C^-$".

MODE-3: If the capacitor is completely discharged, it starts to charge with reverse polarity. Due to reverse polarity of capacitor voltage, the commutating current i_C opposes the load current I_L. As SCR is a unidirectional device, the net current flows through is equal to

$$I_{T1} = I_L - i_C \tag{3.41}$$

Whenever the commutating current i_C is greater than the load current I_L, SCR will be turned off.

MODE-4: As soon as SCR is turned off, capacitor again starts to charge with upper plate positive and lower plate negative. When capacitor is fully charged, SCR operates in the forward blocking state and it will be turned-on when a trigger pulse is applied to SCR. Fig.3.21 shows the voltage and current waveforms of class B commutation. If SCR is turned on by applying a gate pulse and loads current flows though SCR and load for certain specified time duration. After the specified time period, SCR will be turned-off due to *self-commutation.*

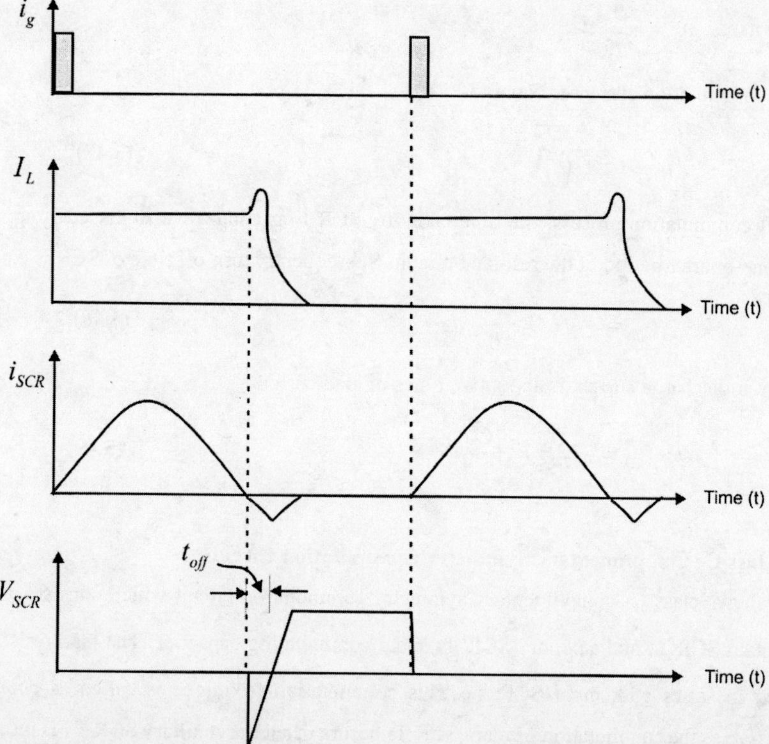

Fig.3.21 Voltage and current waveforms of class B commutation

3.10.2.1 Design of Class B Commutation

The KVL equation for L-C circuit is equal to

$$L\frac{di}{dt}+\frac{1}{C}\int idt = V \qquad (3.42)$$

After differentiating the equation (3.1), we get

$$L\frac{d^2i}{dt^2}+\frac{1}{C}i(t)=\frac{dV}{dt}=0 \qquad (3.43)$$

As $\frac{dV}{dt}=0,\quad L\frac{d^2i}{dt^2}+\frac{1}{C}i(t)=0$ (3.44)

After applying Laplace transform on the equation (3.44), we obtain

$$\left(s^2L+\frac{1}{C}\right)I(s)=0 \qquad (3.45)$$

After solving the equation (3.45), we get

$$i(t)=V\sqrt{\frac{C}{L}}\sin\omega_0 t \qquad (3.46)$$

where, $\omega_0 = \frac{1}{\sqrt{LC}}$

The peak commutation current is equal to

$$I_{c(peak)} = V\sqrt{\frac{C}{L}} \tag{3.47}$$

In class B commutation process, the time taken by SCR to get into reverse biased is approximately equal to one-quarter period of the resonant circuit. Subsequently turn-off time of SCR is equal to

$$t_{off} = \frac{\pi}{2}\sqrt{LC} \tag{3.48}$$

and the peak discharge current is about two times of load current

$$I_{c(peak)} = 2I_L = V\sqrt{\frac{C}{L}} \tag{3.49}$$

3.10.3 Class C- Complimentary Symmetry Commutation Circuit

Fig.3.22 shows class C- complimentary symmetry commutation circuit which consists of two SCRs such as main SCR T_1 and auxiliary SCR T_A and a commutating capacitor. The load resistance R_L is connected in series with main SCR T_1. This commutation technique is called as *complementary commutation* as the commutation of main SCR T_1 occurs when the auxiliary SCR T_A is turned on. This commutation process is also known as complementary impulse commutation. This commutation circuit operates in four different modes as explained below:

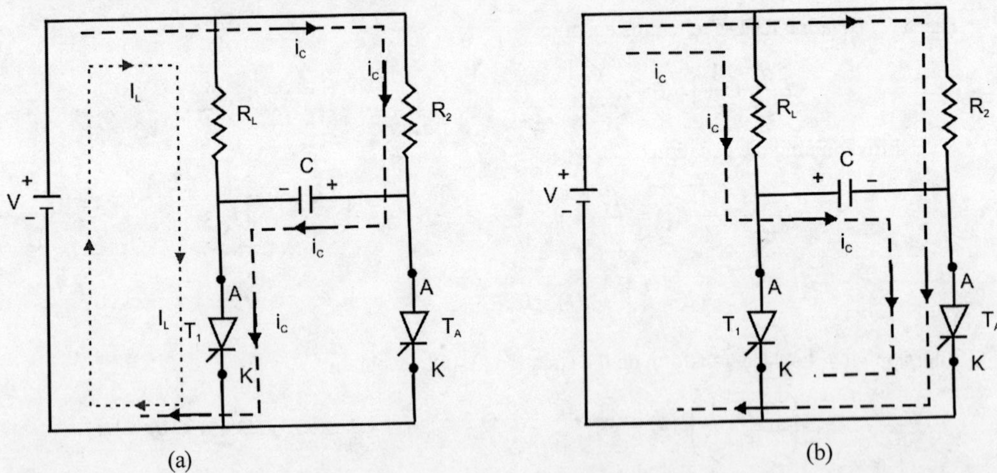

Fig.3.22(a) Class C commutation circuit with T_1 ON and T_A OFF and (b) Class C commutation circuit with T_1 OFF and T_A ON

MODE-1: Initially, both the SCRs T_1 and T_A are in OFF state and the voltage across capacitor is zero. The conditions of T_1 and T_A and capacitor can be represented by "T_1 is in OFF state, T_A is OFF state and $V_C = 0$"

MODE-2: At $t = t_1$, when the triggering pulse is applied to main SCR T_1, SCR T_1 will be turned on and two currents namely load current I_L and capacitor charging current i_C flows through the circuit. Then load current I_L follows though the path "$V^+ - R_L - T_1 - V^-$" and the capacitor charging current flows through the path "$V^+ - R - C^+ - C^- - T_1 - V^-$". During steady state condition, capacitor is fully charged to the supply voltage V with the polarity as shown in Fig.3.22 and the conditions of T_1 and T_A and capacitor can be represented by "T_1 is in ON state, T_A is OFF state and $V_C = V$".

MODE-3: At $t = t_2$, when a triggering pulse is applied to auxiliary SCR T_A, SCR T_A will be turned on. When SCR T_A is turned on and starts conducting, a negative polarity voltage of the capacitor C is applied to cathode of SCR T_1 with respect to anode. Subsequently, SCR T_1 will be reverse biased and turned off immediately. Therefore, the commutation of main SCR T_1 is possible by turning on the auxiliary SCR T_A. After that the capacitor C is charged through the load and its polarity becomes reverse. The charging path of capacitor is "$V^+ - R_L - C^- - C^+ - T_A - V^-$".

At the end of this mode of operation, the conditions of T_1 and T_A and capacitor can be represented by "T_1 is in OFF state, T_A is ON state and $V_C = -V$"

MODE-4: At $t = t_2$, again SCR T_1 is triggered and turned on. Then auxiliary SCR will be turned off immediately as reverse bias voltage is applied across T_A and capacitor starts to charge in reverse direction. At the end of this mode of operation, the conditions of T_1 and T_A and capacitor can be represented by "T_1 is in ON state, T_A is OFF state and $V_C = V$"

The voltage and current waveforms of class C complimentary symmetry commutation circuit is shown in Fig.3.23. The class C commutation is very useful for operating frequency below 1 kHz and this commutation technique is used in Mc Murray Bedford inverter. The characteristics of class c commutation are most reliable commutation.

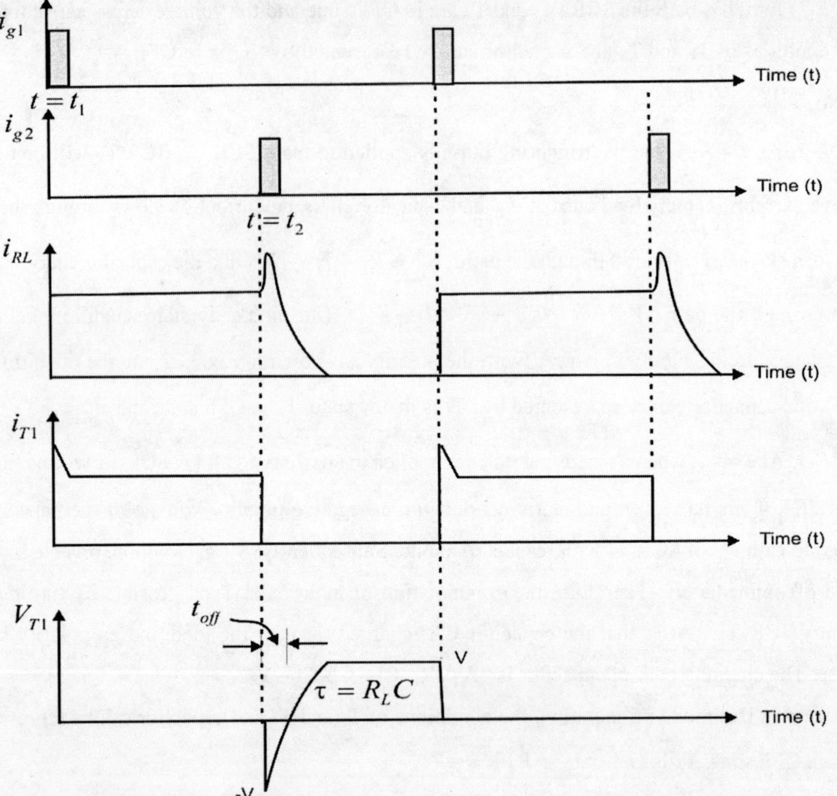

Fig.3.23 Voltage and current waveforms of class C complimentary symmetry commutation circuit

3.10.3.1 Design of Class C Complimentary Symmetry Commutation Circuit

When SCR T_1 conducts and capacitor C is charged to input voltage V through resistance R_L. Whenever a triggering pulse is applied to SCR T_2 and T_2 is turned on, a voltage twice the dc supply voltage is applied to the R_LC series circuit. Subsequently applying the KVL in the closed loop, we get

$$iR_L + \frac{1}{C}\int idt + V_{TA} - V = 0 \tag{3.50}$$

as $V_{TA} = 0$,

$$iR_L + \frac{1}{C}\int idt - V = 0 \tag{3.51}$$

Applying Laplace transform on the equation (3.1), we obtain

$$\left(R_L + \frac{1}{sC}\right)I(s) = \frac{V}{s} \tag{3.52}$$

After applying the inverse Laplace transform of equation (3.1), we get

$$i(t) = \frac{2V}{R_L} e^{-\frac{t}{R_L C}} \tag{3.53}$$

The voltage across SCR T_1 is equal to (3.54)

$$V_{T1} = V - iR_L$$

Therefore, $V_{T1} = V - 2Ve^{-\frac{t}{R_L C}} = V(1 - 2e^{-\frac{t}{R_L C}})$ (3.55)

To turn-off the main SCR T_1, the capacitor voltage should be equal to the voltage across SCR T_1.

Therefore, $V_C = V_{T1} = V(1 - 2e^{-\frac{t}{R_L C}})$ (3.56)

At $t = t_{off}$, $V_C = 0$ and $V(1 - 2e^{-\frac{t_{off}}{R_L C}}) = 0$ (3.57)

Therefore, $t_{off} = 0.6931 R_L C$ (3.58)

So, $C = 1.44 \frac{t_{off}}{R_L}$ (3.59)

The maximum permissible $\frac{dV}{dt}$ across SCR T_1 using commutating components is given by

$$\frac{dV}{dt}\Big|_{\max} > \frac{2V}{R_L C} \tag{3.60}$$

3.10.4 Class D – Auxiliary Commutation:

Fig.3.24 shows the class D -auxiliary commutation circuit which consists of two SCRs such as main SCR T_1 and auxiliary SCR T_A, inductor L, diode D and a commutation capacitor C. The main SCR T_1 and load resistance R_L act as a power circuit but inductor L, diode D and auxiliary SCR T_A are used to form the commutation circuit. This commutation circuit operates in four different mode which are discussed in this section.

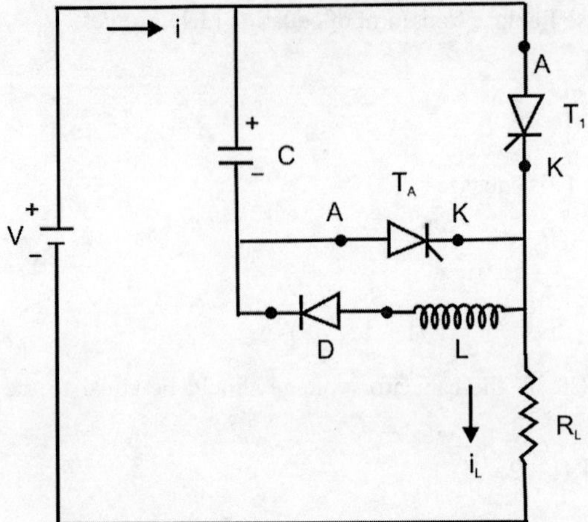

Fig.3.24 Class D - auxiliary commutation circuit

MODE-1: Initially, the dc voltage V is applied to circuit, SCRs T_1 and T_A are in off-state. There is no current flow though dc supply and commutation circuit. The conditions of T_1 and T_A and capacitor can be represented by "T_1 is in OFF state, T_A is OFF state and $V_C = 0$".

MODE-2: When the triggering pulse is applied to auxiliary SCR T_A, SCR T_A will be turned on and capacitor C gets charged. The capacitor charging current flows through the path "$V^+ - C^+ - C^- - T_A - R_L - V^-$". As the voltage across the capacitor C increases gradually, the current flow through SCR T_A decreases slowly. When the capacitor is fully charged to V, the auxiliary SCR T_A gets turned off. In this mode, the conditions of T_1 and T_A and capacitor can be represented by "Initially T_A is in ON state, and at steady state condition; T_1 is in OFF state, T_A is OFF state and $V_C = V$".

MODE-3: When the triggering pulse is applied to main SCR T_1, the current flows in the two different paths: (i) the load current I_L follows though the path "$V^+ - T_1 - R_L - V^-$" and commutation current (capacitor discharging current) flows through the path "$C^+ - T_1 - L - D - C^-$". If the capacitor is fully discharged, its polarity will be reversed. The discharging of capacitor C in reverse direction is not possible due to presence of Diode D. At the end of this mode of operation, T_1 is in ON state, T_A is in OFF state and $V_C = -V$.

MODE-4: In this mode, whenever the SCR T_A is triggered and turned on, capacitor C starts to discharge through the path "$C^+ - T_A - T_1 - .C^-$". Whenever the commutating current (discharging current of capacitor) becomes more than load current I_L, SCR T_1 gets turned off. At the end of mode 3 operation,

"T_1 operates in OFF state and T_A is also in OFF state". Capacitor C will again charge to the supply voltage V with the polarity. As the commutation energy rapidly transfers to the load, high efficiency is possible in class D commutation. This commutation is mostly used in Jones chopper circuit. Voltage and current waveforms of Class D auxiliary commutation are shown in Fig.3.25.

3.10.4.1 Design of Class D Auxiliary Commutation

Design of Commutation capacitor (C)

The value of commutating capacitor (C) depends on (i) maximum load current to be commutated, (ii) turn off time of SCR, t_{off} and (iii) input voltage V.

The turn off time of SCR, t_{off} is available from manufacturer data sheet. The capacitor voltage is changed from –V to 0 during turn off time t_{off} .

As $CV = q = it$, where $i = I_L$ and $t = t_{off}$, $CV = I_L t_{off}$ (3.61)

Therefore, $$C = \frac{I_L t_{off}}{V} \tag{3.62}$$

Design of Commutation inductor (L)

The magnitude of commutating inductor (L) depends on (i) the maximum permissible value of capacitor current is I_C when the main SCR T_1 operates in the ON state and (ii) during the time interval $t_2 - t_1$, the capacitor voltage must be reset to correct polarity for commutating SCR T_1.

As the capacitor current I_C is an oscillatory current in nature and flows through main SCR T_1, L, D and C when SCR T_1 is triggered and turned on. The peak value of current I_C is equal to

$$I_{C(peak)} = \frac{V}{\omega_r L} \tag{3.63}$$

where, $\omega_r = oscillating.frequency = \frac{1}{\sqrt{LC}} rad/\sec$

After substituting the value of ω_r in equation (3.63), we obtain

$$I_{C(peak)} = V\sqrt{\frac{C}{L}} \tag{3.64}$$

The periodic time during oscillation is

$$T_r = \frac{2\pi}{\omega_r} = 2(t_1 - t_2) \tag{3.65}$$

When $I_{L(\max)}$ is the maximum current through the main SCR T_1 for commutation, the capacitor peak current must be equal to the load current.

So, $V\sqrt{\frac{C}{L}} \leq I_{L(\max)}$ (3.66)

Therefore, $L \geq C\left(\frac{V}{I_{L(\max)}}\right)^2$ (3.67)

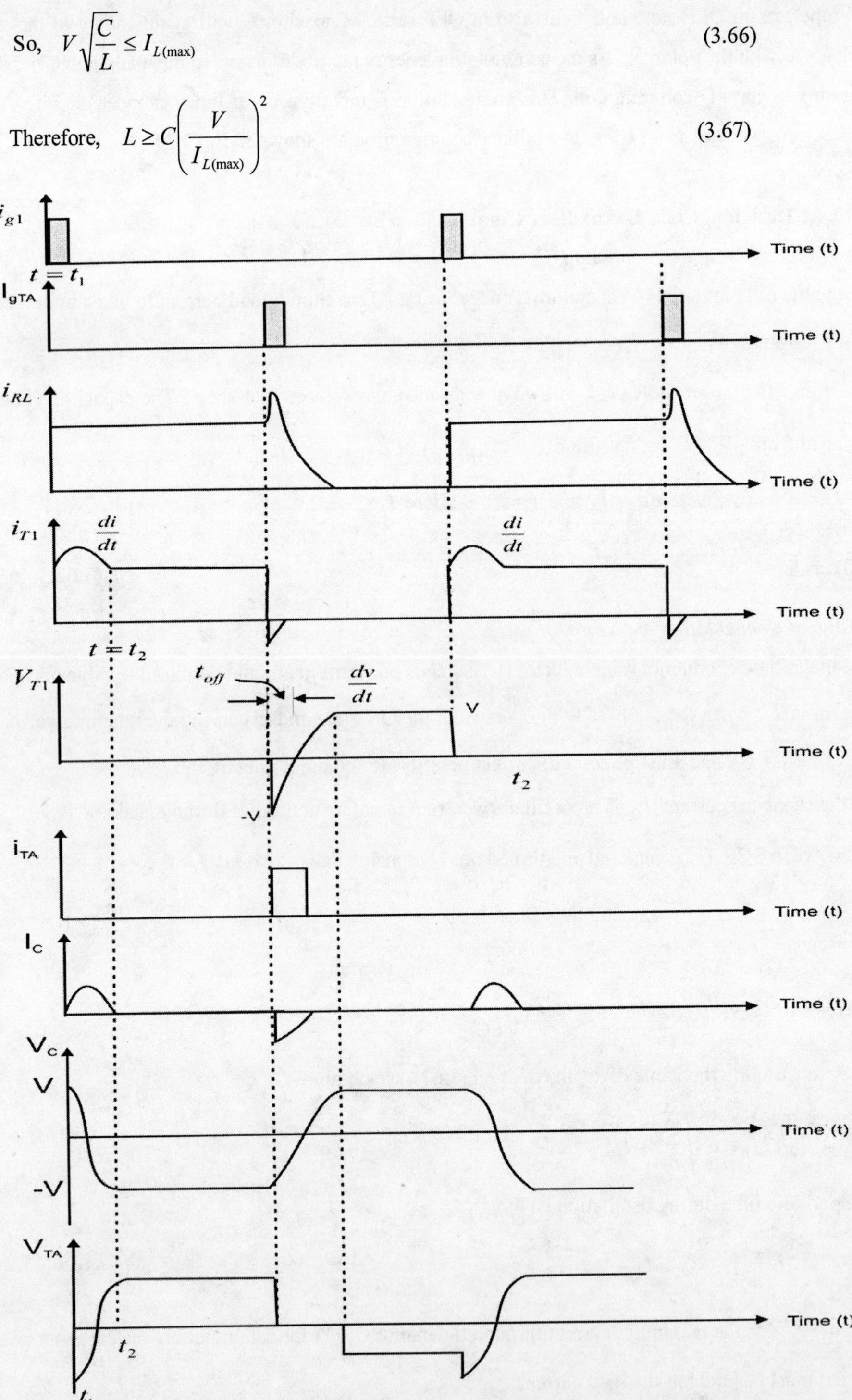

Fig.3.25 Voltage and current waveforms of Class D auxiliary commutation

3.10.5 Class E-External Pulse Commutation

The class E external pulse commutation circuit is shown in Fig.3.26. In this commutation method, the reverse voltage applied to the current carrying SCR from an external pulse source. This commutation is also called as *external pulse commutation.* In this commutation method, the commutating pulse is applied through a pulse transformer which is design in such a way that there should be tight coupling between the primary and secondary windings of pulse transformer. The pulse transformer is also designed with a small air gap so that there will not be any saturation when a pulse voltage is applied to its primary. When the commutation of SCR T_1 is required, the pulse duration equal to or slightly greater than the specified turn-off time of SCR T_1 must be applied. The voltage and current waveforms of class E external pulse commutation is depicted in Fig.3.27. This commutation operates in two different modes.

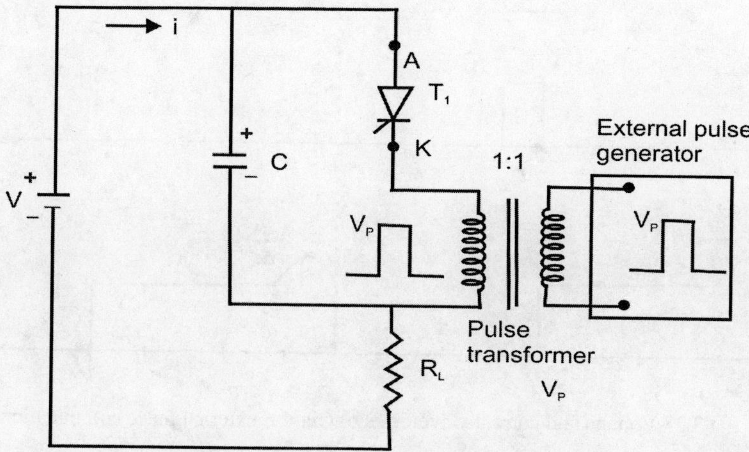

Fig.3.26 Class E external pulse commutation

MODE-1: If the SCR T_1 is triggered and turned on, current starts to flow through the pulse transformer and load resistance R_L. Then current flows through the path "$V^+ - T_1 - \text{Primary.of.Pulse.transformer} - R_L - V^-$"

MODE-2: Whenever an external pulse voltage V_P is applied across the primary of the pulse transformer, a voltage will be induced in the secondary of the pulse transformer. This induced voltage in the secondary appears as reverse voltage (- V_P) across the SCR T_1. After that, SCR T_1 gets turned off. Since the frequency of induced pulse voltage is very high, the capacitor provides almost zero impedance. When SCR T_1 is turned off completely, the load current decays to zero. Before the computation process, the capacitor voltage remains at a value of about 1V. As the minimum energy is required for commutation, both the time ratio and pulse width regulations are easily possible in this commutation. The class E external pulse commutation method is very efficient.

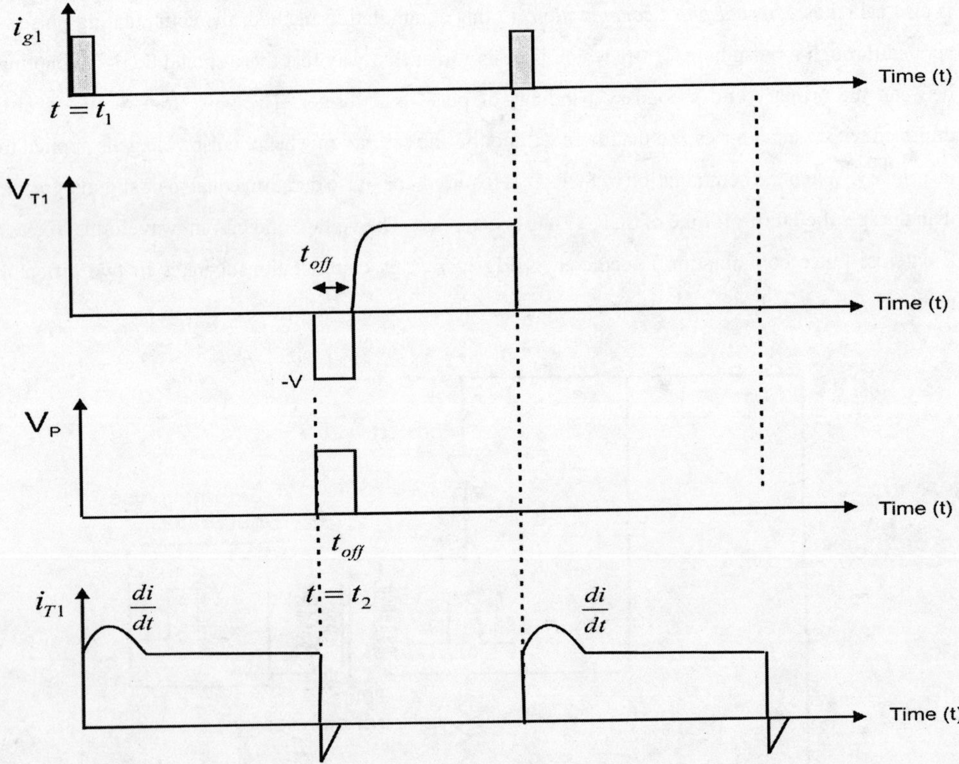

Fig.3.27 Voltage and current waveforms of Class E external pulse commutation

3.10.6 Class F-Line or Natural Commutation

When the supply voltage is ac in any ac circuit, current in the SCR goes through a natural zero and a reverse voltage appears across the SCR and SCR will be turned off. This method of turned-off process of SCR is called as *natural commutation*. During natural commutation, there is no requirement of external circuits for turning off the SCR. For example, class-F commutation is natural commutation. The line or natural commutation is used in single phase and three phase controlled rectifiers, ac voltage controllers and cyclo-converters etc. Fig.3.28 shows the class F commutation and it's voltage and current waveforms are depicted in Fig.3.29..

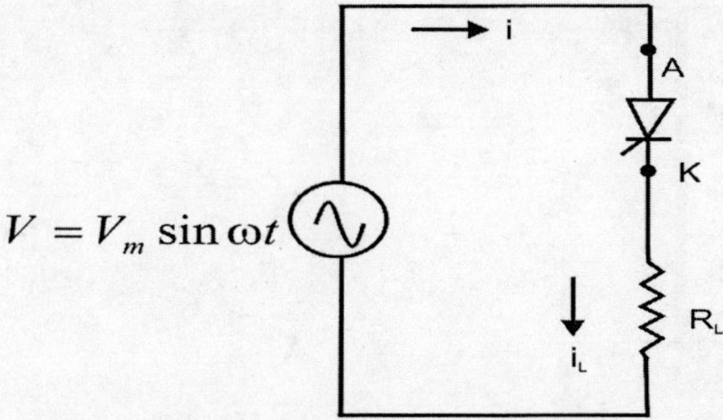

Fig.3.28 Class F line or natural commutation

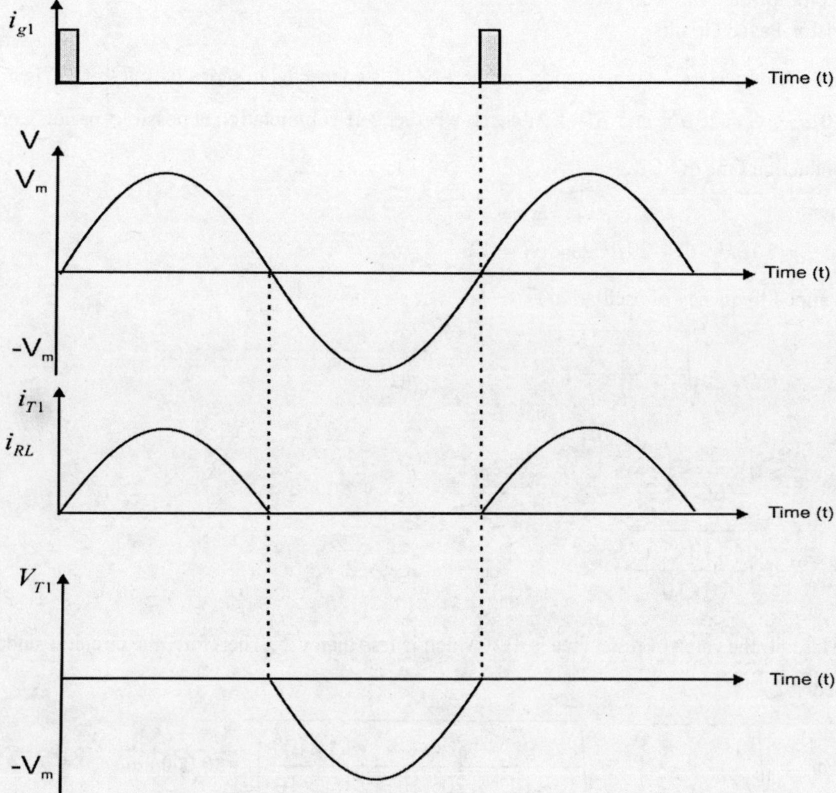

Fig.3.29 Voltage and current waveforms of Class F line or natural commutation

For more on Commutation of Thyristor-Based Circuits

Example 3.5 In class A commutation circuit, a SCR is connected in series with R-L-C. When $L = 10\mu H$, $C = 20\mu F$ and $R = 1\Omega$, check whether self-commutation is possible or not. Compute the conduction time of SCR.

Solution:

Given: $L = 10\mu H$, $C = 20\mu F$ and $R = 1\Omega$

The damped frequency of oscillation is

$$\omega_d = \sqrt{\frac{1}{LC} - \left(\frac{R}{2L}\right)^2}$$

Since $\omega_d > 0$, $\frac{1}{LC} - \left(\frac{R}{2L}\right)^2 > 0$ or $R < \sqrt{\frac{4L}{C}}$

Then $R < \sqrt{\frac{4 \times 10 \times 10^{-6}}{20 \times 10^{-6}}}$ or $R < \sqrt{2}$

In the circuit, the value of resistance is 1Ω which is less than $\sqrt{2}$. Therefore, the circuit is under-damped.

Then $\omega_r = \sqrt{\frac{1}{LC} - \left(\frac{R}{2L}\right)^2} = \sqrt{\frac{1}{10 \times 10^{-6} \times 20 \times 10^{-6}} - \left(\frac{1 \times 10^6}{2 \times 10}\right)^2}$ =50,000 rad/sec

The conduction time of SCR is $t_1 = \frac{\pi}{\omega_r} = \frac{\pi \times 10^6}{50000}\mu s = 62.857\mu s$

Example 3.6 In a class B resonant pulse commutation circuit, $L = 5\mu H$ and $C = 25\mu F$. The initial voltage across capacitor is 210V, determine (a) resonant frequency (b) peak value of resonant current (c) turn-off time of SCR.

Solution:

Given $L = 5\mu H$, $C = 25\mu F$ and V=210V

(a) Resonant frequency is equal to $\omega_0 = \frac{1}{\sqrt{LC}} = \frac{1}{\sqrt{5\times10^{-6}\times25\times10^{-6}}} = 89.445kHz$

(b) Peak value of resonant current is $I_{c(peak)} = V\sqrt{\frac{C}{L}} = 210\sqrt{\frac{25\times10^{-6}}{5\times10^{-6}}}$ =469.56 A

(c) Turn-off time of thyristor is $t_{off} = \frac{\pi}{2}\sqrt{LC} = \frac{\pi}{2}\sqrt{5\times10^{-6}\times25\times10^{-6}} = 17.553\mu s$

Example 3.7 In a Class C commutation, if $V = 200V$, $R_L = 20\Omega$, $R_2 = 100\Omega$ calculate (a) the peak value of current through SCR T_1 and T_2 (b) the value of capacitance C when turn-off time of each SCR is equal to $25\mu s$. Consider that the factor of safety is 2.

Solution:

(a) The peak value of current through SCR T_1 is $V\left(\frac{1}{R_L}+\frac{2}{R_2}\right)$ when SCR T_1 is turned on to commutate at the instant $i_{T2} = 0$, $v_{T2} = -V$, $v_{T1} = -0$ and $i_c = \frac{2V}{R_2}$

$$i_{T1} = V\left(\frac{1}{R_L}+\frac{2}{R_2}\right) = 200\left(\frac{1}{20}+\frac{2}{100}\right) = 14 \text{ A}$$

The peak value of current through T_2 is equal to

$$i_{T2} = V\left(\frac{2}{R_L}+\frac{1}{R_2}\right) = 200\left(\frac{2}{20}+\frac{1}{100}\right) = 40 \text{ A}$$

(b) The value of capacitance C when turn-off time of each thyristor is equal to $25\mu s$ is

$$C = 1.44\frac{t_{off}}{R_L} = 1.44\times\frac{25\times10^{-6}}{20} = 1.8\mu F$$

If factor of safety is 2, the value of C is $C = 1.44\frac{2\times t_{off}}{R_L} = 1.44\times\frac{2\times25\times10^{-6}}{20} = 3.6\mu F$

Example 3.8 In a Class D commutation circuit, $V = 220V$, $L = 25\mu H$ and $C = 50\mu F$. If load current is 100A, determine (a) peak value of current through capacitance, main SCR and auxiliary SCR (b) turn off time of SCR.

Solution:

Given $V = 220V$, $L = 25\mu H$, $C = 50\mu F$ and $I_L = 100$A

(a) An oscillatory current flows through C, T_1, L and L ́and it is expressed by

$$i_c(t) = V\sqrt{\frac{C}{L}}\sin\omega_o t$$

Peak value of current through capacitance is $I_p = V\sqrt{\frac{C}{L}} = 220\sqrt{\frac{50\times10^{-6}}{25\times10^{-6}}}$ =311.12 A

Peak value of current through main SCR T_1 is $I_p + I_L = 311.12 + 100 = 411.12A$

Peak value of current through auxiliary SCR T_A is $I_L = 100A$

(b) As $C = \frac{I_L t_{off}}{V}$, Turn off time of main SCR is equal to

$$t_{off} = C\frac{V}{I_L} = 50\times10^{-6}\times\frac{200}{100} = 100\mu s$$

Turn off time of auxiliary SCR is equal to

$$t_{off} = \frac{\pi}{2\omega_o} = \frac{\pi}{2}\sqrt{LC} = \frac{\pi}{2}\sqrt{25\times10^{-6}\times50\times10^{-6}} = 55.50\mu s \text{ as } \omega_o = \frac{1}{\sqrt{LC}}$$

EXERCISE

MULTIPLE CHOICE QUESTIONS

3.1 If the gate circuit is open, SCR operating state can be changed from forward blocking state to forward conducting state when

(a) The applied voltage is greater than forward break over voltage

(b) The applied voltage is greater than reverse break down voltage

(c) The applied voltage is less than forward break over voltage

(d) The applied voltage is greater than 1.2V

3.2 After turn on a SCR,

(a) the gate pulse should not be removed

(b) the gate pulse should be removed to reduce losses and junction temperature

(c) the gate pulse nay or may not be removed

(d) the gate pulse has no relation with thyristor turn on

3.3 Which of the following is the normal trigger of SCR?

(a) thermal triggering

(b) high forward voltage triggering

(c) $\frac{dv}{dt}$ triggering

(d) gate triggering

3.4 In gate triggering turn on method of SCR,

(a) the gate pulse width must be greater than or equal to turn-on time.

(b) the gate pulse width must be less than or equal to turn-on time.

(c) the gate pulse width must be equal to turn-on time.

(d) the gate pulse width must be less than turn-on time.

3.5 In a RC triggering circuit of SCR, the value of R will be

(a) $R \leq \frac{V - V_{gT} - V_d}{I_{gT}}$ (b) $R \leq \frac{V + V_{gT} - V_d}{I_{gT}}$

(c) $R \leq \frac{V - V_{gT} + V_d}{I_{gT}}$ (d) $R \leq \frac{V + V_{gT} + V_d}{I_{gT}}$

3.6 The oscillation frequency of UJT oscillator is equal to

(a) $f = \frac{1}{R_E C \ln\left(\frac{1}{1+\eta}\right)}$ (b) $f = \frac{1}{R_E C \ln\left(\frac{1}{1-\eta}\right)}$

(c) $f = \frac{1}{R_E \ln\left(\frac{1}{1-\eta}\right)}$ (d) $f = \frac{1}{C \ln\left(\frac{1}{1-\eta}\right)}$

3.7 A DIAC can be used

(a) in triggering circuit of a TRIAC

(b) to increase efficiency of a TRIAC

(c) to decrease efficiency of a TRIAC

(d) for protection of a TRIAC

3.8 The commutation of SCR means

(a) turn–on process

(b) turn-off process

(c) both turn on and turn-off process

(d) breakdown process

3.9 Match List I and List II and select the correct answer

(a) A -3, B-4, C-2, D-1 (b) A -1, B-2, C-4, D-3

(c) A -3, B-4, C-1, D-2 (d) A -1, B-2, C-3, D-4

(A)

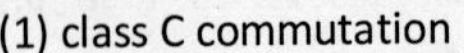
(1) class C commutation

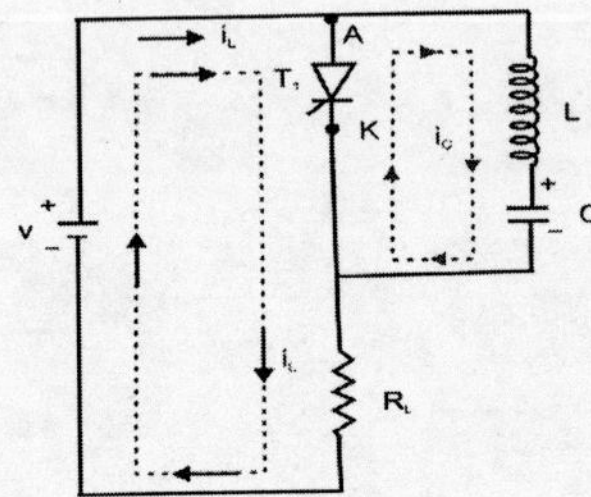

(B)

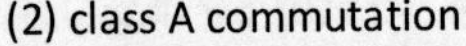
(2) class A commutation

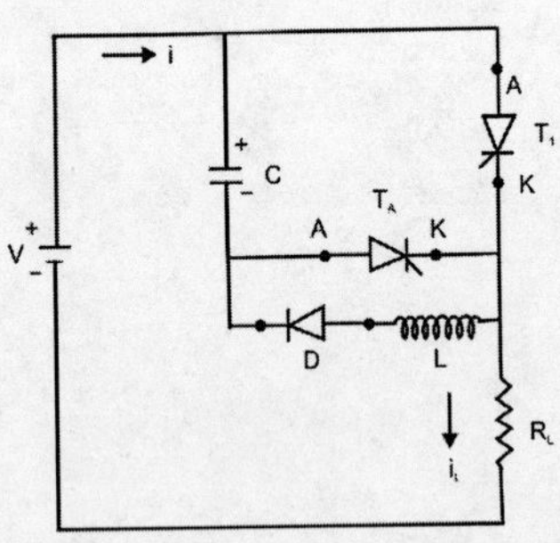

(C) (3) class B commutation

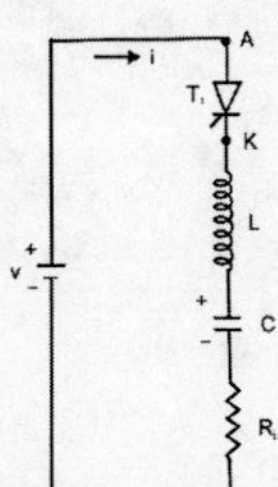

(D) (4) class D commutation

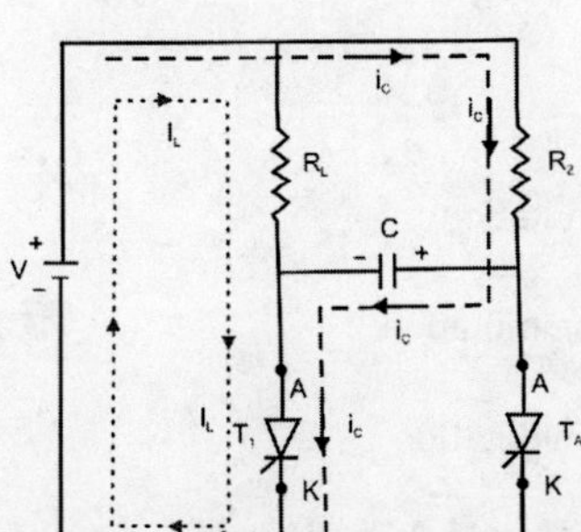

3.10 In a commutation circuit of a SCR, for satisfactory turn of SCR the following condition must be satisfied:

(a) circuit time constant is less than device turn-off time

(b) circuit time constant is greater than device turn-off time

(c) circuit time constant is equal or less than device turn-off time

(d) circuit time constant is equal to device turn-off time

3.11 Match List I and List II and select the correct answer

(a) A -3, B-4, C-2, D-1 (b) A -1, B-2, C-4, D-3

(c) A -3, B-4, C-1, D-2 (d) A -2, B-4, C-1, D-3

(A) Resonant commutation	(1) Class E
(B) Complementary commutation	(2) Class A
(C) External pulse commutation	(3) Class E
(D)Line commutation	(4) Class C

3.12 When the amplitude of the gate pulse to SCR is increased,

(a) the delay time would decrease but the rise time remain un-effected

(b) the delay time would decrease but the rise time would increase

(c) both delay time and rise time would increase

(d) the delay time would increase but the rise time would decrease

3.13 Match List I and List II and select the correct answer

(a) A -1, B-4, C-2, D-3 (b) A -1, B-2, C-4, D-3

(c) A -3, B-4, C-1, D-2 (d) A -3, B-4, C-2, D-1

(A) Resonant commutation	(1) Load commutation
(B) Complementary Impulse commutation	(2) Current commutation
(C) Resonant pulse commutation	(3) Natural commutation
(D)Line commutation	(4) Parallel capacitor commutation

3.14 A pulse transformer is used in a SCR driver circuit

(a) to generate high frequency pulses

(b) to save from dc triggering

(c) to shape the trigger signal

(d) to provide isolation

3.15 In synchronized UJT triggering of a SCR, the voltage across capacitor reaches threshold voltage of UJT thrice in each half cycle. Therefore there are three firing pulses during each half cycle. The firing angle of the SCR can be controlled

(a) thrice in each half cycle

(b) twice in each half cycle

(c) Once in each half cycle

(d) none of these

3.16 The function of a zener diode which is connected in an UJT trigger circuit for SCR is,

(a) to expedite the generation of triggering pulses

(b) to provide a variable voltage to UJT as the source voltage changes

(c) to provide delay the generation of triggering pulses

(d) to provide a constant voltage to UJT to prevent unreliable firing

3.17 In reliable gate triggering of SCR, usually we use

(a) slight over triggering

(b) very hard triggering

(c) very soft triggering

(d) none of these

3.18 The most commonly used gate triggering signal for SCR is

(a) a short duration pulse

(b) a steady dc signal

(c) a high-frequency pulse train

(d) a low frequency pulse train

3.19 When an UJT relaxation oscillator circuit is used for triggering a SCR, the wave shape of the voltage which is generated by UJT circuit is a

(a) saw-tooth wave

(b) sine wave

(c) square wave

(d) trapezoidal wave

3.20 UJT is used for relaxation oscillator due to

(a) negative resistance part of V-I characteristics

(b) positive resistance part of V-I characteristics

(c) peak-point potential

(d) valley-point potential

Answers of Multiple Choice Questions :

3.1 (a) 3.2 (b) 3.3 (d) 3.4 (a) 3.5 (a) 3.6 (b) 3.7 (a) 3.8 (b) 3.9 (a) 3.10 (b)

3.11 (d) 3.12 (a) 3.13 (a) 3.14 (d) 3.15 (c) 3.16 (d) 3.17 (a) 3.18 (c) 3.19 (a) 3.20 (a)

REVIEW QUESTIONS

Short Answer Type Questions

3.21 What are the necessary conditions for turning on of SCR?

3.22 What is gate triggering of thyristor ?

3.33 What is latching current of SCR ?

3.34 What is holding current of thyristor ?

3.35 What is relaxation oscillator?

3.36 What are the different turnings on methods of SCR?

3.37 What is the nature of gate current waveform?

3.38 What is the limitation of a resistance triggering circuit?

3.39 What is commutation?

3.40 What are the types of commutation?

3.41 What are the advantages and disadvantages of self-commutation with respect to other commutations?

3.42 What are the advantages of RC triggering over R triggering circuit?

3.43 What are the applications of DIAC?

3.44 Is it possible to obtain a firing angle greater than $90°$ using R triggering method?

3.45 Why pulse train gating is preferred over pulse gating?

Long Answer Type Questions

3.46 Describe the different methods of triggering of SCR

3.47 Explain the different triggering signals of a thyristor.

3.48 Draw a resistance triggering circuit and explain its operation.

3.49 Draw a RC triggering circuit and explain its operation.

3.50 Write short notes on the following

(a) UJT triggering circuit (b) Synchronized UJT triggering circuit

3.51 Compare UJT triggering circuit with R and RC firing circuit.

3.52 Explain any one commutation circuit with a diagram and waveforms.

3.53 Distinguish between (a) natural commutation and forced commutation

(b) voltage commutation and current commutation

3.54 Draw the self-commutation circuit and discuss in detail.

3.55 Explain the different features of a thyristor firing circuit.

3.56 Draw the trigger circuit for a triac using a diac and explain its operation with waveforms.

3.57 Describe a gate trigger circuit for a single phase half wave and full wave controlled rectifier

Problems

3.58 In resistance triggering circuit as shown in Fig.3.3, $I_{g(\min)} = 0.12mA$ and $V_{g(\min)} = 0.6V$.

If the peak amplitude of input voltage is 220V, find the trigger angle α for $R_1 = 100k\Omega$ and $R_{\min} = 10k\Omega$

3.59 A 230V, 50Hz ac supply is connected to a resistance capacitance triggering circuit. If the resistance R is variable from $2k\Omega$ to $20k\Omega$, $V_{gT} = 2V$ and $C = 0.47\mu F$, what is the minimum and maximum firing angle?

3.60 The frequency of relaxation oscillator can be varied by changing the value of charging resistance R. Compute the maximum and minimum values of R and their corresponding firing frequencies.

Assume that $\eta = 0.65$, $I_p = 0.65mA$, $V_p = 10V$, $I_v = 2.0mA$, $V_v = 2V$, $V_{BB} = 20V$, and $C = 0.047\mu F$

3.61 In class A commutation circuit, a SCR is connected in series with R-L-C. When $L = 10\mu H$, $C = 20\mu F$ and $R = 1\Omega$, check whether self-commutation is possible or not. Compute the conduction time of SCR.

3.62 In a class B resonant pulse commutation circuit, $L = 5\mu H$ and $C = 25\mu F$. The initial voltage across capacitor is 200V, determine (a) resonant frequency (b) peak value of resonant current (c) turn-off time of SCR.

3.63 In a Class C commutation, if V=220, $R_L = 20\Omega$, $R_2 = 100\Omega$ calculate (a) the peak value of current through SCR T_1 and T_2 (b) the value of capacitance C when turn-off time of each SCR is equal to $25\mu s$. Consider that the factor of safety is 2.

3.64 In a Class D commutation circuit, V=230V, $L = 25\mu H$ and $C = 50\mu F$. If load current is 100A, determine (a) peak value of current through capacitance, main SCR and auxiliary SCR (b) turn off time of SCR.

PRACTICAL

1. Experiment on test the variation of R, C in R and RC triggering circuits on firing angle of SCR

Aim:

To study the different triggering circuits for SCR such as (i) Resistance (R) triggering circuit and (ii) Resistor-capacitor (RC) triggering circuit

Apparatus:

AC power supply 230V, 50 Hz; Transformer 0-12V, 500mA; Digital voltmeter 0-30V, CRO 20 MHz; SCR-SN 104/TYN604/TYN 612, Diode D1 4007, Diode D2 4007, Capacitor C 0.22µF, Resistance $R_1=R_L=100k\Omega$,5W , $R_4=R_G=100\Omega$,0.5W, $R_2=100\Omega$,0.5W, $R_3=100k\Omega$,0.5W potentiometer.

Circuit Diagram:

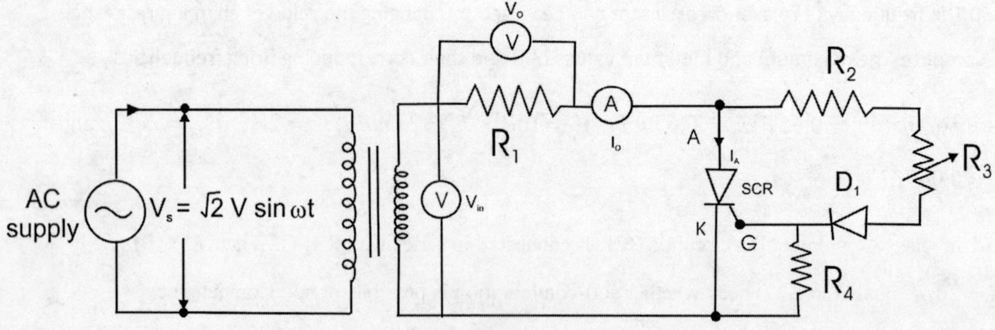

Fig.3.30 Circuit diagram for Resistance (R) triggering circuit of SCR.

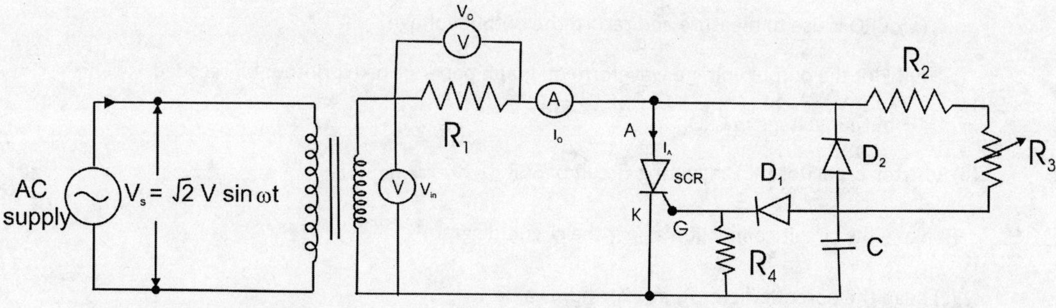

Fig.3.31 Circuit diagram for Resistor-capacitor (RC) triggering circuit of SCR.

Theory:

SCR can be turned-on by applying gate signal at any phase angle with respect to applied 230V, 50 Hz ac voltage when SCR is forward biased. Basically firing angle of SCR is the phase angle of ac voltage at which SCR is turned-on. The different of gate triggering are (i) Resistance (R) triggering circuit and (ii) Resistor-capacitor (RC) triggering circuit. The operating principle of resistance (R) triggering circuit is explained in section 3.4 and resistor-capacitor (RC) triggering circuit is explained in section 3.5. These are most commonly used triggering method of SCR as these circuits are most simplest and economical type of triggering. In R triggering circuit firing angle can be varied by varying the gate current by R_3=100kΩ and is limited to 90˚. In RC triggering circuit, firing angle can be varied by varying the gate current by R_3=100kΩ and it provides the firing angle control from 0 to 180˚.

Precautions:

1. Initially connection to 230V, 50 Hz AC supply carefully.
2. The applied voltage and current should not be more than the maximum rating of SCR.
3. Reading of observation should be error free.

Procedure

(a) Resistance (R) triggering circuit of SCR

(i) Make the circuit connection as per the circuit diagram

(ii) Keep the potentiometer R2 at maximum value

(iii) Switch on ac power supply

(iv)The resistance of potentiometer R2 decreases in steps and measure the output voltage using voltmeter.

(v) CRO is use to measure and record the output voltage

(vi) Plot the output voltage waveform on graph paper or take printout of recoded output voltage in CRO

(b) Resistor-capacitor (RC) triggering circuit of SCR

(i) Make the circuit connection as per the circuit diagram

(ii) Keep the potentiometer R_2 at maximum value

(iii) Switch on ac power supply

(iv) The resistance of potentiometer R_2 decreases in steps and measure the output voltage using voltmeter.

(v) CRO is use to measure and record the output voltage

(vi) Plot the output voltage waveform on graph paper or take printout of recoded output voltage in CRO

Observation Table:

Table 3.1 Observation table for Resistance (R) triggering circuit of SCR

Sr.No.	Firing angle $\alpha°$	DC output voltage in Volts

Table 3.2 Observation table for Resistance-Capacitor (RC) triggering circuit of SCR

Sr.No.	Firing angle $\alpha°$	DC output voltage in Volts

Related Questions

1. If R3 is shorted, state the effect on output voltage
2. If Diode D1 is open circuit, state the effect on output voltage
3. Write the range of firing angle of R, RC triggering circuit.

2. Experiment on test effect of variation of R, C in UJT triggering technique

Aim:

To study the performance and waveforms of UJT triggering or SCR

Apparatus:

Regulated DC power supply 0-30V, UJT-2N2646; Capacitor C 10μF; Digital voltmeter 0-30V, CRO 20 MHz; Resistance R_1=10kΩ,0.5W , R_2=500Ω,0.5W, R_3=150Ω,0.5W, R_4=27Ω,0.5W potentiometer

Circuit Diagram:

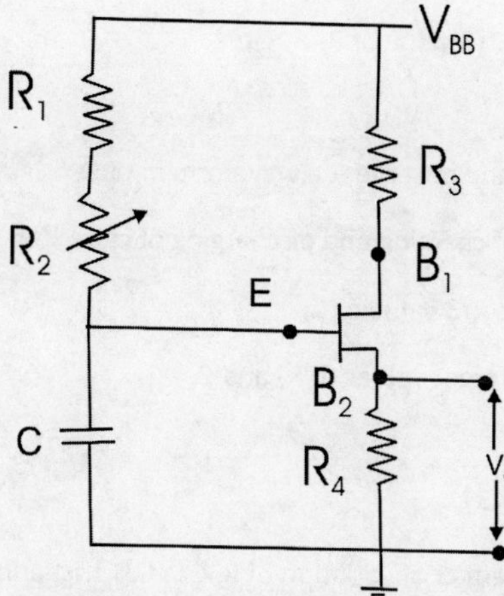

Fig.3.32 Circuit diagram to test effect of variation of R, C in UJT triggering technique.

Theory:

Unijunction Transistor (UJT) is a semiconductor device which has three terminals Emitter E and two bases B_1 and B_2. The base of UJT is formed by lightly doped n-type silicon bar. Two ohomic contacts B1 and B2 are attached at its ends. The emitter (E) is p-type and it is highly doped. If the emitter is open circuit, the resistance between B_1 and B_2 is called interbase resistance. The operating principle of UJT triggering is explained in section 3.6. The rate of charging of capacitor is varied as the R_C pot is varied and firing angle is controlled by varying the potentiometer RC. UJT triggering circuit is used in power control using SCR and Triac.

Precautions:

1. Initially all the knobs of the power supplies are at zero value.
2. The applied voltage and current should not be more than the maximum rating of UJT.
3. Reading of observation should be error free.

Procedure

Case-I

(i) Make the circuit connection as per the circuit diagram

(ii) Switch on regulated dc power supply

(ii) Increase power supply voltage V_{BB} gradually to increase I_B

(iv) Resistance R_2 is in maximum value

(v) C is fixed. When resistance R_2 decreases in steps to observe the voltage across B_1 and C_1.

(vi) Measure the time period of charging and discharging of capacitor

(vii) Repeat step (v) and (vi) up to 2/3 value of R_2

Case -II

(i) R_2 is fixed and C_1 is varied and observe the waveform in CRO

(ii) Measure the time period of charging and discharging of capacitor

(iii) Repeat step (i) and (ii) upto 2/3 value of C_1

(iv) Plot the waveforms for different values of R_2 and C_1

Observation Table:

Table 3.3 Observation table for test effect of variation of R, C in UJT triggering technique

Sr.No.	R_2	C_1	Measured time period T	Calculate time period T

Related Questions

1. Why UJT triggering circuit is better compared to R and RC triggering circuit?

2. Write the specification of UJT from manufacturer data sheet

3. Compare measured time period and calculated time period.

3. Experiment on perform the operation of Class-A, B, C turn-off circuits

Aim:

To study the operation of class-A, B, C turn-off circuits of SCR observing waveforms in CRO

Apparatus:

Regulated DC power supply 0-30V, SCR-SN 104/TYN604/TYN 612, Dual Trace CRO 20 MHz; Resistance 1kΩ,0.5W, Inductance 5mH, Capacitor C 1µF; UJT base SCR triggering circuit.

Circuit Diagram:

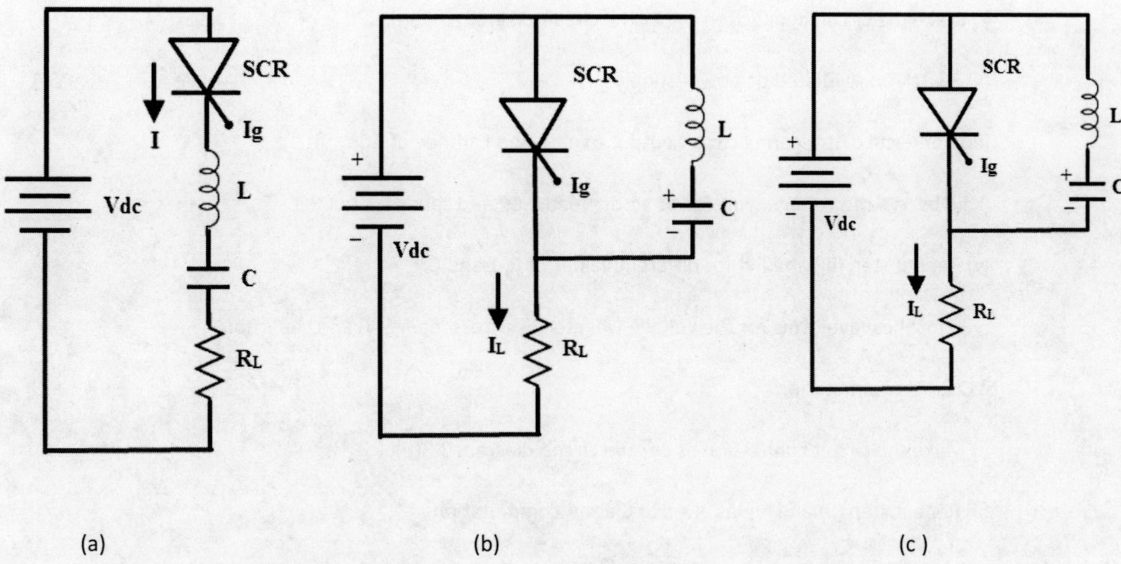

Fig.3.33 Circuit diagram (a) Class-A (b) Class-B and (c) Class-C commutation circuits

Theory:

The commutation circuits are required to turn-off SCR. The following method are used for commuting SCR (i) when the anode current is reduced below the holding current and (ii) by applying reversing the anode voltage across the conducting thyristor. In Class-A, Class-B, Class-C commutation, SCR can be turned off by reducing the anode current below the holding current. The detail operating principle of

Class-A, Class-B and Class-C commutation are explained in section 3.10.1, 3.10.2 and3.10.3 respectively.

Precautions:

1. Initially all the knobs of the power supplies are at zero value.
2. The applied voltage and current should not be more than the maximum rating of UJT.
3. Reading of observation should be error free.

Procedure

(a) Class-A commutation

(i) Make the circuit connection as per the circuit diagram 3.33(a).

(ii) Switch on regulated dc power supply

(iii) Connect UJT triggering circuit output to gate and cathode of SCR

(iv) Observe the voltage waveforms across load, SCR and capacitor in CRO

(v) Repeat step (iii) and (iv) for different values of R, L and C

(vi) Plot the waveforms for the voltage waveforms across load, SCR and capacitor

(b) Class-B commutation

(i) Make the circuit connection as per the circuit diagram 3.3(b).

(ii) Repeat step (ii) and (v) as given in Class-A commutation

(c) Class-C commutation

(i) Make the circuit connection as per the circuit diagram 3.3(c).

(ii) Switch on regulated dc power supply

(iii) Connect UJT triggering circuit output to gate and cathode of SCRs (T_1 and T_2)

(iv) Observe the voltage waveforms across load, SCR and capacitor in CRO

(v) Plot the waveforms for the voltage waveforms across load, SCR and capacitor

VIVA Questions Related to Class-A, Class-B and Class-C Commutation

(i) What happened if inductor is disconnected in Class-B commutation circuit?

(ii) What is the effect on Class-C commutation circuit when capacitor is short-circuit?

(iii) Explain the waveforms for the voltage waveforms across load, SCR and capacitor

4. Experiment on perform the operation of Class-D, E, F turn-off circuits

Aim:

To study the operation of class-D, E, F turn-off circuits of SCR observing waveforms in CRO

Apparatus:

Regulated DC power supply 0-30V, 230 V, 50 Hz AC power supply, SCR-SN 104/TYN604/TYN 612, Dual Trace CRO 20 MHz; Resistance 1kΩ,0.5W, Inductance 5mH, Capacitor C 1μF; UJT base SCR triggering circuit.

Circuit Diagram:

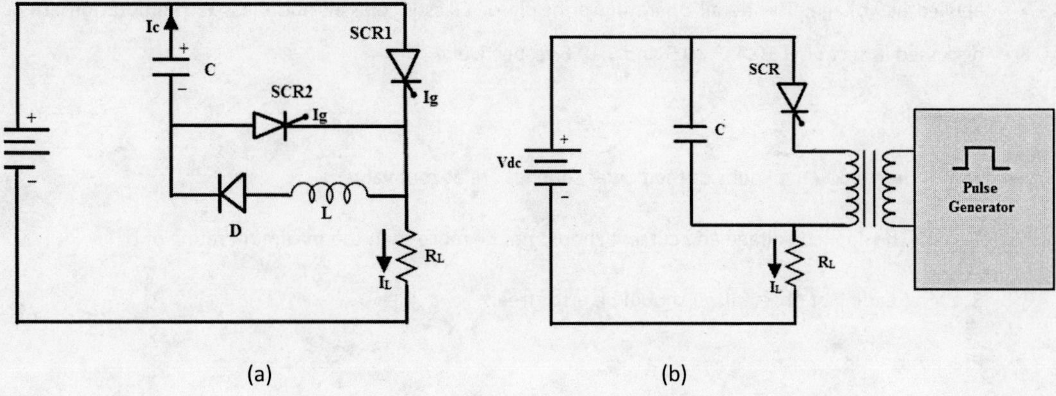

(a) (b)

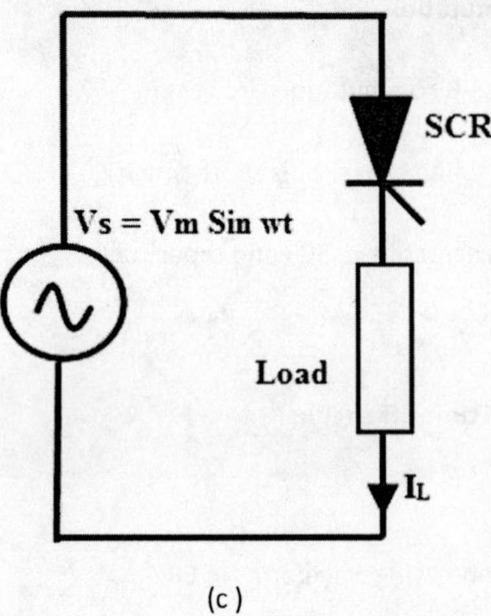

(c)

Fig.3.34 Circuit diagram (a) Class-D (b) Class-E and (c) Class-F commutation circuits

Theory:

In Class-D and Class-E commutation, SCR can be turned off by applying reverse voltage across SCR and external pulse respectively. In Class-F commutation, SCR is turned off due to negative half cycle of applied ac voltage. The detail operating principle of Class-D, Class-E and Class-F commutation are discussed in section 3.10.4, 3.10.5 and 3.10.6 respectively.

Precautions:

1. Initially all the knobs of the power supplies are at zero value.
2. The applied voltage and current should not be more than the maximum rating of UJT.
3. Reading of observation should be error free.

Procedure

(i) Make the circuit connection as per the circuit diagram

(ii) Switch on regulated dc power supply/ac power supply

(iii) Connect UJT triggering circuit output to gate and cathode of SCR

(iv) Observe the voltage waveforms across load, SCR in CRO

(v) Plot the waveforms for the voltage waveforms across load and SCR

VIVA Questions Related to Class-A, Class-B and Class-C Commutation

1. What happened in Class-D commutation circuit if trigger pulse is disconnected to auxiliary SCR?

2. What is the effect on Class-E commutation circuit when the pulse width of external pulse is smaller than turn-off of SCR ?

3. Explain the waveforms for the voltage waveforms across load and SCR in class-D, E, and F turn-off circuits

KNOW MORE

The basic requirement for successful firing of SCRs in any converter circuit is that the current supplied to the gate should be of adequate amplitude and sufficiently short rise time and the current must flows for adequate duration. These conditions can be achieved properly by gate triggering. Therefore, the gate triggering method is most commonly used to turn-on SCRs in different converter circuits. In this module, resistance-capacitance triggering circuits and synchronized UJT triggering circuit using pulse transformer and opto-coupler, are already discussed elaborately. To study the gate triggering circuits for single phase full wave controlled rectifier using analog and digital devices, design and implementation of firing circuit for single-phase converter and microprocessor/microcontroller based firing circuits for SCRs, please scan the following QR code.

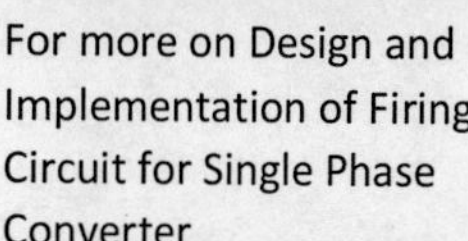
For more on Design and Implementation of Firing Circuit for Single Phase Converter

For more on Microcontroller Based Advanced Triggering Circuit for Converters/Inverters

References:

1. Ramamoorty M., An Introduction to Thyristor and their applications, East-West Press Pvt. Ltd, New Delhi

2. Sugandhi, Rajendra Kumar and Sugandhi, Krishna Kumar, Thyristors: Theory and Applications, New Age International (P) ltd. publishers, New Delhi

3. Bhattacharya, S.K., Fundamental of Power Electronics, Vikas Publishing House Pvt. Ltd. Noida

4. Jain & Alok, Power Electronics and its Applications, Penram International Publishing(India) Pvt. Ltd, Mumbai

5. Rashid, Muhammad, Power Electronics Circuits Devices and Applications, Pearson Education India, Noida

6. Sing, M.D. and Khanchandani, K.B., Power Electronics, Tata McGraw Hill Publishing Co. Ltd, New Delhi

7. Zbar, Paul B., Industrial Electronics: A Text-Lab manual, McGraw Hill Publishing Co. Ltd, New Delhi

8. Grafham D.R., SCR Manual, General Electric Co.

Dynamic QR Code for Further Reading

-QR codes embedded in the unit.

4 Phase Controlled Rectifiers

UNIT SPECIFIES

Through this unit we have discussed the following aspects:

- Phase control: firing angle and conduction angle
- Single phase half controlled, full controlled and midpoint–controlled rectifier with R and RL load
- Circuit diagram, working principle, input-output waveforms of phase controlled rectifiers
- Effect of freewheeling diode in phase controlled rectifiers
- Different configurations of bridge controlled rectifiers: full bridge, half bridge with common anode, common cathode and SCRs in one arm and diodes in another arm

The practical applications of the above topics are discussed for generating further curiosity and creativity as well as improving problem solving capacity of learners.

Not only a large number of multiple choice questions are incorporated in this unit but also questions of short and long answer types marked in two categories following lower and higher order of Bloom's taxonomy, assignments through a number of numerical problems. A list of references and suggested readings are illustrated in this unit so that learners can go through them for practice and get detailed knowledge. It is also be noted that some QR codes have been incorporated in different sections of this unit for getting more information on various topics of interest.

After discussion the content related to theory, there is a "Know More" section at end of this unit which is related to laboratory experiments. This section has been carefully designed so that the supplementary information provided in this section becomes beneficial for the earners. Usually, this section highlights the initial activity, examples of some interesting facts, analogy, and history of the development of the thyristor family devices to focusing the salient observations and finding, timelines starting from the development of the concerned topics up to the recent time, applications of thyristor in our day-to-day life, and industrial applications of thyristor in power electronics converter circuits.

RATIONALE

This unit on "phase controlled rectifiers" can be able to provide detail information regarding phase control: firing angle and conduction angle, single phase half controlled, full controlled and midpoint–controlled rectifier with R and RL load, circuit diagram, working principle, and input-output waveforms of phase controlled rectifiers. The equations of dc output voltage, performance parameters of converters, and the effect of freewheeling diode in phase controlled rectifiers are also incorporated in this unit. After study this unit, student can able to get knowledge about different configurations of bridge controlled rectifiers: full bridge, half bridge with common anode, common cathode and SCRs in one arm and diodes in another arm.

PRE-REQUISITES

Semiconductor physics

Analog and digital electronics

Electrical circuit theory

Electrical Machines and power system

UNIT OUTCOMES

List of outcomes of this unit is as follows:

U4-O1: To define phase control of controlled rectifiers

U4-O2: To define firing angle and conduction angle of SCR

U4-O3: To draw the circuit diagram of single phase half controlled, full controlled and mid-point controlled rectifier with R and RL load

U4-O4: To discuss the working principle of single phase controlled rectifiers

U4-O5: To understand the effect of freewheeling diode in phase controlled rectifiers

U4-O6: To explain operating principle of different configurations of bridge controlled rectifiers: full bridge, half bridge with common anode, common cathode and SCRs in one arm and diodes in another arm

Unit-4 Outcomes	EXPECTED MAPPING WITH COURSE OUTCOMES (*1-Weak Correlation; 2- Medium Correlation; 3-Strong Correlation*)				
	CO-1	CO-2	CO-3	CO-4	CO-5
U4-O1	2	2	2	3	2
U4-O2	1	2	3	2	1
U4-O3	2	2	2	3	2
U4-O4	2	2	2	3	3
U4-O5	1	2	3	2	1
U4-O6	2	2	2	3	3

4.1 Introduction

Rectifiers are converters which convert ac to dc. In a rectifier circuit, the average output voltage is constant for a specified load and a fixed input voltage. When a variable input voltage is applied to uncontrolled rectifier through an auto transformer, the output voltage will also vary. Using uncontrolled rectifier with auto transformer, the variable output voltage can be obtained very simply, but these circuits have some demerits such as large size, heavy weight and high cost of transformers. As a result, phase controlled rectifiers are used in place of half wave and full wave rectifiers using auto transformer. Single phase controlled rectifiers are extensively used in many industrial applications such as electric traction systems, paper mills, steel rolling mills, textile mills, magnet power supply and electro-mechanical devices etc. SCRs are also used as switches in phase controlled converter.

4.2 Classification of Single Phase Controlled Rectifiers

A controllable output voltage is able to obtain by controlling the firing angle "α" of controlled rectifier. So variable voltage can be available at output terminals of a controlled rectifier. The single phase controlled converters are classified as single phase half wave (1-pulse) controlled converter, and single phase full wave (2-pulse) controlled converter. The classification of single phase controlled rectifiers is shown in Fig.4.1.

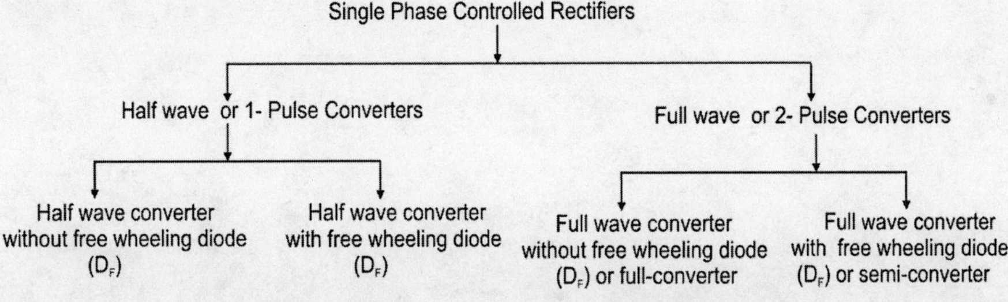

Fig.4.1 Classification of single phase controlled rectifiers

4.3 Single Phase Half-Controlled Rectifiers with R Load

The single phase half controlled rectifier with resistive (R) load is shown in Fig.4.2. The thyristor T_1 will conduct only when the anode is positive with respect to cathode and a positive gate pulse is applied. Otherwise it can operate in forward blocking state and the load current does not flow though the thyristor. In positive half cycle of input voltage, thyristor T_1 is forward biased. Whenever the thyristor T_1 is fired at $\omega t = \alpha$, it is turned on and starts to conduct at $\omega t = \alpha$ and the load current flows through T_1 and R load. Then thyristor T_1 continue it's conduction up to $\omega t = \pi$ and conduction angle is ($\pi - \alpha$).

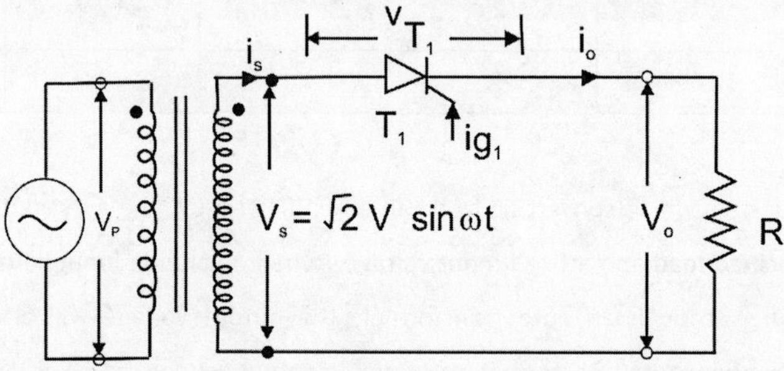

Fig.4.2 Single phase half controlled rectifier with R load

For the negative half cycle of input voltage, thyristor T_1 is reverse biased. Therefore, thyristor T_1 will be turned off due to natural commutation at $\omega t = \pi$ and the current flow through load becomes zero. So the average (dc) output voltage can be controlled by varying the firing angle α. Fig.4.3 shows the input voltage (V_s), triggering pulse of T_1 (i_{g1}), output voltage (V_O), output current (I_O) and V_{T1}, voltage across T_1 of half wave controlled rectifier with R load. The output voltage waveform is same as load current waveform due to zero phase difference in resistive load as depicted in Fig.4.3. In single phase half-controlled rectifier, the dc output voltage is always positive and current is also positive. Therefore, single phase half wave controlled rectifier with R load operates in first quadrant and this controlled rectifier is also called as *half wave converter*.

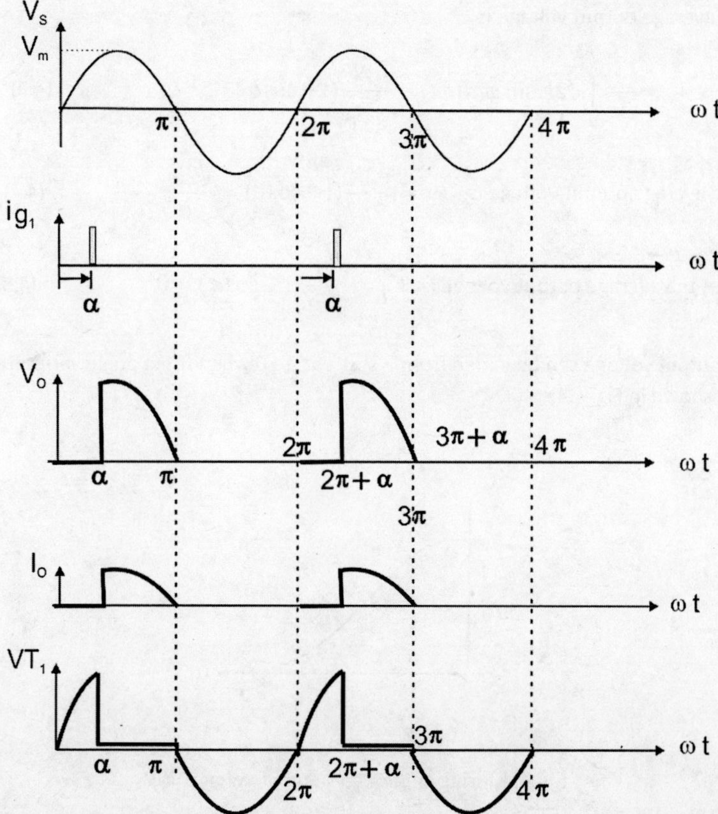

Fig.4.3 Input voltage (Vs), triggering pulse of T_1 (i_{g1}), output voltage (Vo), output current (Io) and V_{T1} voltage across T_1 for half wave controlled rectifier with R load

As α is the firing angle of thyristor T_1, T_1 operates in off-state from $\omega t = 0$ to $\omega t = \alpha$ and $\omega t = \pi$ to $\omega t = 2\pi$. The device T_1 operates in on-state from $\omega t = \alpha$ to $\omega t = \pi$. Consequently the output voltage across load is zero from $\omega t = 0$ to $\omega t = \alpha$ and $\omega t = \pi$ to $\omega t = 2\pi$ and the output voltage across load is equal to input voltage during $\omega t = \alpha$ to $\omega t = \pi$. Hence the conduction angle of thyristor T_1 is ($\pi - \alpha$). Therefore, the output voltage across load is available during positive half cycle only.

The output voltage across the load is equal to

$$v_O = V_m \sin \omega t \qquad \text{for} \quad \alpha \le \omega t \le \pi \qquad (4.1)$$

$$= 0 \qquad \text{for} \quad 0 \le \omega t \langle \alpha \quad \text{and} \quad \pi \le \omega t \le 2\pi \qquad (4.2)$$

The value of average output voltage is

$$V_O = \frac{1}{2\pi}\int_{\alpha}^{\pi}\sqrt{2}V \sin\omega t d(\omega t) = \frac{\sqrt{2}V}{2\pi}(1+\cos\alpha) \tag{4.3}$$

If $\alpha = 0$, the average output voltage is $V_O = \frac{\sqrt{2}V}{2\pi}(1+\cos 0) = \frac{\sqrt{2}V}{\pi}$ (4.4)

When $\alpha = \pi$, the average output voltage is $V_O = \frac{\sqrt{2}V}{2\pi}(1+\cos\pi) = 0$ (4.5)

Thus the dc output voltage can be varied from $\sqrt{2}V/\pi$ to 0 when firing angle varies from $\alpha = 0$ to $\alpha = \pi$ as shown in Fig. 4.4.

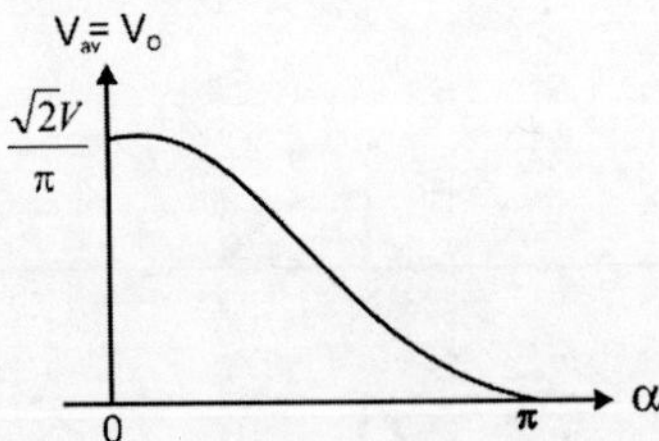

Fig.4.4 Firing angle α vs output voltage V_O

The average current is $I_{dc} = I_o = V_o / R = \sqrt{2}V(1+\cos\alpha)/2\pi R$ (4.6)

Since both the current and voltage are positive, the converter operates in first quadrant as shown in Fig.4.5.

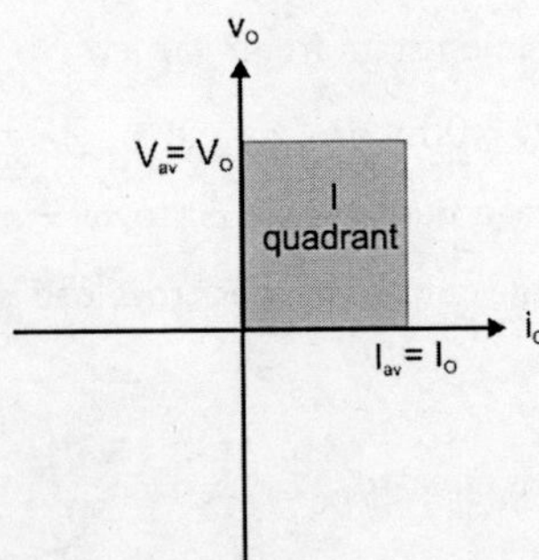

Fig.4.5 First quadrant operation of single half wave controlled rectifier with R load

The dc output power is equal to

$$P_{dc} = V_{dc} \times I_{dc} = V_o \times I_o = \frac{\sqrt{2}V}{2\pi}(1+\cos\alpha) \times \frac{\sqrt{2}V}{2R\pi}(1+\cos\alpha) \tag{4.7}$$

$$= \frac{V^2}{2\pi^2 R}(1+\cos\alpha)^2 \quad (4.8)$$

The rms value of output voltage across load is

$$V_{rms} = \left[\frac{1}{2\pi}\int_{\alpha}^{\pi}(\sqrt{2}V\sin\omega t)^2 d\omega t\right]^{\frac{1}{2}} \quad (4.9)$$

$$= \frac{V}{\sqrt{2}}\left[\frac{1}{\pi}(\pi-\alpha+\frac{1}{2}\sin 2\alpha)\right]^{\frac{1}{2}} \quad (4.10)$$

The rms value of load current is equal to

$$I_{rms} = \frac{V_{rms}}{R} = \frac{V}{\sqrt{2}R}\left[\frac{1}{\pi}(\pi-\alpha+\frac{1}{2}\sin 2\alpha)\right]^{\frac{1}{2}} \quad (4.11)$$

The ac output power is expressed as

$$P_{ac} = V_{rms} \times I_{rms} \quad (4.12)$$

$$= \frac{V}{\sqrt{2}}\left[\frac{1}{\pi}(\pi-\alpha+\frac{1}{2}\sin 2\alpha)\right]^{\frac{1}{2}} \times \frac{V}{\sqrt{2}R}\left[\frac{1}{\pi}(\pi-\alpha+\frac{1}{2}\sin 2\alpha)\right]^{\frac{1}{2}} \quad (4.13)$$

$$= \frac{V^2}{2R}\left[\frac{1}{\pi}(\pi-\alpha+\frac{1}{2}\sin 2\alpha)\right] \quad (4.14)$$

The from factor is equal to

$$FF = \frac{I_{rms}}{I_{av}} = \frac{V}{\sqrt{2}R}\left[\frac{1}{\pi}(\pi-\alpha+\frac{1}{2}\sin 2\alpha)\right]^{\frac{1}{2}} / \frac{\sqrt{2}V}{2R\pi}(1+\cos\alpha) \quad (4.15)$$

$$= \frac{1}{(1+\cos\alpha)}\left[\pi(\pi-\alpha+\frac{1}{2}\sin 2\alpha)\right]^{\frac{1}{2}} \quad (4.16)$$

The ripple factor of load current is

$$RF = \sqrt{FF^2-1} = \frac{1}{(1+\cos\alpha)}\left[\pi(\pi-\alpha+\frac{1}{2}\sin 2\alpha)-(1+\cos\alpha)\right]^{\frac{1}{2}} \quad (4.17)$$

The volt ampere rating of transformer is

$$\text{VA}=VI_{rms} = \frac{V^2}{\sqrt{2}R}\left[\frac{1}{\pi}(\pi-\alpha+\frac{1}{2}\sin 2\alpha)\right]^{\frac{1}{2}} \quad (4.18)$$

Transformer utilization factor is expressed as

$$TUF = \frac{Power.delivered.to.load}{Input.VA} = \frac{P_{dc}}{VA} \tag{4.19}$$

$$= \frac{V_{dc}I_{dc}}{VI_{rms}} = \frac{V_o I_o}{VI_{rms}} = \frac{(1+\cos\alpha)^2}{\pi\left[2\pi(\pi-\alpha+\frac{1}{2}\sin 2\alpha)\right]^{\frac{1}{2}}} \tag{4.20}$$

4.4 Single Phase Half-Controlled Rectifiers with R-L Load

Fig. 4.6 shows a single phase half wave rectifier with R-L load. In the positive half-cycle of supply voltage, thyristor T_1 is forward biased and a gate pulse is applied at $\omega t = \alpha$. Then Thyristor T_1 will be turned on and input voltage is applied across the load. Due to presence of inductive load, output current increases gradually from zero to maximum value and then it starts to decrease. At $\omega t = \pi$, the polarity of supply voltage has been changed from positive to negative but load current continue to flow through thyristor T_1 due to inductive load.

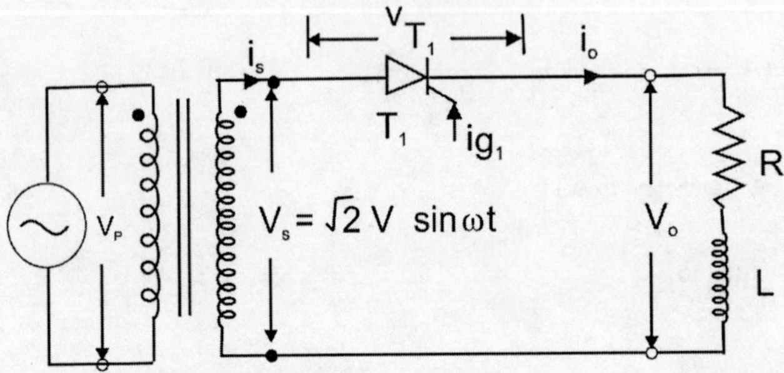

Fig.4.6 Single phase half controlled rectifier with R-L load

At $\omega t = \pi$, thyristor T_1 is subjected to reverse anode voltage and the device T_1 will not be turned off as load current i_o is greater than the holding current. Consequently, thyristor T_1 remains in conduction. At $\omega t = \beta$, load current i_o becomes zero and thyristor T_1 will be turned off as it is already reverse biased. During $\omega t = \beta$ to $\omega t = 2\pi + \alpha$, the output voltage across load is zero and current i_o is also zero. Since the thyristor T_1 starts conduction at $\omega t = \alpha$ and it is turned off at $\omega t = \beta$, the conduction period of T_1 is $(\beta - \alpha)$. Hence, β is known as *extinction angle* and the *conduction angle* is equal to $(\beta - \alpha)$.

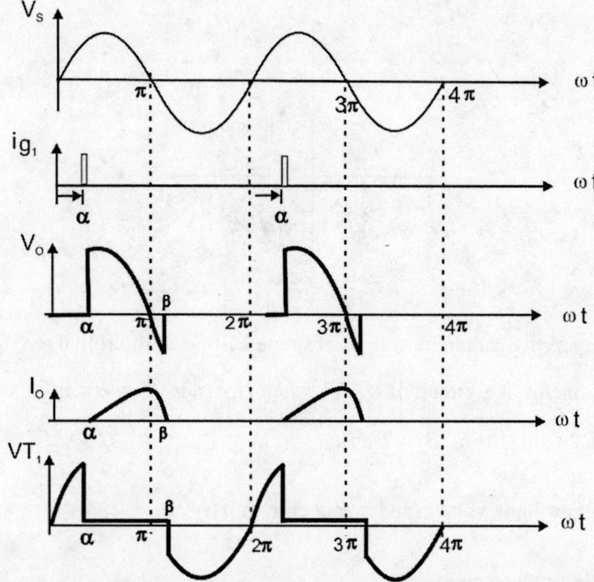

Fig.4.7 Input voltage (V_s), triggering pulse of T_1 (i_{g1}), output voltage (V_o), output current (I_o) and V_{T1} voltage across T_1 for single phase half wave controlled rectifier with R-L load

The output voltage across the load can be expressed as

$$v_O = V_m \sin \omega t \qquad \text{for} \quad \alpha \leq \omega t \leq \beta \qquad (4.21)$$

$$= 0 \qquad \text{for} \quad 0 \leq \omega t \langle \alpha \quad \text{and} \quad \beta \leq \omega t \leq 2\pi \qquad (4.22)$$

The waveforms of input voltage V_s, triggering pulse of T_1 (i_{g1}), output voltage V_o, output current I_o and V_{T1} voltage across T_1 for half wave controlled rectifier with R-L load are shown in Fig.4.7. In the time from $\omega t = \alpha$ to $\omega t = \pi$, input voltage and output current are positive. Therefore power flows from the supply to load. This operating mode of single phase half wave controlled rectifier with R-L load is known as *rectification mode* and this controlled rectifier operates in first quadrant. During the time period from $\omega t = \pi$ to $\omega t = \beta$, input voltage is negative, output current is positive and power flows in reverse direction from the load to supply. This mode of operation of single phase half wave controlled rectifier with R-L load is called as *inversion mode* and this controlled rectifier operates in fourth (IV) quadrant. Based on the firing angle α, the dc output voltage will be either positive or negative. So, the single phase half wave controlled rectifier with R-L load operates in two quadrants as shown in Fig.4.8. This controlled rectifier is called *two quadrant converter*.

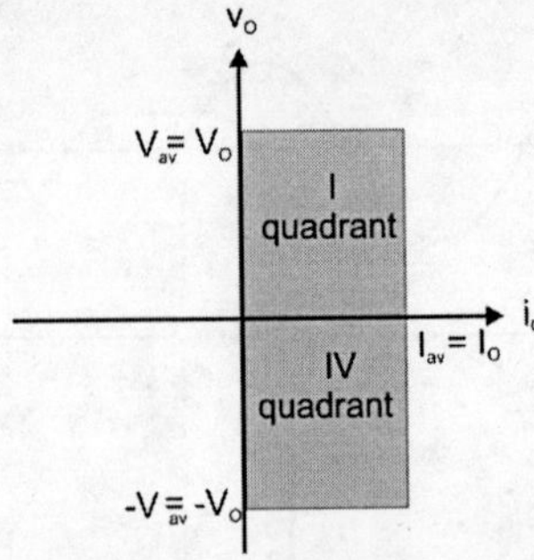

Fig.4.8 First and fourth quadrant operation of single half wave controlled rectifier with R-L load

The voltage equation for the circuit [Fig.4.6] when thyristor T_1 operates in conduction state and $\alpha \le \omega t \le \beta$, is $\sqrt{2}V \sin \omega t = L.di_o / dt + Ri_o$ (4.23)

where, V is the rms input voltage and output current is i_o

The output current i_o can be expressed as

$$i_o(t) = i_s + i_t = \frac{\sqrt{2}V}{Z}\sin(\omega t - \phi) + Ae^{-\frac{R}{L}t} \tag{4.24}$$

The current i_o has two components such as steady state component (i_s) and transient component (i_t). The steady state component of $i_o(t)$ is $i_s(t) = (\sqrt{2}V / Z)\sin(\omega t - \phi)$ and the transient component of $i_o(t)$ is $i_t(t) = Ae^{-\frac{R}{L}t}$ (4.25)

At $\omega t = \alpha$, current i_o=0, the equation (4.24) is expressed as

$$0 = \frac{\sqrt{2}V}{Z}\sin(\alpha - \phi) + Ae^{-\frac{R\alpha}{L\omega}} \tag{4.26}$$

From the equation (4.26), we obtain $A = -\frac{\sqrt{2}V}{Z}\sin(\alpha - \phi) \times e^{\frac{R\alpha}{L\omega}}$ (4.27)

After substituting the value of A in equation (4.24), we get

$$i_o(t) = \frac{\sqrt{2}V}{Z}\left[\sin(\omega t - \phi) - \sin(\alpha - \phi)e^{\frac{R\alpha}{L\omega}}e^{-\frac{R}{L}t}\right] \tag{4.28}$$

The equation (4.28) can be written as

Finally we get $i_o(t) = \frac{\sqrt{2}V}{Z}\left[\sin(\omega t - \phi) - \sin(\alpha - \phi)e^{-\frac{\omega t - \alpha}{\tan\phi}}\right]$ for $\alpha \le \omega t \le \beta$ (4.29) were, $Z = \sqrt{R^2 + (\omega L)^2}$, and $\tan\phi = \dfrac{\omega L}{R}$

As per output current waveform, i_o=0 at $\omega t = \beta$.

$$\text{So, } 0 = \frac{\sqrt{2}V}{Z}\left[\sin(\beta - \phi) - \sin(\alpha - \phi)e^{-\frac{\beta-\alpha}{\tan\phi}}\right] \tag{4.30}$$

From the above equation, we obtain, $\sin(\beta - \phi) = \sin(\alpha - \phi)e^{-\frac{\beta-\alpha}{\tan\phi}}$ (4.31)

The equation (4.31) will be solved to find out the value of β .

As average output voltage across inductor (L) is zero, average output voltage across R-L is equal to the average voltage across R. The average output voltage can be computed from equation (4.32)

$$V_{av} = V_O = \frac{1}{2\pi}\int_{\alpha}^{\beta}\sqrt{2}V\sin\omega t d(\omega t) = \frac{\sqrt{2}V}{2\pi}(\cos\alpha - \cos\beta) \tag{4.32}$$

If $\beta = 0$, the average output voltage is maximum. When the value of β increases, the average output voltage decreases.

The rms value of output voltage is

$$V_{rms} = \left[\frac{1}{2\pi}\int_{\alpha}^{\beta}(\sqrt{2}V\sin\omega t)^2 d\omega t\right]^{\frac{1}{2}} \tag{4.33}$$

$$= \frac{V}{\sqrt{2}\sqrt{\pi}}\left[(\beta - \alpha) - \frac{1}{2}(\sin 2\beta - \sin 2\alpha)\right]^{\frac{1}{2}} \tag{4.34}$$

The form factor is

$$FF = \frac{V_{rms}}{V_{av}} = \frac{V}{\sqrt{2}\sqrt{\pi}}\left[(\beta - \alpha) - \frac{1}{2}(\sin 2\beta - \sin 2\alpha)\right]^{\frac{1}{2}} / \frac{\sqrt{2}V}{2\pi}(\cos\alpha - \cos\beta) \tag{4.35}$$

$$= \frac{\sqrt{\pi}\left[(\beta - \alpha) - \frac{1}{2}(\sin 2\beta - \sin 2\alpha)\right]^{\frac{1}{2}}}{\cos\alpha - \cos\beta} \tag{4.36}$$

For more on Single Phase Controlled Rectifier

Example 4.1 A single phase 220V, 1.5 kW heater is connected a half wave controlled rectifier and it is supplied from a 220V, 50Hz ac supply. Calculate the power absorbed by the heater if the firing angle is (a) $\alpha = 60°$ and (b) $\alpha = 45°$

Solution:

Given: $V = 220V$, $\alpha = 60°$ =1.046 rad and $\alpha = 45°$ =0.785 rad

The resistance of heater is $R = V^2/W = 220^2/1500 = 32.26\Omega$ as $W = 1.5kW = 1500 Watt$

(a) The rms value of output voltage at $\alpha = 60°$ is

$$V_{rms} = \frac{V}{\sqrt{2}}\left[\frac{1}{\pi}(\pi - \alpha + \frac{1}{2}\sin 2\alpha)\right]^{\frac{1}{2}} = \frac{220}{\sqrt{2}}\left[\frac{1}{\pi}(\pi - 1.046 + \frac{1}{2}\sin(2\times 60))\right]^{\frac{1}{2}} = 139.56 \text{ V}$$

The power absorbed by the heater at firing angle $\alpha = 60°$ is

$$P_{at.\alpha=60°} = V_{rms}^2/R = 139.56^2/32.26 \text{ Watt} = 603.75 \text{ Watt}$$

(b) The rms value of output voltage at $\alpha = 45°$ is

$$V_{rms} = \frac{V}{\sqrt{2}}\left[\frac{1}{\pi}(\pi - \alpha + \frac{1}{2}\sin 2\alpha)\right]^{\frac{1}{2}} = \frac{220}{\sqrt{2}}\left[\frac{1}{\pi}(\pi - 0.785 + \frac{1}{2}\sin(2\times 45))\right]^{\frac{1}{2}} = 148.35 \text{ V}$$

The power absorbed by the heater at firing angle $\alpha = 45°$ is

$$P_{at.\alpha=45°} = V_{rms}^2/R = 148.35^2/32.26 \text{ Watt} = 682.19 \text{ Watt}$$

Example 4.2 A single phase half wave controlled rectifier is connected across R-L load and feeds from a 230V, 50Hz ac supply. If R=20Ω and L=0.05H, Compute (a) the firing angle to ensure no transient current and (b) the firing angle for the maximum transient.

Solution:

Given V=230V, $f = 50$Hz, R=20Ω and L=0.05H

The current flows through load is

$$i(t) = \frac{\sqrt{2}V}{Z}\left[\sin(\omega t - \phi) - \sin(\alpha - \phi)e^{-\frac{\omega t - \alpha}{\tan\phi}}\right] \quad \text{for } \alpha \le \omega t \le \beta$$

$$\text{Where, } Z = \sqrt{R^2 + (\omega L)^2}\text{, and } \tan\phi = \frac{\omega L}{R}$$

The transient current is $i_t(t) = -\frac{\sqrt{2}V}{Z}\left[\sin(\alpha - \phi)e^{-\frac{\omega t - \alpha}{\tan\phi}}\right]$

(a) When there is no transient current

$$i_t(t) = 0 = -\frac{\sqrt{2}V}{Z}\left[\sin(\alpha - \phi)e^{-\frac{\omega t - \alpha}{\tan\phi}}\right]$$

Or, $\sin(\alpha - \phi) = 0$

Then the firing angle to ensure no transient current is

$$\alpha = \phi = \tan^{-1}\frac{\omega L}{R} = \tan^{-1}\frac{2\pi f L}{R} = \tan^{-1}\frac{2\pi \times 50 \times 0.05}{20} = 38.14°$$

(b) For maximum transient current,

$$\sin(\alpha - \phi) = 1 = \sin 90°$$

So, $\alpha = 90° + \phi = 90° + 38.14° = 128.14°$

As a result, the maximum transient occurs at firing angle $\alpha = 128.14°$

Example 4.3 The voltage across secondary winding of a single phase transformer is 220V, 50 Hz and the transformer deliver power to resistive load of $R = 2\Omega$ through a single phase half wave controlled rectifier. When the firing angle of thyristor is $90°$, determine (a) average dc output voltage (b) average dc output current (c) rms output voltage (d) rms output current (e) form factor (f) voltage ripple factor (g) rectification efficiency (h) transformer utilization factor (i) PIV of thyristor.

Solution: Given V=220, f=50 Hz, $R = 2\Omega$, $\alpha = 90°$

(a) the average output voltage is

$$V_O = \frac{\sqrt{2}V}{2\pi}(1+\cos\alpha) = \frac{\sqrt{2}\times 220}{2\pi}(1+\cos 90)$$ =49.54 V

(b) average dc output current is $I_o = \frac{V_o}{R} = \frac{49.54}{2}A = 24.77A$

(c) the rms value of output voltage is equal to

$$V_{rms} = \frac{V}{\sqrt{2}}\left[\frac{1}{\pi}(\pi - \alpha + \frac{1}{2}\sin 2\alpha)\right]^{\frac{1}{2}} = \frac{200}{\sqrt{2}}\left[\frac{1}{\pi}(\pi - \frac{\pi}{2} + \frac{1}{2}\sin(2\times 90))\right]^{\frac{1}{2}}$$ =110 V

(d) rms output current $I_{rms} = \frac{V_{rms}}{R} = \frac{110}{2} = 55A$

(e) the form factor is equal to

$$FF = \frac{V_{rms}}{V_{av}} = \frac{110}{49.54}$$ =2.22

(f) the voltage ripple factor of load current is

$$RF = \sqrt{FF^2 - 1} = \sqrt{(2.22)^2 - 1}$$ =1.982

(e) The dc output power is equal to

$P_{dc} = V_{dc} \times I_{dc} = V_o \times I_o = 49.54 \times 24.77$ Watt=1227.10 Watt

The ac output power is equal to

$P_{ac} = V_{rms} \times I_{rms} = 110 \times 55$ =6050 Watt

Rectification efficiency= $\frac{P_{dc}}{P_{ac}} \times 100\% = \frac{1227.10}{6050} \times 100\%$ =20.28%

(f)The volt ampere rating of transformer is

VA= $VI_{rms} = 220 \times 55$ VA

Transformer utilization factor is

$$TUF = \frac{Power.delivered.to.load}{Input.VA} = \frac{P_{dc}}{VA} = \frac{1227.10}{220 \times 55}$$ =0.1014

(g) The peak inverse voltage of thyristor is

PIV= $\sqrt{2}V = \sqrt{2} \times 220$ V= 311.12 V

4.5 Single Phase Half-Controlled Rectifiers With Free Wheeling Diode

In single phase half wave controlled rectifier with R-L load, output voltage will be negative during $\omega t = \pi$ to $\omega t = \beta$. Consequently, the average output voltage decreases with increasing β and the performances of controlled rectifier are reduced. To improve controlled rectifier performance, a freewheeling diode will be connected across R-L load. The circuit diagram of single phase half wave controlled converter with R-L load and freewheeling diode D_F is shown in Fig.4.9. Since free wheeling diode D_F is connected across load, the controlled rectifier operates as half controlled converter. This converter operates in two operating modes such as continuous mode and discontinuous mode.

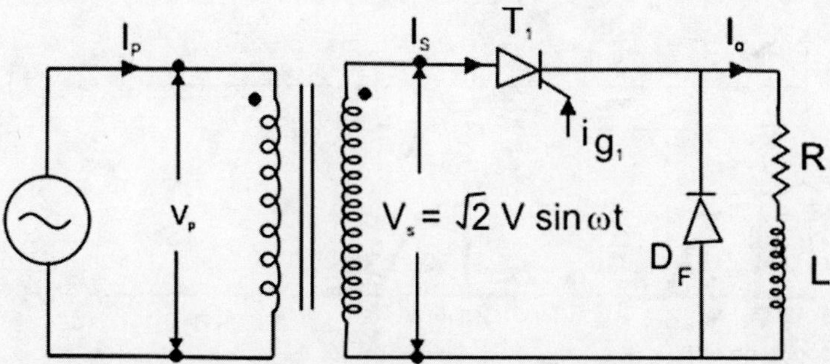

Fig.4.9 Single phase half controlled rectifier with R-L load and freewheeling diode D_F

4.5.1 Discontinuous mode of operation

In the positive half-cycle of supply voltage, thyristor T_1 is forward biased and firing pulse is applied at $\omega t = \alpha$. Thereafter, thyristor T_1 operates in on-state and input voltage is applied across R-L load and current starts to flow through load. Consequently thyristor T_1 conducts from $\omega t = \alpha$ to $\omega t = \pi$. Subsequent to $\omega t = \pi$, the polarity of input voltage is reversed, although the load current still flows due to inductance (L). In this time, the amplitude of load current decreases with time. At $\omega t > \pi$, diode D_F is forward biased and starts conduction. After that the decaying inductive current flows though load and diode D_F upto β for discontinuous mode of operation. Since thyristor T_1 is reverse biased, it will be turned off at $\omega t = \pi$. The waveforms of input voltage, triggering pulse of T_1, output voltage, output current and voltage across T_1 for half wave controlled rectifier with R-L load and freewheeling diode during *discontinuous mode operation* are shown in Fig.4.10.

The dc output voltage of controlled rectifier is

$$V_{av} = V_O = \frac{1}{2\pi}\int_{\alpha}^{\pi} \sqrt{2}V \sin\omega t d(\omega t) = \frac{\sqrt{2}V}{2\pi}(1+\cos\alpha) \tag{4.37}$$

The dc output current is equal to

$$I_{av} = \frac{V_O}{R} = \frac{\sqrt{2}V}{2\pi R}(1+\cos\alpha) \tag{4.38}$$

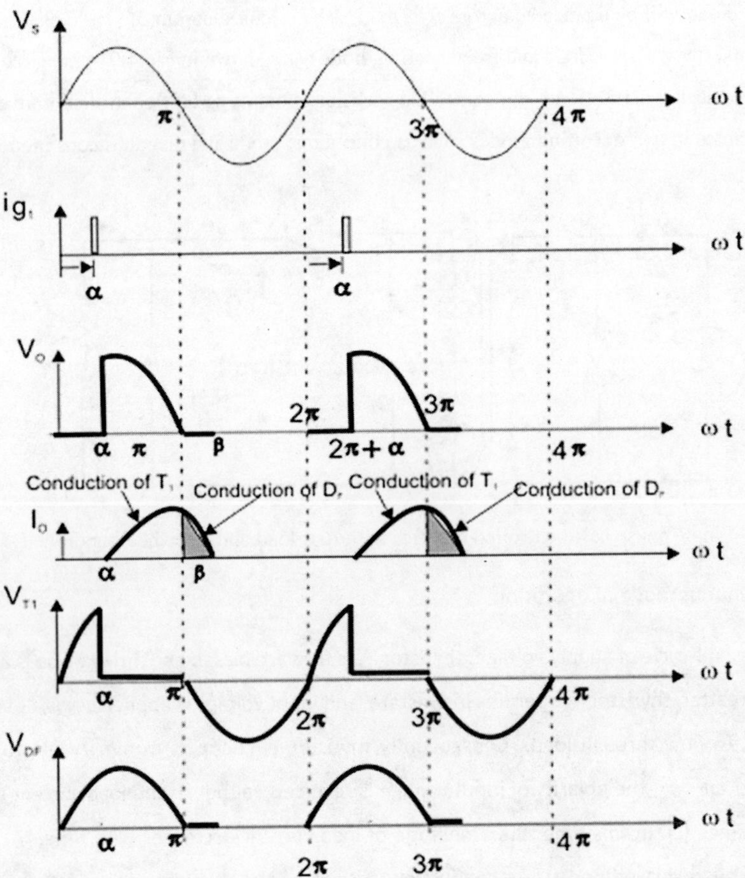

Fig.4.10 Supply voltage (V_s), triggering pulse of T_1 (i_{g1}), output voltage (V_o), output current (I_o) and voltage across T_1 for single phase half wave controlled rectifier with R-L load and freewheeling diode (discontinuous mode)

4.5.2 Continuous mode of operation

The continuous mode of operation of single phase half wave controlled rectifier with R-L load and freewheeling diode has two sub parts namely

(i) *MODE I : Conduction of T_1* and (ii) *MODE II : Conduction of D_F.*

The supply voltage (V_s), triggering pulse of T_1 (i_{g1}), output voltage (V_o), output current (I_o) and voltage across T_1 for single phase half wave controlled rectifier with R-L load and freewheeling diode is shown in Fig.4.11

MODE I : Conduction of T_1:

During continuous mode of operation, the voltage equation for the circuit [Fig.4.9] is

$$\sqrt{2}V\sin\omega t = L\frac{di_o}{dt} + Ri_o \qquad \text{for } \alpha \le \omega t \le \pi \qquad (4.39)$$

Where, V is the rms input voltage and output current is i_o

The output current i_o is equal to

$$i_o(t) = i_s + i_t = \frac{\sqrt{2}V}{Z}\sin(\omega t - \phi) + Ae^{-\frac{R}{L}t} \qquad (4.40)$$

The load current i_o has two components: (i) steady state component (i_s) and transient component (i_t). The steady state component current is $i_s(t) == \frac{\sqrt{2}V}{Z}\sin(\omega t - \phi)$ and the transient component current is $i_t(t) = Ae^{-\frac{R}{L}t}$

When $\omega t = \alpha$, load current i_o=I_o, the equation (4.40) can be expressed as

$$I_o = \frac{\sqrt{2}V}{Z}\sin(\alpha - \phi) + Ae^{-\frac{R\alpha}{L\omega}} \qquad (4.41)$$

Therefore, $A = \{I_o - \frac{\sqrt{2}V}{Z}\sin(\alpha - \phi)\} \times e^{\frac{R\alpha}{L\omega}}$ (4.42)

When the value of A is substituted in equation (4.40), we obtain

$$i_o(t) = \frac{\sqrt{2}V}{Z}\sin(\omega t - \phi) + \left[I_o - \frac{\sqrt{2}V}{Z}\sin(\alpha - \phi)\right] \times e^{\frac{R\alpha}{L\omega}} \times e^{-\frac{R}{L}t} \qquad (4.43)$$

So, $i_o(t) = \frac{\sqrt{2}V}{Z}\sin(\omega t - \phi) + \left[I_o - \frac{\sqrt{2}V}{Z}\sin(\alpha - \phi)\right] \times e^{-\frac{R}{L\omega}(\omega t - \alpha)}$ for $\alpha \le \omega t \le \pi$ (4.44)

MODE II : Conduction of D_F:

The voltage equation during conduction of D_F is equal to

$$0 = L\frac{di_o}{dt} + Ri_o \qquad \text{for } \pi \le \omega t \le 2\pi + \alpha \qquad (4.45)$$

The solution of the equation (4.45) is

$$i_o(t) = Ae^{-\frac{R}{L}t} \qquad (4.46)$$

When $\omega t = \pi$, $t = \dfrac{\pi}{\omega}$ and load current i_o =I_{o1}, the equation (4.46) can be rewritten as

$$I_{o1} = Ae^{-\frac{R\pi}{L\omega}} \tag{4.47}$$

When the value of A is substituted in equation (4.46), we get

$$i_o(t) = I_{o1}e^{\frac{R\pi}{L\omega}}e^{-\frac{R}{L}t} = I_{o1}e^{-\frac{R}{L}(\omega t-\pi)} \quad \text{for } \pi \le \omega t \le 2\pi + \alpha \tag{4.48}$$

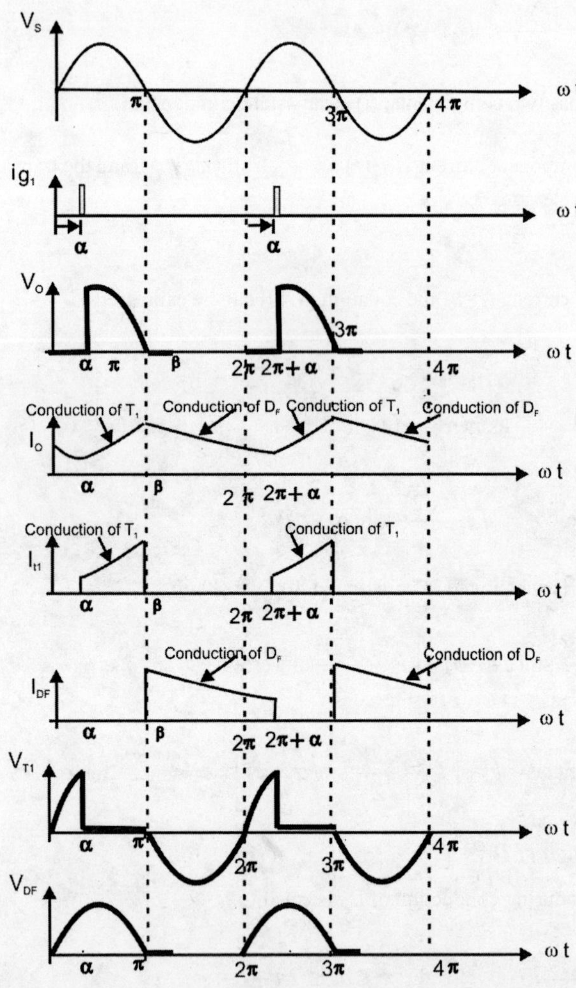

Fig.4.11 Supply voltage (V_s), triggering pulse of T_1 (i_{g1}), output voltage (V_o), output current (I_o) and voltage across T_1 for single phase half wave controlled rectifier with R-L load and freewheeling diode (continuous mode)

During the positive half cycle of supply voltage, a voltage is induced across the inductance and its value is $L\frac{di}{dt}$. When $\frac{di}{dt}$ changes its sign, the polarity of induced voltage is changed and free wheeling diode D_F starts to conduct as soon as the induced voltage reaches a sufficient amplitude. Subsequently inductance will discharge its stored energy into load resistance.

In this converter circuit, the power flow from input to load takes place only when the thyristor is conducting. Due to free wheeling diode D_F, no power will be returned to source during negative half cycle of power supply. If diod D_F is absent, the stored energy in the load inductance is returned to supply through thyristor. Therefore, the ratio of recative power flow from supply to the total power consumed in load is less in single phase half wave controlled rectifier with R-L load and free wheeling diode.

Hence, $\frac{Reactive.Power}{Power.Consumed} = \frac{VI\sin\phi}{VI} = \sin\phi$ is less. As $\sin\phi$ is less, $\cos\phi$ is more and power factor of single phase half wave controlled rectifier with R-L load and freewheeling diode is more. Consequently, the freewheeling diode helps to improve power factor of this converter.

The advantages of free wheeling diode in single phase half wave controlled rectifier with R-L load are (i) average output voltage increases, (ii) input power factor improves, (iii) load current wave form is improved, (iv) performance of controlled rectifier is better and (v) since the stored energy in inductance L is transferred to R when the free-wheeling diode conducts, the converter efficiency is improved.

4.6 Single Phase Full Wave Midpoint-Controlled Rectifiers

Fig.4.12 shows the single phase full wave midpoint controlled rectifiers using centre tap transformer. This circuit diagram is same as the single phase full wave uncontrolled rectifier, but thyristors are used in place of diodes. These converters are called as two pulse converters as two set triggering pulses are used to trigger thyristors. When the ac supply voltage is applied across the primary winding of transformer, the center tap of secondary winding is used as ground or zero reference voltage. The voltage between the center tap of transformer and either end of the secondary winding is half of the secondary winding voltage.

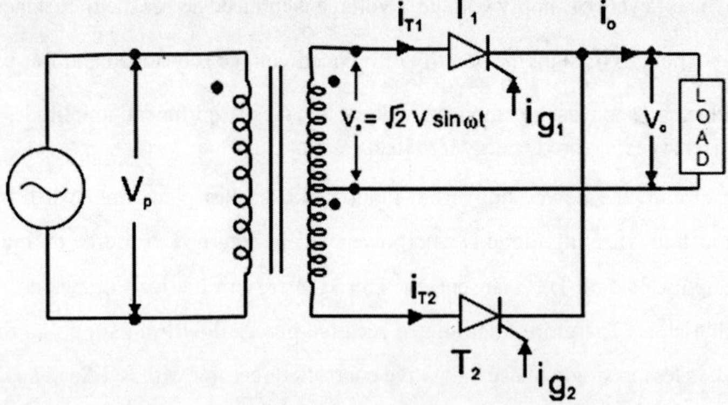

Fig.4.12 Single phase full wave midpoint controlled rectifier

4.6.1 Single Phase Full Wave Midpoint Controlled Rectifier With R Load

In the positive half cycle of supply voltage, thyristor T_1 is forward biased and thyristor T_2 is reverse biased. Whenever a gate pulse is applied to thyristor T_1 at $\omega t = \alpha$, T_1 starts to conduct and it operates in on-state and thyristor T_2 is in off-state and current flows through load resistance R and T_1 as shown in Fig.4.13.

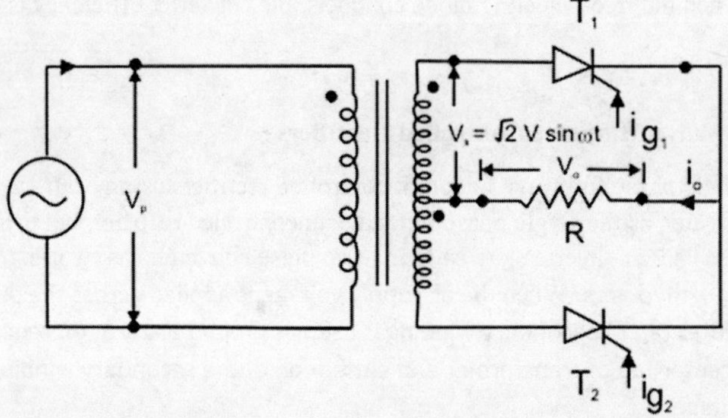

Fig.4.13 Single phase full wave midpoint controlled rectifier with R load

During negative half cycle of supply voltage, the polarities of secondary winding have been reversed. Then thyristor T_2 is forward biased and thyristor T_1 is reverse biased. As firing pulse is applied at $\omega t = \pi + \alpha$, thyristor T_2 operates in on-state and thyristor T_1 is in off-state and current flows through load resistance R and thyristor T_2. Therefore, the current flows through load in the same direction for positive and negative half cycle of supply voltage. Thus, the output voltage across load resistance is full wave rectified dc voltage. Since thyristor T_1 conducts from $\omega t = \alpha$ to $\omega t = \pi$ and thyristor T_2

conducts from $\omega t = \pi + \alpha$ to $\omega t = 2\pi$, the input voltage, triggering pulses, output voltage, output current, voltage across thyristors are shown in Fig.4.14.

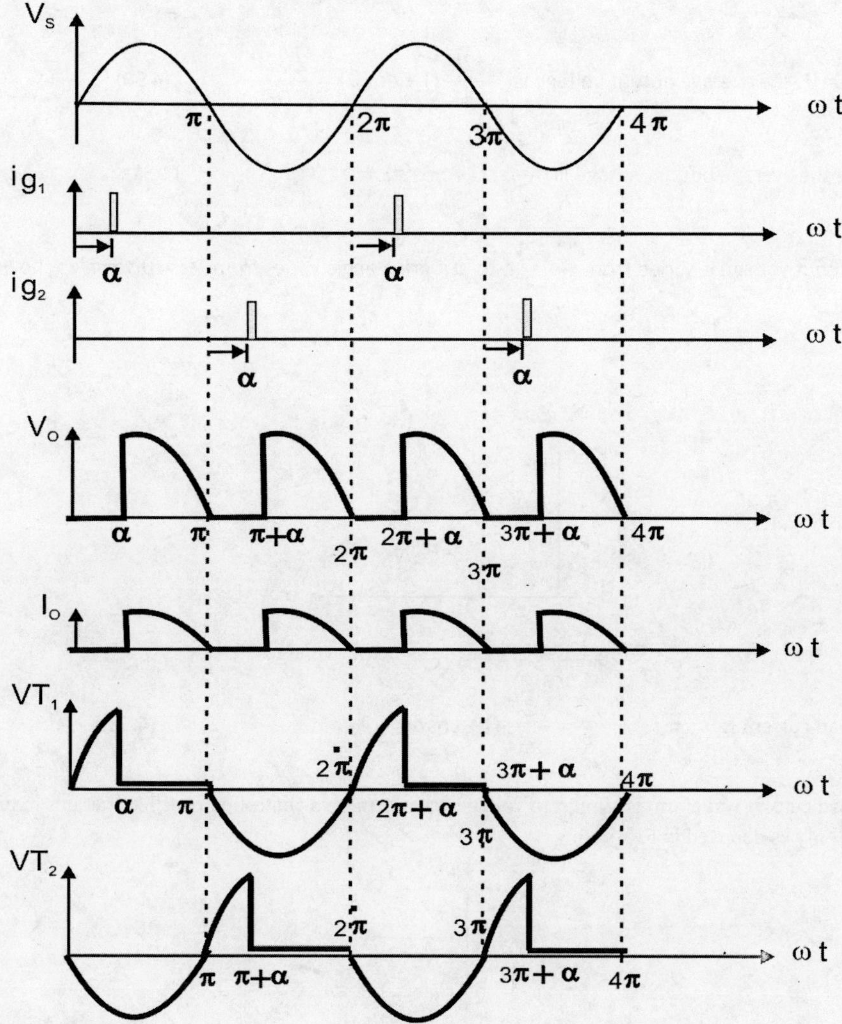

Fig.4.14 Supply voltages (V_s), gate pulse, output voltage (V_o), voltage across thyristor and load current (I_o) waveforms at firing angle α for single phase full wave controlled rectifier with R load

The average or dc output voltage is

$$V_o = V_{av} = \frac{1}{\pi}\int_{\alpha}^{\pi}\sqrt{2}V \sin \omega t.d\omega t = \frac{\sqrt{2}V}{\pi}(1+\cos\alpha) \quad (4.49)$$

When $\alpha = 0$, the average output voltage is $\frac{\sqrt{2}V}{\pi}(1+\cos 0) = \frac{2\sqrt{2}V}{\pi}$ (4.50)

If $\alpha = \pi$, the average output voltage is $\frac{\sqrt{2}V}{2\pi}(1+\cos\pi) = 0$ (4.51)

The dc output voltage is varied from $\frac{2\sqrt{2}V}{\pi}$ to 0 if firing angle varies from $\alpha = 0$ to π as shown in Fig.4.15.

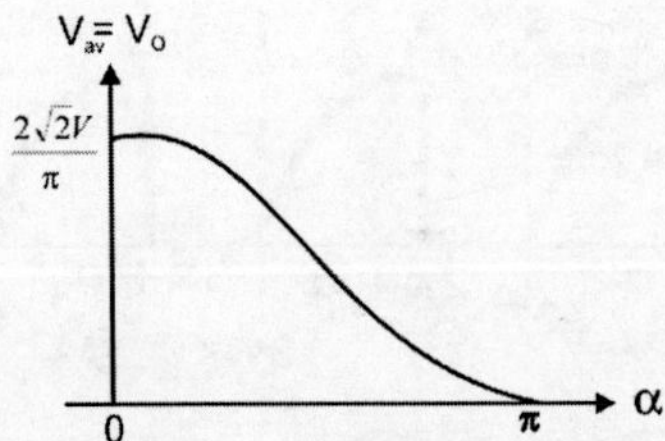

Fig.4.15 Firing angle vs output voltage

The dc load current is $I_o = I_{av} = \frac{V_o}{R} = \frac{\sqrt{2}V}{\pi R}(1+\cos\alpha)$ (4.52)

As the load current waveform is similar to voltage waveform, this controlled rectifier operates in first quadrant only as depicted in Fig.4.16.

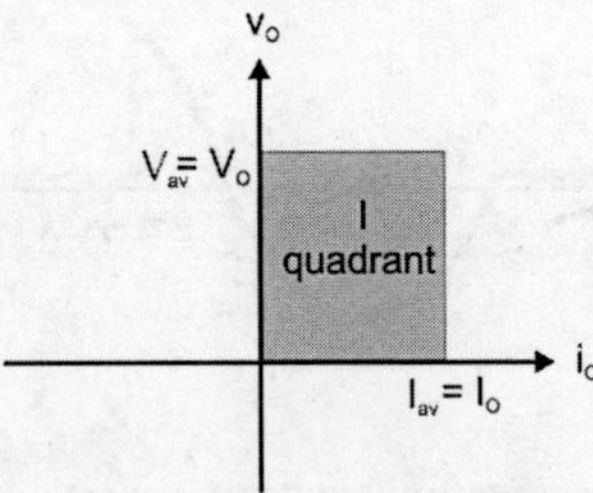

Fig.4.16 First quadrant operation of single phase full wave mid point controlled rectifier with R load

The rms output voltage is

$$V_{rms} = \left[\frac{1}{\pi} \int_{\alpha}^{\pi} (\sqrt{2}V \sin \omega t)^2 \, d\omega t \right]^{\frac{1}{2}} = \sqrt{2}V \left[\frac{1}{2\pi} (\pi - \alpha + \frac{1}{2} \sin 2\alpha) \right]^{\frac{1}{2}} \quad (4.53)$$

The rms output current is equal to

$$I_{rms} = \frac{V_{rms}}{R} = \frac{\sqrt{2}V}{R} \left[\frac{1}{2\pi} (\pi - \alpha + \frac{1}{2} \sin 2\alpha) \right]^{\frac{1}{2}} \quad (4.54)$$

The form factor is

$$FF = \frac{V_{rms}}{V_{av}} = \sqrt{2}V \left[\frac{1}{2\pi} (\pi - \alpha + \frac{1}{2} \sin 2\alpha) \right]^{\frac{1}{2}} / \frac{\sqrt{2}V}{\pi} (1 + \cos\alpha) \quad (4.55)$$

The load ripple factor is equal to $RF = \sqrt{FF^2 - 1}$ (4.56)

The dc output power is

$$P_{dc} = V_{av} I_{av} = V_o I_o \quad (4.57)$$

$$= \frac{\sqrt{2}V}{\pi} (1 + \cos\alpha) \times \frac{\sqrt{2}V}{\pi R} (1 + \cos\alpha) = \frac{2V^2}{\pi^2 R} (1 + \cos\alpha)^2 \quad (4.58)$$

The ac output power is $P_{ac} = V_{rms} I_{rms} = \frac{V^2}{\pi R} \left[(\pi - \alpha + \frac{1}{2} \sin 2\alpha) \right]$ (4.59)

The rms value of transformer secondary current is

$$I_{rms-transformer} = \frac{I_{rms}}{2} = \frac{\sqrt{2}V}{2R} \left[\frac{1}{2\pi} (\pi - \alpha + \frac{1}{2} \sin 2\alpha) \right]^{\frac{1}{2}} \quad (4.60)$$

The VA rating of transformer is

$VA = 2V_{rms} I_{rms-transformer}$

$$= \frac{\sqrt{2}V^2}{R} \left[\frac{1}{2\pi} (\pi - \alpha + \frac{1}{2} \sin 2\alpha) \right]^{\frac{1}{2}} \quad (4.61)$$

The transformer utilization factor (TUF) is

$$TUF = \frac{P_{dc}}{VA} = \frac{2(1 + \cos\alpha)^2}{\pi [\pi (\pi - \alpha + \frac{1}{2} \sin 2\alpha)]^{\frac{1}{2}}} \quad (4.62)$$

Example 4.4 A single phase full-wave midpoint controlled rectifier is fed from a 220V, 50Hz ac source and it is connected a resistive load R=20Ω. Calculate (a) the output voltage, (b) form factor, (c) ripple factor, (d) efficiency, (e) transformer utilization factor at $\alpha = 30°$. Assume turn ratio of transformer is 1:1

Solution

Given : $\alpha = 30° = 30 \times \frac{\pi}{180}$ = 0.523rad

As turn ratio of transformer is 1:1, the rms voltage between center tap and other terminal of transformer secondary is V=220/2=110V.

(a) The dc output voltage is

$$V_o = V_{av} = \frac{1}{\pi}\int_{\alpha}^{\pi} \sqrt{2}V \sin \omega t.d\omega t = \frac{\sqrt{2}V}{\pi}(1+\cos\alpha) = \frac{\sqrt{2}\times 110}{\pi}(1+\cos 30°) = 92.42 \text{ V}$$

(b) The rms output voltage is

$$V_{rms} = \left[\frac{1}{\pi}\int_{\alpha}^{\pi} (\sqrt{2}V \sin \omega t)^2 d\omega t\right]^{\frac{1}{2}} = \sqrt{2}V\left[\frac{1}{2\pi}(\pi - \alpha + \frac{1}{2}\sin 2\alpha)\right]^{\frac{1}{2}}$$

$$= \sqrt{2}\times 110\left[\frac{1}{2\pi}(\pi - 0.523 + \frac{1}{2}\sin(2\times 30))\right]^{\frac{1}{2}} = 108.39 \text{ V}$$

Form factor is

$$FF = \frac{V_{rms}}{V_{av}} = \frac{108.39}{92.42} = 1.172$$

(c) The load ripple factor is $RF = \sqrt{FF^2 - 1} = \sqrt{(1.172)^2 - 1}$ =0.6112

(d) The dc output current is $I_o = \frac{V_o}{R} = \frac{92.42}{20} = 4.621$ A

RMS output current $I_{rms} = \frac{V_{rms}}{R} = \frac{108.39}{20} = 5.419$ A

The dc output power is $P_{dc} = V_o I_o = 92.42 \times 4.621$ Watt=427.07 Watt

The ac output power is $P_{ac} = V_{rms} I_{rms} = 108.39 \times 5.419$ Watt=587.365 Watt

Rectifier efficiency is $\eta = \frac{P_{dc}}{P_{ac}} \times 100\% = \frac{427.07}{597.368} \times 100\%$ =71.49 %

(e) The rms value of transformer secondary current is

$$I_{rms-transformer} = \frac{I_{rms}}{2} = \frac{5.419}{2} \text{ A=2.709A}$$

VA rating of transformer is

VA =2 $VI_{rms-transformer}$ = $2 \times 110 \times 2.709$ = 595.98 VA

Transformer utilization factor (TUF) is

$$TUF = \frac{P_{dc}}{VA} = \frac{427.07}{595.98} = 0.7165$$

4.6.2 Single phase full wave controlled rectifier using centre tap transformer with R-L Load

The single phase full wave midpoint controlled rectifier using centre tap transformer at R-L load is shown in Fig.4.17. During the positive half cycle of supply voltage, thyristor T_1 is forward biased and thyristor T_2 is reverse biased. If a gate pulse is applied to the thyristor T_1 at $\omega t = \alpha$, T_1 operates in on-state and thyristor T_2 is in off-state and current flows through R-L load and T_1.

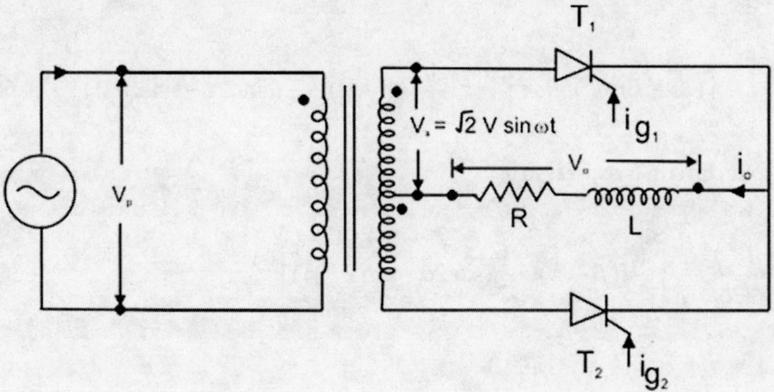

Fig.4.17 Single phase full wave midpoint controlled rectifier with R-L load

In the negative half cycle of supply voltage, the polarities of secondary winding have been reversed. Although the voltage across thyristor T_1 is negative, thyristor T_1 will not be turned-off due to inductive load. If a firing pulse is applied to thyristor T_2 at $\omega t = \pi + \alpha$, then thyristor T_2 conducts and thyristor T_1 will be in off-state and current flows through R-L load and T_2. The load current may be discontinuous or continuous depending upon the inductive load. This controlled rectifier operates in continuous or discontinuous mode. If inductance value is very large, the load current will be continuous and each thyristor conducts for 180° duration. In a complete cycle, Thyristor T_1 conducts from $\omega t = \alpha$ to

$\omega t = \pi + \alpha$ and Thyristor T_2 conducts from $\omega t = \pi + \alpha$ to $\omega t = 2\pi + \alpha$. Voltage and current waveforms for continuous mode operation are shown in Fig.4.19.

When inductance value is low, the load current will be discontinuous and each thyristor conducts for less than 180° duration. For a complete cycle, Thyristor T_1 conducts from $\omega t = \alpha$ to $\omega t = \beta$ and Thyristor T_2 conducts from $\omega t = \pi + \alpha$ to $\omega t = \pi + \beta$. Voltage and current waveforms for discontinuous mode of operation are depicted in Fig.4.18.

4.6.2.1 Discontinuous Load Current

In discontinuous load current, each thyristor conducts for less than 180° duration and the average output voltage is

$$V_o = V_{av} = \frac{1}{\pi}\int_{\alpha}^{\beta} \sqrt{2}V \sin \omega t.d\omega t = \frac{\sqrt{2}V}{\pi}(\cos\alpha - \cos\beta) \tag{4.63}$$

The dc load current is

$$I_o = I_{av} == \frac{\sqrt{2}V}{\pi R}(\cos\alpha - \cos\beta) \tag{4.64}$$

The rms output voltage is equal to

$$V_{rms} = \left[\frac{1}{\pi}\int_{\alpha}^{\beta} (\sqrt{2}V \sin \omega t)^2 \, d\omega t\right]^{\frac{1}{2}} = V\left[\frac{1}{\pi}\{(\beta - \alpha) + \frac{1}{2}(\sin 2\alpha - \sin 2\beta)\}\right]^{\frac{1}{2}} \tag{4.65}$$

The rms output current is expressed as

$$I_{rms} = \frac{V_{rms}}{R} = \frac{V}{R}\left[\frac{1}{\pi}\{(\beta - \alpha) + \frac{1}{2}(\sin 2\alpha - \sin 2\beta)\}\right]^{\frac{1}{2}} \tag{4.66}$$

Form factor is $FF = \dfrac{V_{rms}}{V_{av}}$ (4.67)

Load ripple factor is $RF = \sqrt{FF^2 - 1}$ (4.68)

The dc output power is $P_{dc} = V_{av} I_{av} = V_o I_o = \dfrac{2V^2}{\pi^2 R}(\cos\alpha - \cos\beta)^2$ (4.69)

The ac output power is $P_{ac} = V_{rms} I_{rms}$ (4.70)

The rms value of transformer secondary current is $I_{rms-transformer} = I_{rms}/2$ (4.71)

VA rating of transformer is $VA = 2V I_{rms\text{-}transformer}$ (4.72)

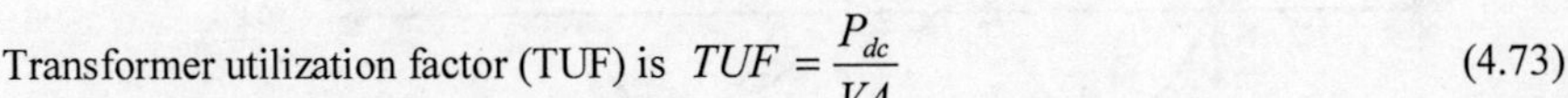

Transformer utilization factor (TUF) is $TUF = \dfrac{P_{dc}}{VA}$ (4.73)

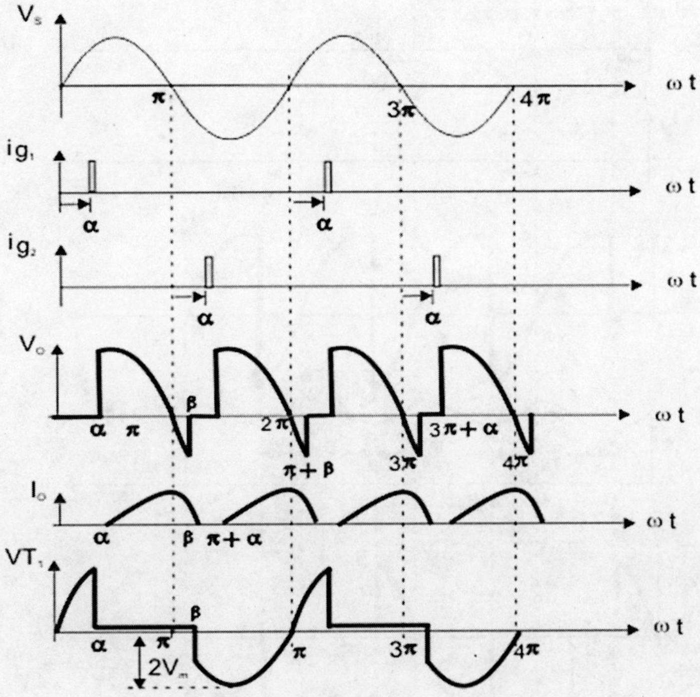

Fig.4.18 Voltage and current waveforms of single phase full wave midpoint controlled rectifier with R-L load in discontinuous load current

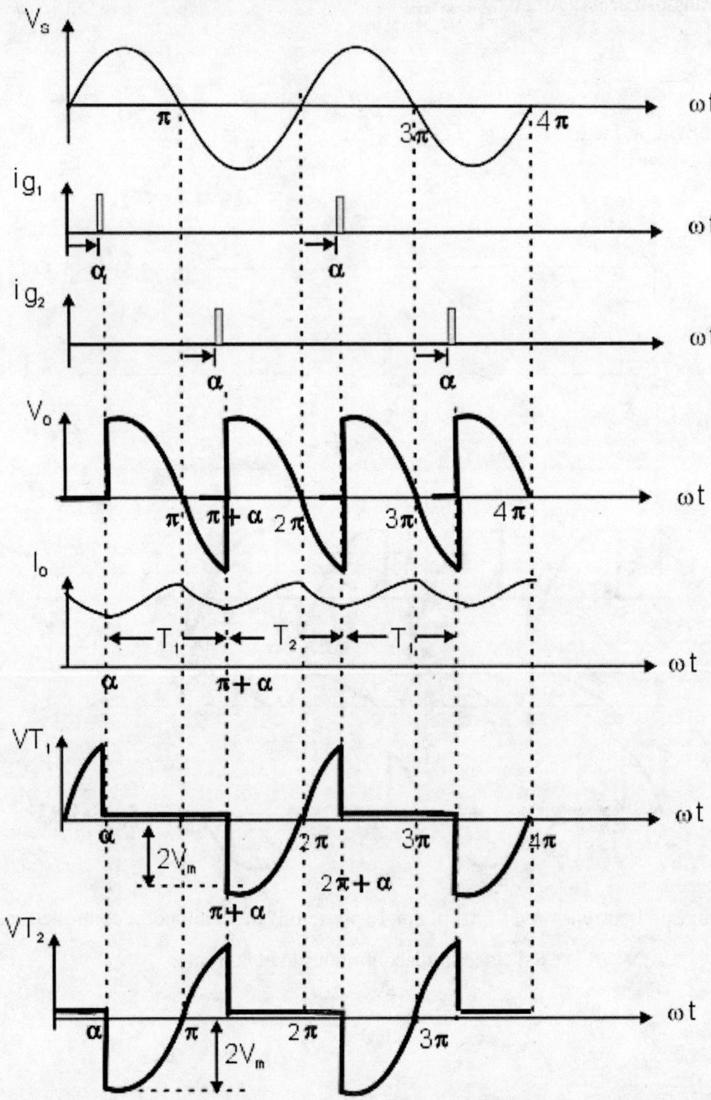

Fig.4.19 Voltage and current waveforms of single phase full wave midpoint controlled rectifier with R-L load in continuous load current

4.6.2.2 Continuous Load Current

During continuous mode operation, each thyristor conducts for 180° duration and the average output voltage is

$$V_o = V_{av} = \frac{1}{\pi}\int_{\alpha}^{\pi+\alpha} \sqrt{2}V \sin\omega t.d\omega t = \frac{2\sqrt{2}V}{\pi}\cos\alpha \quad (4.74)$$

If $\alpha = 0$, average output voltage is $\frac{2\sqrt{2}V}{\pi}$. When $\alpha = \pi/2$, average output voltage is 0 and if $\alpha = \pi$, average output voltage is -$\frac{2\sqrt{2}V}{\pi}$. So, the dc output voltage can be varied from $\frac{2\sqrt{2}V}{\pi}$ to -$\frac{2\sqrt{2}V}{\pi}$ when firing angle varies from $\alpha = 0$ to π as shown in Fig.4.20.

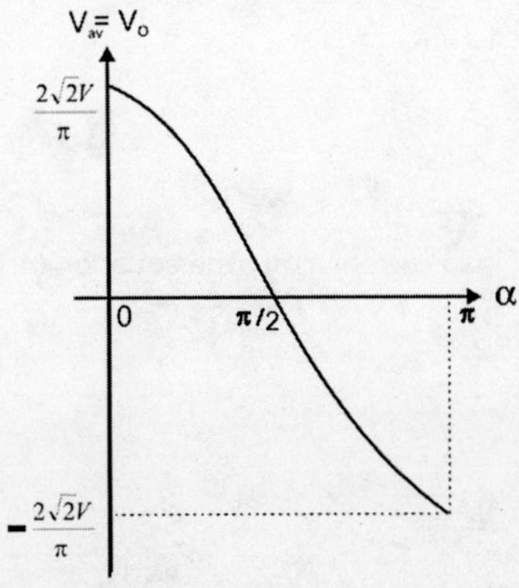

Fig.4.20 Firing angle vs output voltage of single phase full wave midpoint controlled rectifier with R-L load in continuous load current

As the load current is always positive and output voltage is varied from $\frac{2\sqrt{2}V}{\pi}$ to -$\frac{2\sqrt{2}V}{\pi}$, this controlled rectifier operates in first and fourth quadrant as depicted in Fig.4.21.

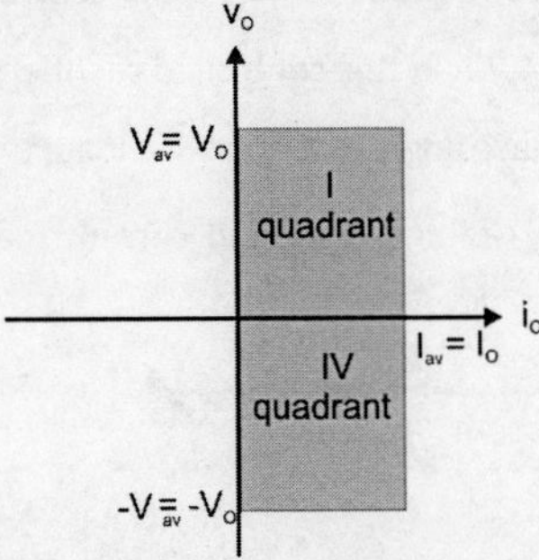

Fig.4.21 First and fourth quadrant operation of single phase full wave midpoint controlled rectifier with R-L load

The rms output voltage is

$$V_{rms} = \left[\frac{1}{\pi}\int_{\alpha}^{\pi+\alpha}(\sqrt{2}V\sin\omega t)^2 d\omega t\right]^{\frac{1}{2}} = \left[\frac{V^2}{\pi}\left| \omega t - \frac{1}{2}\sin 2\omega t\right|_{\alpha}^{\pi+\alpha}\right]^{\frac{1}{2}} = V \quad (4.75)$$

The voltage equation for the circuit [Fig.4.17] is

$$\sqrt{2}V\sin\omega t = L\frac{di_o}{dt} + Ri_o \qquad \text{for } \alpha \le \omega t \le \pi + \alpha \qquad (4.76)$$

Where, V is the rms input voltage and output current is i_o

The output current i_o is expressed as

$$i_o(t) = \frac{\sqrt{2}V}{Z}\sin(\omega t - \phi) + Ae^{-\frac{R}{L}t} \qquad (4.77)$$

When $\omega t = \alpha$ and $\omega t = \pi + \alpha$, load current i_o =I$_o$, the equation (4.77) can be written as

$$I_o = \frac{\sqrt{2}V}{Z}\sin(\alpha - \phi) + Ae^{-\frac{R\alpha}{L\omega}} \qquad (4.78)$$

and $I_o = \frac{\sqrt{2}V}{Z}\sin(\pi + \alpha - \phi) + Ae^{-\frac{R(\pi+\alpha)}{L\ \ \omega}}$ (4.79)

After solving the equation (4.77), we obtain

$$i_o(t) = \frac{\sqrt{2}V}{Z}\sin(\omega t - \phi) + \frac{\frac{\sqrt{2}V}{Z}2\sin(\phi - \alpha)}{1 - e^{-\frac{\pi}{\tan\phi}}}e^{-\frac{(\omega t - \alpha)}{\tan\phi}} \qquad (4.80)$$

4.6.3 Single phase full wave midpoint controlled rectifier with R-L Load and Free-wheeling Diode

The single phase full wave midpoint controlled rectifier using centre tap transformer with R-L load and free wheeling diode is shown in Fig.4.22. In this controlled rectifier, the conduction period of thyristor depends upon firing angle α and phase angle ϕ. Due to R-L load, the conduction of thyristor continues in the negative half cycle of supply voltage. The load current may be continuous or discontinuous depending on load impedance.

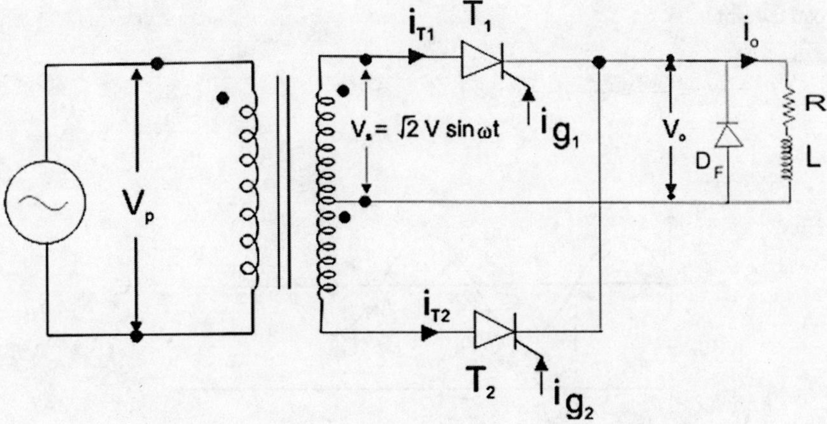

Fig.4.22 Single phase full wave midpoint controlled rectifier with R-L load and free wheeling diode

While a free wheeling diode D_F is connected across R-L load of single phase full wave midpoint controlled rectifier using centre tap transformer, this converter works as a *semi-converter*. During the positive half cycle of supply voltage, thyristor T_1 is forward biased and firing pulse is applied at $\omega t = \alpha$. Subsequently thyristor T_1 is turned on at $\omega t = \alpha$ and conducts upto $\omega t = \pi$. When $\omega t = \pi$, the diode D_F becomes forward biased as cathode is negative with respect to anode and conducts. Therefore, load current flows through diode D_F and R-L load from $\omega t = \pi$ to $\omega t = \beta$ for discontinuous load current. At $\omega t = \beta$, diode D_F is turned-off as load current reduces to zero.

During the negative half cycle of supply voltage, thyristor T_2 is forward biased and firing pulse is applied at $\omega t = \pi + \alpha$. After that thyristor T_2 is turned on at $\omega t = \pi + \alpha$ and conducts up to $\omega t = 2\pi$. At $\omega t = 2\pi$, the diode D_F becomes forward biased and conducts. So, load current flows through diode D_F and R-L load from $\omega t = 2\pi$ to $\omega t = \pi + \beta$ for discontinuous load current. At $\omega t = \pi + \beta$, diode D_F is turned off as load current reduces to zero. The voltage and current waveforms in discontinuous load current are shown in Fig.2.23 and the voltage and current waveforms in continuous load current are depicted in Fig.4.24.

The dc output voltage is equal to

$$V_o = V_{av} = \frac{1}{\pi}\int_{\alpha}^{\pi} \sqrt{2}V \sin \omega t.d\omega t = \frac{\sqrt{2}V}{\pi}(1+\cos\alpha) \qquad (4.81)$$

The dc load current is

$$I_o = I_{av} = \frac{V_o}{R} = \frac{\sqrt{2}V}{\pi R}(1+\cos\alpha) \tag{4.82}$$

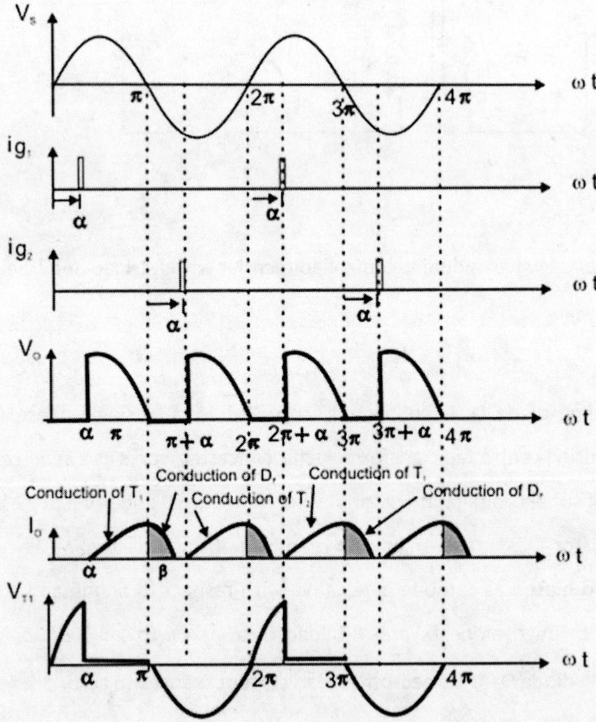

Fig.4.23 Voltage and current waveforms of single phase full wave midpoint controlled rectifier with R-L load and free wheeling diode in discontinuous load current

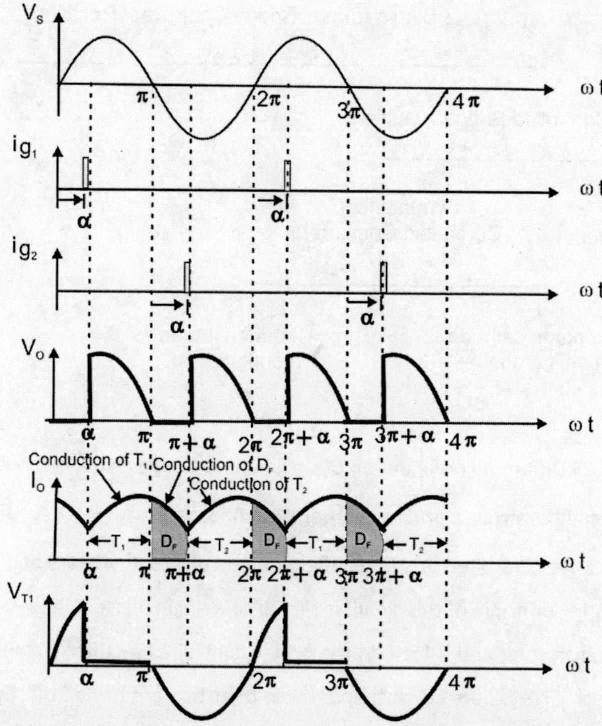

Fig.4.24 Voltage and current waveforms of Single phase full wave midpoint controlled rectifier with R-L load and free wheeling diode in continuous load current

4.7 Different Configurations of Bridge Controlled Rectifiers

Single phase full wave controlled bridge rectifiers are extensively used in various industrial applications. This controlled rectifier circuit provides unidirectional current flow during both positive and negative half cycle of input voltage and current flow also depends on the firing angle of thyristors. Therefore this converter can operate as a controlled rectifier or an inverter. As a single phase full wave controlled bridge rectifier consists of four thyristors, the gate drive circuit will be complex and expensive. The form factor and input power factor of this converter is relatively poor. To improve the output voltage and current form factor, a free wheeling diode D_F is connected across load. When the output voltage becomes negative, the diodes D_F across load becomes forward bias and clamp the load voltage to zero. So, this converter circuit will not be able to operate in the *inverter mode*. If two thyristors of a single phase fully controlled rectifier are replaced by two diodes, the fully controlled converter becomes half controlled converters. The different configuration of single phase bridge controlled rectifiers is shown in Fig.4.25.

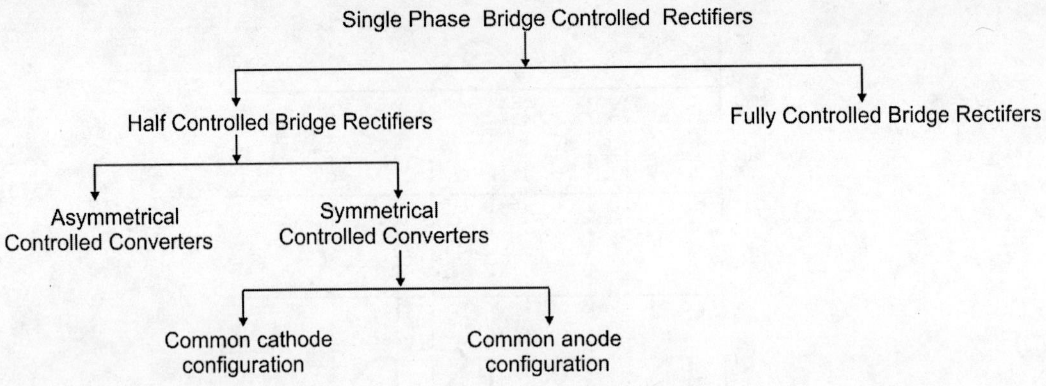

Fig.4.25 Different configuration of single phase bridge controlled rectifier

4.7.1 Single phase fully controlled bridge rectifier With R load

A single phase full wave controlled bridge rectifier consists of four thyristors and R load as shown in Fig.4.26. The operating principle of this circuit is similar to a single phase full wave controlled rectifier using center tap transformer and R load. In the bridge controlled rectifier circuit, diagonally pair of thyristors "T_1 & T_2" or "T_3&T_4" are conduct and these thyristors are turned off by line commutation simultaneously.

During positive half cycle of supply voltage, thyristors T_1 and T_2 are forward biased. When any firing pulse is applied to T_1 and T_2 simultaneously at $\omega t = \alpha$, thyristors T_1 and T_2 will be turned on and conduct during $\alpha \le \omega t \le \pi$. Subsequently current flows through T_1, T_2 and R as shown in Fig.4.27.

In the negative half cycle of input voltage, thyristors T_3 and T_4 are forward biased. Whenever any firing pulse is applied to T_3 and T_4 simultaneously at $\omega t = \pi + \alpha$, thyristors T_3 and T_4 can be turned on and conduct during $\pi + \alpha \le \omega t \le 2\pi$. Then current flows through T_3 and T_4 and R as depicted in Fig.4.27.

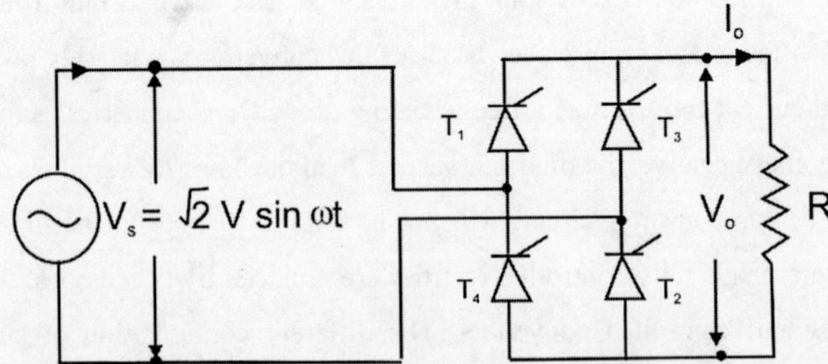

Fig.4.26 Single phase full wave controlled bridge rectifier with R load

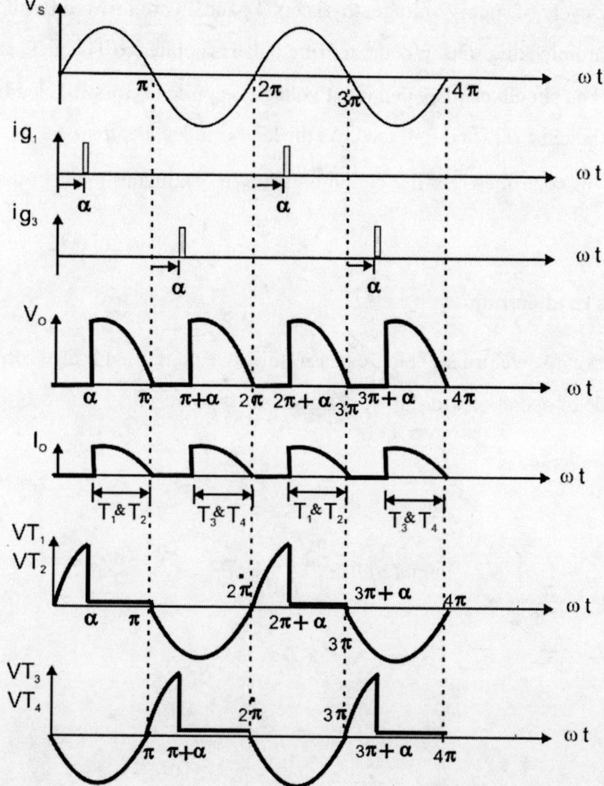

Fig.4.27 voltage and current waveforms of single phase full wave controlled bridge rectifier with R load

4.7.2 Single phase fully controlled bridge rectifier With R-L load

Fig.4.28 shows a single phase full wave controlled bridge rectifier with R-L load. During negative half-cycle of supply voltage, thyristors T_1 and T_2 are forward biased and operates in the forward blocking sate. If the firing pulse is applied to T_1 and T_2 at $\omega t = \alpha$, both thyristors T_1 and T_2 will be turned on simultaneously and input voltage is applied across R-L load. Then, the load current flows though T_1, T_2 and R-L load. Due to inductive load, thyristors T_1 and T_2 conduct beyond $\omega t = \pi$.

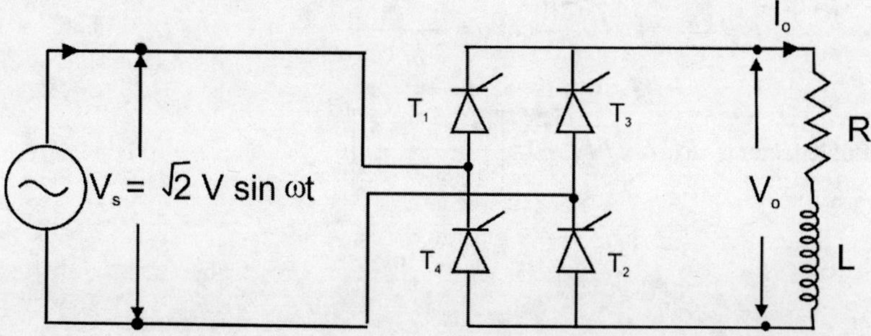

Fig.4.28 Single phase full wave controlled bridge rectifier with R-L load

In the negative half-cycle of supply voltage, thyristors T_3 and T_4 are forward biased and these devices operate in the forward blocking sate. When the firing pulse is applied to T_3 and T_4 at $\omega t = \pi + \alpha$, both thyristors are turned on simultaneously and input voltage is applied across R-L load. Consequently, the load current flows though T_3, T_4 and R-L load. As the load is inductive, thyristor T_3 and T_4 will conduct beyond $\omega t = 2\pi$. This controlled rectifier circuit operates in continuous mode or discontinuous mode.

4.7.2.1 Continuous Load Current

The voltage and current waveforms of single phase full wave controlled bridge rectifier with R-L load for continuous mode of operation are shown in Fig.4.29.

The average output voltage is

$$V_o = V_{av} = \frac{1}{\pi}\int_{\alpha}^{\pi+\alpha} \sqrt{2}V \sin \omega t . d\omega t = \frac{2\sqrt{2}V}{\pi}\cos\alpha \qquad (4.83)$$

The dc load current is

$$I_o = I_{av} = \frac{V_o}{R} = \frac{2\sqrt{2}V}{\pi R}\cos\alpha \qquad (4.84)$$

The rms output voltage is

$$V_{rms} = \left[\frac{1}{\pi}\int_{\alpha}^{\pi+\alpha} (\sqrt{2}V \sin \omega t)^2 d\omega t\right]^{\frac{1}{2}} = \left[\frac{V^2}{2\pi}\int_{\alpha}^{\pi+\alpha} (1-\cos 2\omega t).d\omega t\right]^{\frac{1}{2}} = V \qquad (4.85)$$

Form factor is

$$FF = \frac{V_{rms}}{V_{av}} = V / \frac{2\sqrt{2}V}{\pi}\cos\alpha = \frac{\pi}{2\sqrt{2}\cos\alpha} \qquad (4.86)$$

The load ripple factor is $RF = \sqrt{FF^2 - 1} = \left[\frac{\pi^2}{8\cos^2\alpha} - 1\right]^{1/2}$ (4.87)

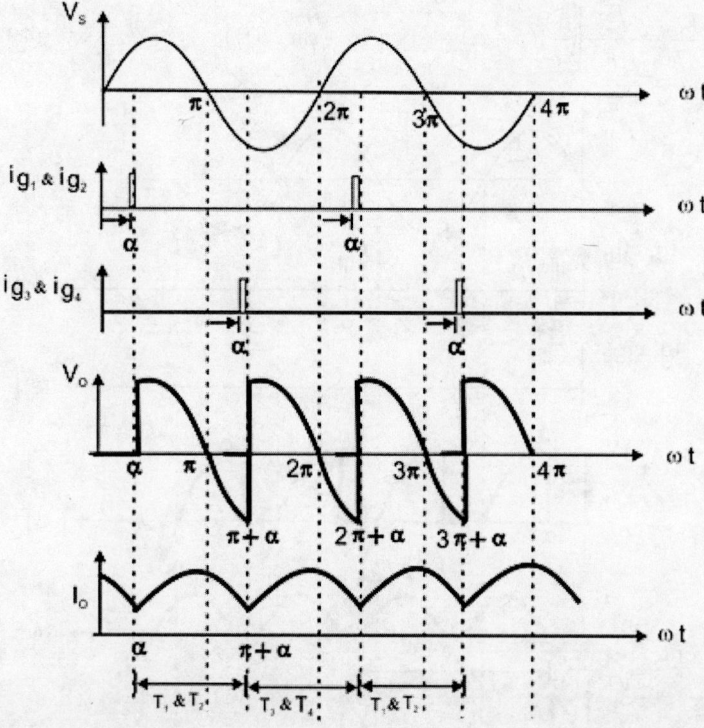

Fig.4.29 The voltage and current waveforms of single phase full wave controlled bridge rectifier with R-L load for continuous load current

4.7.2.2 Discontinuous Load Current

For discontinuous load current, the voltage and current waveforms of single phase full wave controlled bridge rectifier with R-L load are shown in Fig.4.30.

The average output voltage is

$$V_o = V_{av} = \frac{1}{\pi}\int_{\alpha}^{\beta} \sqrt{2}V \sin \omega t.d\omega t = \frac{\sqrt{2}V}{\pi}(\cos\alpha - \cos\beta) \qquad (4.88)$$

The average or dc load current is

$$I_o = I_{av} = \frac{V_o}{R} = \frac{\sqrt{2}V}{\pi R}(\cos\alpha - \cos\beta) \qquad (4.89)$$

The rms output voltage is

$$V_{rms} = \left[\frac{1}{\pi}\int_{\alpha}^{\beta} (\sqrt{2}V \sin \omega t)^2 \, d\omega t\right]^{\frac{1}{2}} = \sqrt{2}V\left[\frac{1}{\pi}\{(\beta - \alpha) + \frac{1}{2}(\sin 2\alpha - \sin 2\beta)\}\right]^{\frac{1}{2}} \qquad (4.90)$$

The rms output current is

$$I_{rms} = \frac{V_{rms}}{R} = \frac{\sqrt{2}V}{R}\left[\frac{1}{\pi}\{(\beta-\alpha)+\frac{1}{2}(\sin 2\alpha - \sin 2\beta)\}\right]^{\frac{1}{2}} \tag{4.91}$$

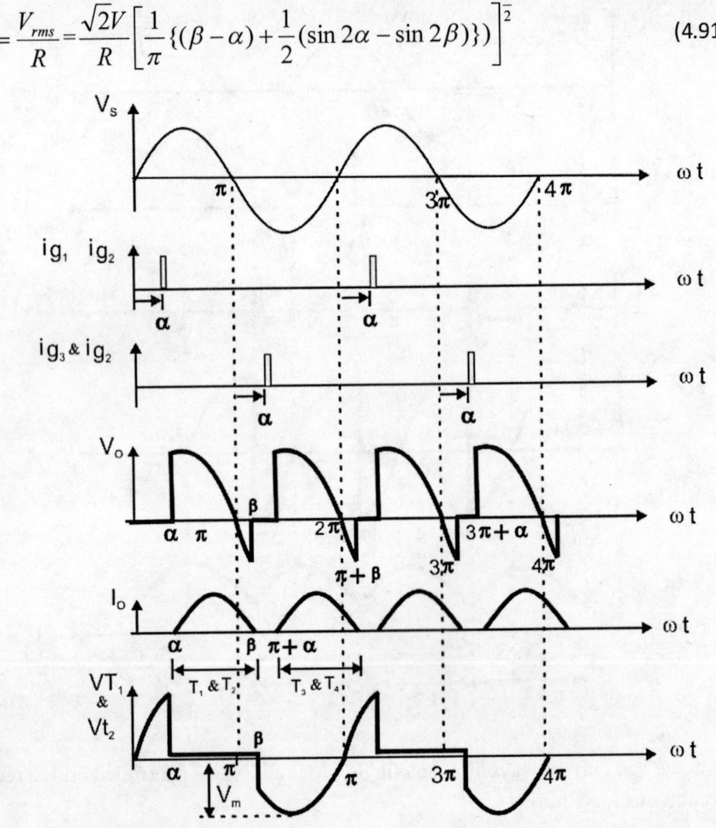

Fig.4.30 The voltage and current waveforms of single phase full wave controlled bridge rectifier with R-L load for discontinuous load current

4.7.3 Single Phase Fully Controlled Bridge Rectifier With Free Wheeling Diode D_F and R-L Load

A single phase full wave controlled bridge rectifier which consists of four thyristors (T_1,T_2,T_3 and T_4), R-L load and free wheeling diode D_F as shown in Fig.4.31. In the negative half-cycle of supply voltage, thyristors T_1 and T_2 are forward biased and the firing pulse is applied to T_1 and T_2 at $\omega t = \alpha$. Subsequently both thyristors T_1 and T_2 are turned on at the same time and supply voltage is applied across R-L load. Then, the load current flows through T_1, T_2 and R-L load. Since load is inductive, the load current continuously flows after $\omega t = \pi$. The load current can flow through D_F as thyristors T_1 and T_2 are turned off due to reverse bias after $\omega t = \pi$ and thyristor current is below the holding current.

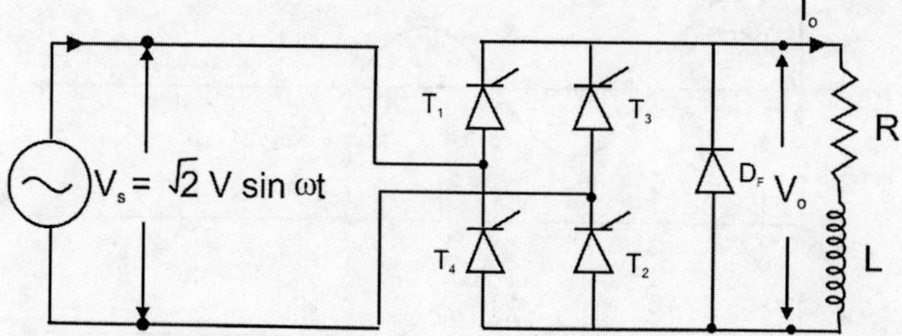

Fig.4.31 Single phase full wave controlled bridge rectifier with R-L load and free wheeling diode D_F

During negative half-cycle of supply voltage, thyristors T_3 and T_4 are forward biased and these devices operate in the forward blocking sate. Whenever firing pulse is applied to T_3 and T_4 at $\omega t = \pi + \alpha$, both thyristors T_3 and T_4 are turned on simultaneously and input voltage is applied to R-L load. So, the load current flows through T_3, T_4 and R-L load. Due to presence of inductive load, the load current continuously flows after $\omega t = 2\pi$. Then the load current flows through D_F as thyristors T_3 and T_4 are turned off due to reverse bias after $\omega t = 2\pi$ and the current flow through thyristor is less than the holding current. This controlled rectifier circuit operates in continuous mode and discontinuous mode. The voltage and current waveforms of single phase full wave controlled bridge rectifier with R-L load and free wheeling diode D_F for continuous load current and for discontinuous load current are shown in Fig.4.32 and Fig.4.33 respectively.

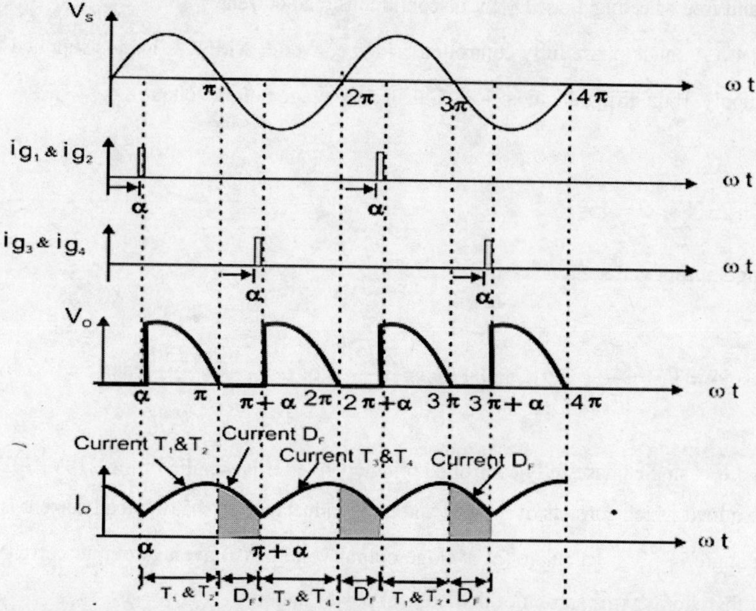

Fig.4.32 The voltage and current waveformsof single phase full wave controlled bridge rectifier with R-L load and free wheeling diode D_F for continuous load current

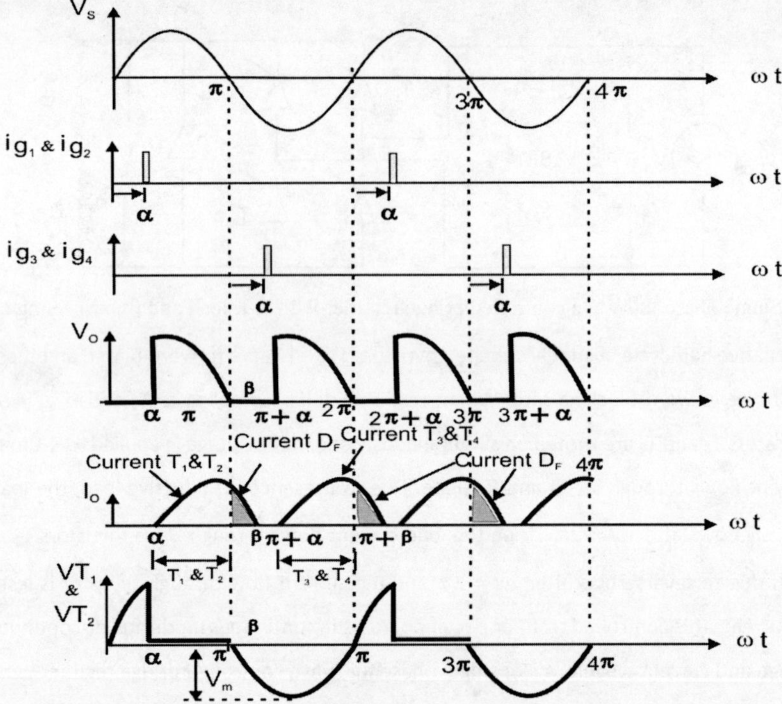

Fig.4.33 The voltage and current waveforms of single phase full wave controlled bridge rectifier with R-L load and free wheeling diode D_F for discontinuous load current

Example 4.5 A single phase fully controlled bridge converter with R-L load is supplied from 230V, 50Hz ac supply. If the firing angle is $45°$, Calculate average output voltage

Solution

Given: $V = 230V$, $\alpha = 45°$

The average output voltage is

$$V_o = V_{av} = \frac{1}{\pi}\int_{\alpha}^{\pi+\alpha} \sqrt{2}V \sin \omega t.d\omega t = \frac{2\sqrt{2}V}{\pi}\cos\alpha = \frac{2\sqrt{2}\times 230}{\pi}\cos 45 = 146.46\text{V}$$

Example 4.6 A single phase fully controlled bridge converter is supplied from 210V, 50Hz ac supply and fed to a load which consists of R=10Ω and large inductance so that the load current is constant. If the firing angle is $45°$, calculate (a) average output voltage, (b) average output current (c) average current of thyristor (d) rms current of thyristor (e) power factor.

Solution

Given: $V = 210V$, f = 50 Hz, $R = 10\Omega$ A and $\alpha = 45°$

(a)At firing angle α the average output voltage is

$$V_o = V_{av} = \frac{1}{\pi}\int_{\alpha}^{\pi+\alpha} \sqrt{2}V \sin \omega t.d\omega t = \frac{2\sqrt{2}V}{\pi}\cos\alpha$$

$$= \frac{2\sqrt{2}\times 210}{\pi}\cos 45 = 133.73 \text{ V}$$

(b) The average current is $I_o = \frac{V_o}{R} = \frac{133.73}{10}$ = 13.373 A

(c) The average current of thyristor is $I_{Tav} = \frac{I_o}{2} = \frac{13.373}{2} = 6.68\,\text{A}$

(d) The rms current of thyristor is $I_{Trms} = \frac{I_o}{\sqrt{2}} = \frac{13.373}{\sqrt{2}} = 9.45A\,\text{A}$

(e) rms value of source current $I_{rms} = I_o = 13.373A$

DC output is $P_{dc} = V_o I_o = 133.73 \times 13.373 = 1788.37$ Watt

Power factor $\cos\phi = \frac{V_o I_o}{VI_{rms}} = \frac{133.73\times 13.373}{210\times 13.373} = 0.636$ lag

4.7.4 Single Phase Half Bridge Controlled Rectifier With R Load

If two thyristors of full converter are replaced diodes as depicted in Fig.4.26, the modified controlled rectifier circuit is called as half bridge controlled rectifier. Different types of half bridge controlled rectifier are (i) symmetrical configurations (common cathode [Fig.4.34] or common anode [Fig.4.35]) and (ii) asymmetrical configurations (thyristors in one arm and diodes in another arm) [Fig.4.36]. The common cathode symmetrical type half bridge converter is most commonly used as a single firing circuit can be used to turn on both thyristors T_1 and T_3 as shown in Fig.4.34.

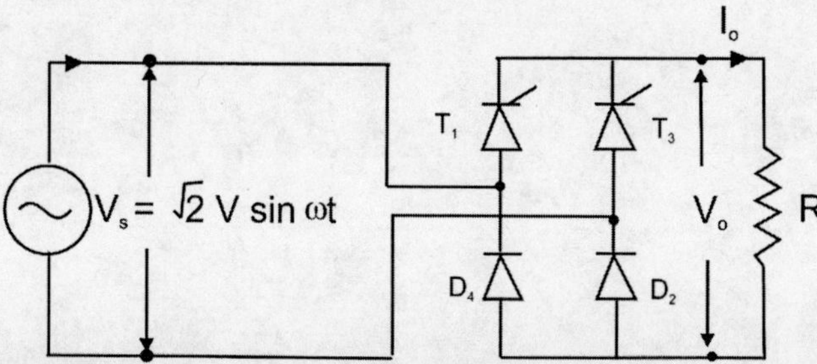

Fig.4.34 Single phase half controlled bridge controlled rectifier common-cathode with R load

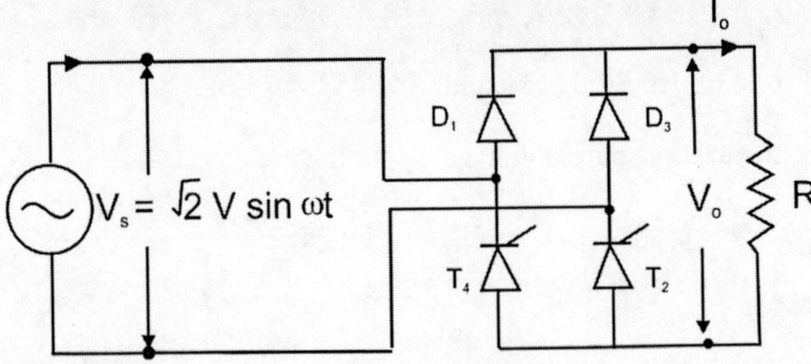

Fig.4.35 Single phase half controlled bridge controlled rectifier common-anode with R load

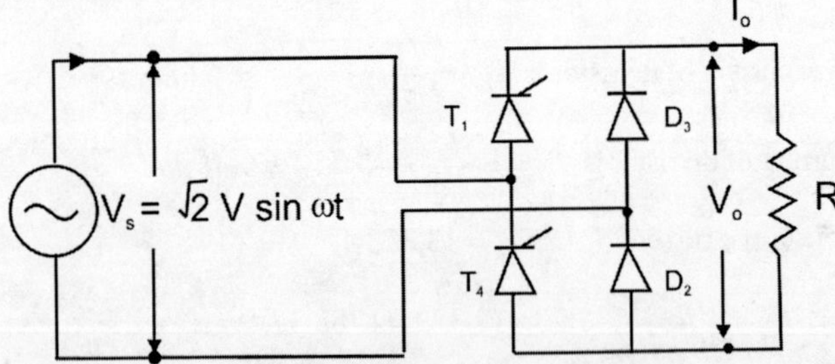

Fig.4.36 Single phase asymmetrical half controlled bridge controlled rectifier with R load

During positive half cycle of input voltage, thyristor T_1 is forward biased and operates in forward bloacking state. When a firing pulse is applied at $\omega t = \alpha$, the thyristor T_1 operates in on-state and supply voltage is connected across resistance. Subsequently current flows through T_1 and D_2. At $\omega t = \pi$, thyristor T_1 is reverse biased and it is turned-off.

In negative half cycle of input voltage, thyristor T_3 is forward biased. If a firing pulse is applied at $\omega t = \pi + \alpha$, the thyristor T_3 becomes ON and supply voltage is applied across resistance. Then current flows through T_3 and D_4. At $\omega t = 2\pi$, the thyristor T_3 is reverse biased and it is turned- off. The voltage and current wave forms of single phase half bridge controlled rectifier are shown in Fig.6.37.

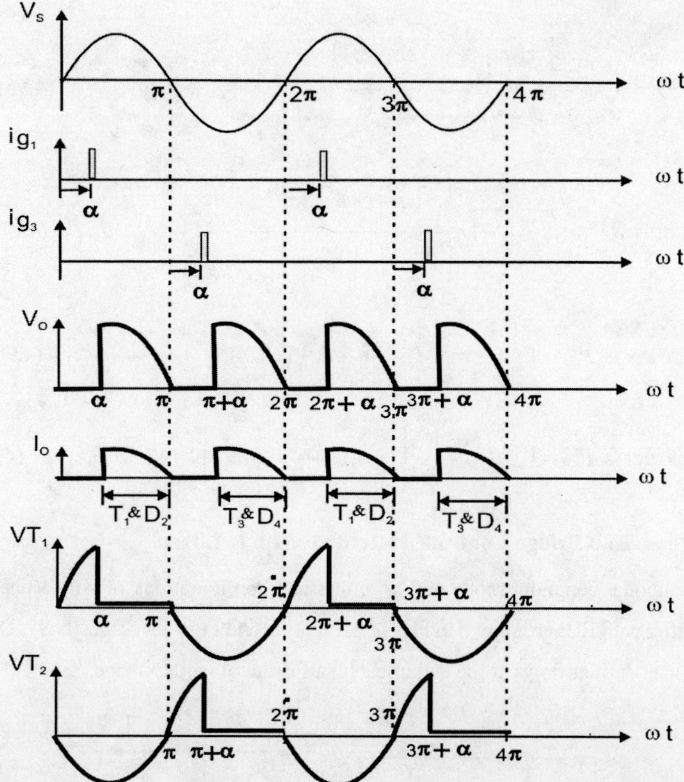

Fig.4.37 Voltage and current waveforms of single phase half bridge controlled rectifier with R load

The average output voltage is

$$V_o = V_{av} = \frac{1}{\pi}\int_{\alpha}^{\pi} \sqrt{2}V \sin \omega t.d\omega t = \frac{\sqrt{2}V}{\pi}(1+\cos\alpha) \tag{4.92}$$

The dc load current is

$$I_o = I_{av} = \frac{V_o}{R} = \frac{\sqrt{2}V}{\pi R}(1+\cos\alpha) \tag{4.93}$$

The rms output voltage is

$$V_{rms} = \left[\frac{1}{\pi}\int_{\alpha}^{\pi}(\sqrt{2}V \sin \omega t)^2 \, d\omega t\right]^{\frac{1}{2}} = V\left[\frac{1}{\pi}(\pi - \alpha + \frac{1}{2}\sin 2\alpha)\right]^{\frac{1}{2}} \tag{4.94}$$

The rms output current is $I_{rms} = \frac{V_{rms}}{R} = \frac{V}{R}\left[\frac{1}{\pi}(\pi - \alpha + \frac{1}{2}\sin 2\alpha)\right]^{\frac{1}{2}}$ (4.95)

Form factor is $FF = \frac{V_{rms}}{V_{av}} = \frac{\left[\pi(\pi - \alpha + \frac{1}{2}\sin 2\alpha)\right]^{\frac{1}{2}}}{\sqrt{2}(1+\cos\alpha)}$ (4.96)

Load ripple factor is $RF = \sqrt{FF^2 - 1} = \left[\frac{\pi(\pi - \alpha + \frac{1}{2}\sin 2\alpha)}{2(1+\cos\alpha)^2} - 1\right]^{\frac{1}{2}}$ (4.97)

The dc output power is $P_{dc} = V_{av} I_{av} = V_o I_o = \frac{2V^2}{\pi^2 R}(1+\cos\alpha)^2$ (4.98)

The ac output power is $P_{ac} = V_{rms} I_{rms} = \frac{V^2}{\pi R}\left[(\pi - \alpha + \frac{1}{2}\sin 2\alpha)\right]$ (4.99)

4.7.5 Single Phase Half Bridge Controlled Rectifier With R-L Load

The common cathode, common anode and asymmetrical configurations of single phase half bridge controlled rectifier with R-L are shown in Fig.4.38, Fig.4.39 and Fig.4.40 respectively. Due to presence of inductance, load current flows at the end of each half cycle of input voltage.

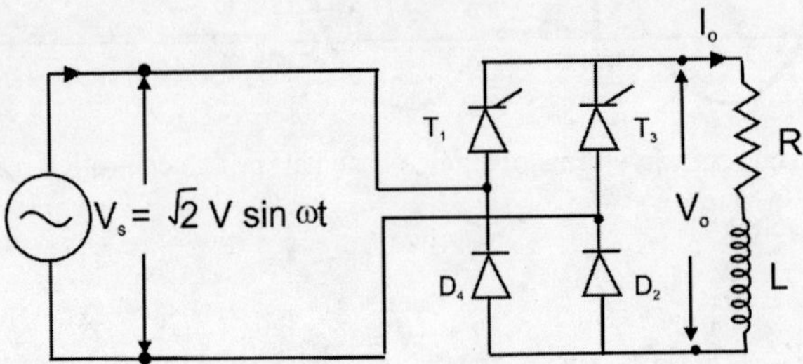

Fig.4.38 Single phase half controlled bridge controlled rectifier common-cathode with R-L load

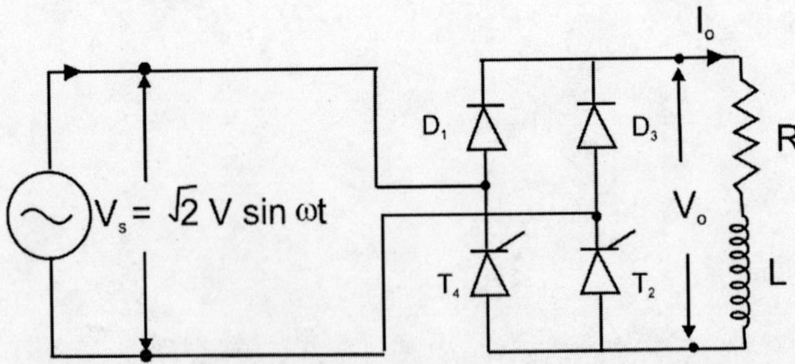

Fig.4.39 Single phase half controlled bridge controlled rectifier common-anode with R-L load

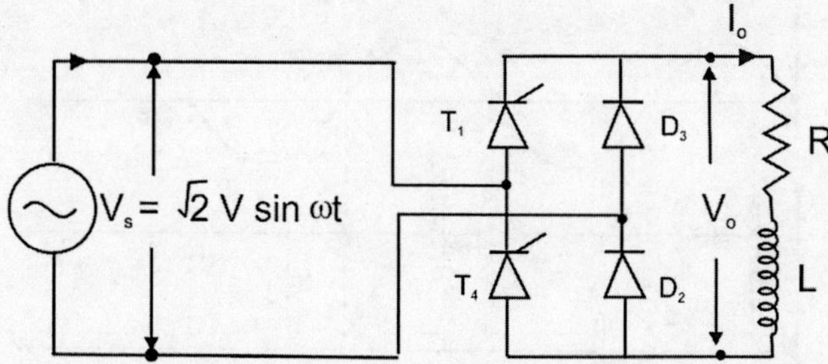

Fig.4.40 Single phase asymmetrical half controlled bridge controlled rectifier with R-L load

In the positive half cycle of input voltage, Thyristor T_1 conducts and load current flows through T_1 and D_2 from $\omega t = \alpha$ to π. After $\omega t = \pi$, the input voltage is negative and passes through zero crossing and diode D_4 comes into conduction commutating diode D_2.. Subsequently load current flows through T_1 and diode D_4 and the amplitude of load current decays exponentially. As soon as thyristor T_3 is fired in the negative half cycle of supply voltage, thyristor T_1 is to be turned off. After that load current flows through T_3 and D_4 from $\omega t = \pi + \alpha$ to 2π.

After $\omega t = 2\pi$, the input voltage is positive and passes through zero crossing and diode D_2 becomes into conduction commutating diode D_4. Then the load current flows though T_2 and D_2 and it again decays exponentially. There after the cycle operation will be repetitive after firing T_1. This controlled rectifier circuit can able to operate in continuous or discontinuous mode. Voltage and current waveforms for continuous and discontinuous load current mode of operation are shown in Fig.4.41 and Fig.4.42 respectively.

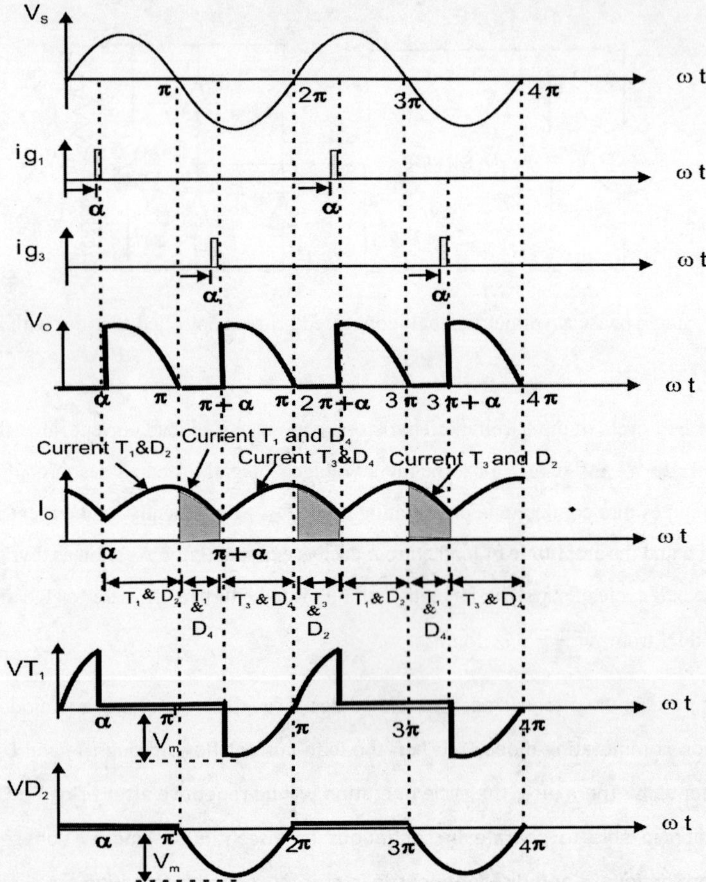

Fig.4.41 Voltage and current waveforms of single phase half controlled bridge controlled rectifier with R-L load for continuous load current

The dc output voltage is

$$V_o = V_{av} = \frac{1}{\pi}\int_{\alpha}^{\pi}\sqrt{2}V\sin\omega t.d\omega t = \frac{\sqrt{2}V}{\pi}(1+\cos\alpha) \tag{4.100}$$

The dc load current is

$$I_o = I_{av} = \frac{V_o}{R} = \frac{\sqrt{2}V}{\pi R}(1+\cos\alpha) \tag{4.101}$$

The rms output voltage is

$$V_{rms} = \left[\frac{1}{\pi}\int_{\alpha}^{\pi}(\sqrt{2}V\sin\omega t)^2 d\omega t\right]^{\frac{1}{2}} = \sqrt{2}V\left[\frac{1}{2\pi}(\pi-\alpha+\frac{1}{2}\sin 2\alpha)\right]^{\frac{1}{2}} \tag{4.102}$$

The rms output current is $I_{rms} = \frac{V_{rms}}{R} = \frac{\sqrt{2}V}{R}\left[\frac{1}{2\pi}(\pi - \alpha + \frac{1}{2}\sin 2\alpha)\right]^{\frac{1}{2}}$ (4.103)

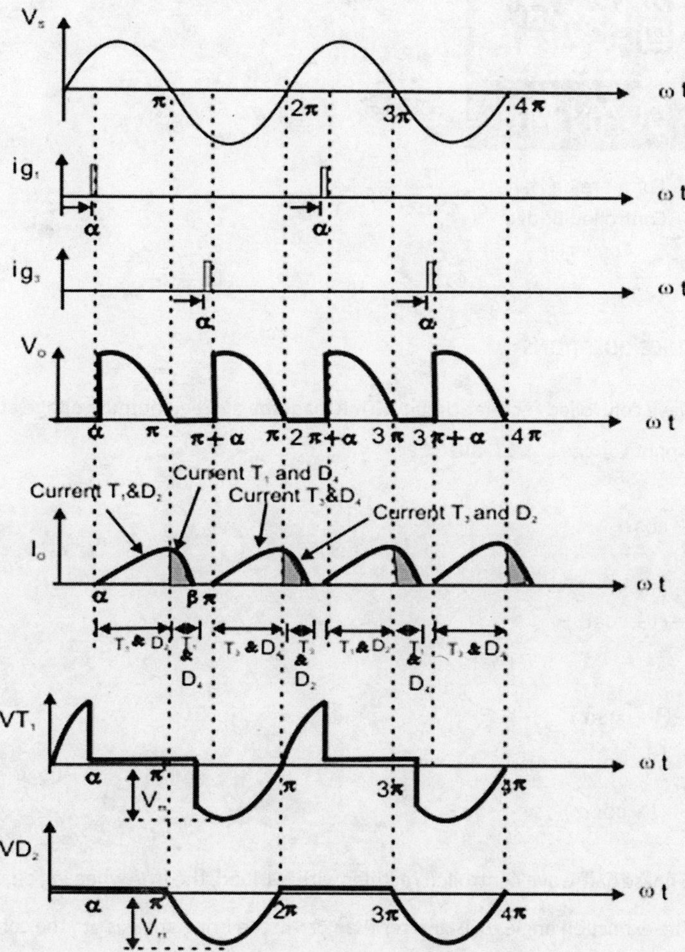

Fig.4.42 Voltage and current waveforms of single phase half controlled bridge controlled rectifier with R-L load for discontinuous load current

For more on Half Controlled Bridge Converter

EXERCISE

MULTIPLE CHOICE QUESTIONS

4.1 In a half-wave controlled rectifier circuit with R load, the average output voltage at firing angle α is ____ when input voltage is $\sqrt{2}V\sin\omega t$

(a) $V_O = \frac{\sqrt{2}V}{\pi}\cos\alpha$

(b) $V_O = \frac{\sqrt{2}V}{2\pi}(1-\cos\alpha)$

(c) $V_O = \frac{\sqrt{2}V}{2\pi}(1+\cos\alpha)$

(d) $V_O = \frac{\sqrt{2}V}{\pi}(1+\cos\alpha)$

4.2 In a single phase half-wave controlled rectifier with R-L load, the free wheeling diode is connected across load. The extinction angle β is greater than π. If the firing angle is α, the conduction period of thyristor and free wheeling diode are

(a) $\alpha \leq \omega t \leq \pi$ and $\pi \leq \omega t \leq \beta$

(b) $\pi \leq \omega t \leq \beta$ and $\alpha \leq \omega t \leq \pi$

(c) $0 \leq \omega t \leq \pi$ and $\pi \leq \omega t \leq \beta$

(d) $\alpha \leq \omega t \leq \pi$ and $2\pi \leq \omega t \leq \beta$

4.3 In a single phase full bridge controlled rectifier, the average dc output voltage is equal to

(a) $V_o = \frac{1}{\pi}\int_{\alpha}^{\pi+\alpha} \sqrt{2}V \sin \omega t.d\omega t$

(b) $V_o = \frac{1}{\pi}\int_{\alpha}^{\pi} \sqrt{2}V \sin \omega t.d\omega t$

(c) $V_o = \frac{1}{\pi}\int_{0}^{\pi+\alpha} \sqrt{2}V \sin \omega t.d\omega t$

(d) $V_o = \frac{1}{\pi}\int_{\alpha}^{2\pi+\alpha} \sqrt{2}V \sin \omega t.d\omega t$

4.4 In a single phase semi-converter, the average dc output voltage is equal to

(a) $V_o = \frac{1}{\pi}\int_{0}^{\pi+\alpha} \sqrt{2}V \sin \omega t.d\omega t$

(b) $V_o = \frac{1}{\pi}\int_{\alpha}^{2\pi+\alpha} \sqrt{2}V \sin \omega t.d\omega t$

(c) $V_o = \frac{1}{\pi}\int_{\alpha}^{\pi+\alpha} \sqrt{2}V \sin \omega t.d\omega t$

(d) $V_o = \frac{1}{\pi}\int_{\alpha}^{\pi} \sqrt{2}V \sin \omega t.d\omega t$

4.5 If a single phase full wave controlled rectifier operates in continuous conduction mode, each SCR conducts for ______duration

(a) π (b) 2π (c) $\pi - \alpha$ (d) $\pi + \alpha$

4.6 If a single phase full controlled rectifier operates in discontinuous conduction mode and extinction angle β is greater than π, each SCR conducts for ______duration

(a) π (b) $\beta - \alpha$ (c) $\pi - \alpha$ (d) $\pi + \alpha$

4.7. If a single phase semi converter operates in continuous conduction mode, each SCR conducts for ______duration

(a) π (b) α (c) $\pi - \alpha$ (d) $\pi + \alpha$

4.8 A freewheeling diode is connected across R-L load to provide

(a) reduce utilization factor

(b) first turn on

(c) slow turn off

(d) power factor improvement

4.9 A single phase full-wave bridge controlled rectifier operates as an inverter when the firing angle in the range_______

(a) $0 \leq \alpha \leq 90°$

(b) $90° < \alpha \leq 180°$

(c) $30 \leq \alpha \leq 90°$

(d) $60° < \alpha \leq 180°$

4.10 Which rectifier requires two diodes and two SCRs?

(a) half-wave controlled rectifier

(b) full-wave controlled bridge rectifier

(c) half controlled bridge rectifier

(d) semi-converter

4.11 Which rectifier requires four diodes?

(a) half-wave controlled rectifier

(b) full-wave controlled bridge rectifier

(c) Half controlled bridge rectifier

(d) Semi-converter

4.12 Which circuit requires one SCR?

(a) half-wave controlled rectifier

(b) full-wave controlled rectifier circuit using center tap transformer

(c) half-controlled bridge rectifier

(d) semi-converter

4.13 In a controlled rectifier, the load current depends on

(a) firing angle and type of load

(b) firing angle only

(c) type of load only

(d) none of these

4.14 In a single phase half wave rectifier circuit with R-L load and a free wheeling diode across the load and extinction angle β is greater than π. If the firing angle of SCR is α, then SCR conducts for____ duration and free wheeling diode conducts for ______ duration.

(a) $\pi-\alpha$, β

(b) $\alpha-\pi$, $\pi-\beta$

(c) $\pi-\alpha$, $\pi-\beta$

(d) $\pi-\alpha$, $\beta-\pi$

4.15 In a single phase half controlled rectifier with resistive load and firing angle α, the load current is zero for ______ duration and non-zero for ____________ duration.

(a) $\pi-\alpha,\alpha$ (b) $\alpha,\pi-\alpha$ (c) $\pi+\alpha,\alpha$ (d) $\alpha,\pi+\alpha$

REVIEW QUESTIONS

Short Answer Type Questions

4.16 What is phase angle controlled technique of converter?

4.17 Write the applications of phase controlled rectifiers.

4.18 What is a semiconverter?

4.19 What is a full converter?

4.20 Write the advantages of freewheeling diode in single phase half wave controlled rectifier with R-L load.

4.21 Compare between semi converter and full converter

4.22 Compare between half controlled converter and full controlled converter.

4.23 How a single phase full converter operates as an inverter.

4.24 What is the principle of phase control?

4.25 What is extinction-angle control of converter?

Long Answer Type Questions

4.26 (a) Draw the circuit diagram of single phase half-wave controlled rectifier with R load. Describe its working principle.

(b) Draw the voltage and current waveforms. Determine the following parameters

(i) dc output voltage (ii) Average d.c. load current (iii) rms output voltage

(iv) rms load current *(v)* Ripple factor (vi) regulation (vii) efficiency

4.27 Explain the operation of single phase angle full controlled rectifier. Derive the expression for average dc output voltage.

4.28 A single phase full wave controlled rectifier circuit feeds power to a resistive load. Draw the waveforms for input voltage, output voltage, load current and voltage across SCR for a specified firing angle α.

4.29 (a) Draw the circuit diagram of a single phase half-wave controlled rectifier with RL load.

(b) Prove that average dc output voltage is $V_O = \frac{\sqrt{2}V}{2\pi}(\cos\alpha - \cos\beta)$ where V is rms input voltage, α is firing angle and β is extinction angle.

(c) Derive the expression for rms output voltage.

4.30 Determine the output voltage of a single phase half wave controlled converter with RL load and freewheeling diode at the firing angle α.

4.31 Draw the circuit diagram of single phase full wave controlled rectifier using centre tap transformer with R load and determine

(a) dc output voltage (b) average d.c. load current (c) rms output voltage

(d) rms load current *(e)* Ripple factor and (g) efficiency

4.32 (a) Draw the circuit diagram of single phase full wave controlled rectifier using centre tap transformer with RL load and freewheeling diode.

(b) Prove that the average output voltage at firing angle α is $V_o = \frac{\sqrt{2}V}{\pi}(1+\cos\alpha)$

4.33 (a) Draw the circuit diagram of single phase bridge converter with R load. Explain its working principle.

(b) Draw the voltage and current waveforms. Determine the following parameters

(i) dc output voltage (ii) Average d.c. load current and (iii) rms output voltage

4.34 (a) Draw the circuit diagram of single phase half-controlled bridge rectifier with R load. Discuss its working principle.

(b) Draw the voltage and current waveforms. Determine the following parameters

(i) dc output voltage (ii) Average d.c. load current and (iii) rms output voltage

Problems

4.35 A single phase 210V, 1.5 kW heater is connected a half wave controlled rectifier and it is supplied from a 210V, 50Hz ac supply. Calculate the power absorbed by the heater if the firing angle is (a) $\alpha = 60°$ and (b) $\alpha = 45°$

4.36 A single phase half wave controlled rectifier is connected across R-L load and feeds from a 220V, 50Hz ac supply. If R=20Ω and L=0.05H, Compute (a) the firing angle to ensure no transient current and (b) the firing angle for the maximum transient.

4.37 The voltage across secondary winding of a single phase transformer is 210V, 50 Hz and the transformer deliver power to resistive load of $R = 2\Omega$ through a single phase half wave controlled rectifier. When the firing angle of thyristor is $90°$, determine (a) average dc output voltage (b) average dc output current (c) rms output voltage (d) rms output current (e) form factor (f) voltage ripple factor (g) rectification efficiency (h) transformer utilization factor and (i) PIV of thyristor.

4.38 A single phase full-wave midpoint controlled rectifier is fed from a 230V, 50Hz ac source and it is connected a resistive load R=20Ω. Calculate (a) the output voltage, (b) form factor, (c) ripple factor, (d) efficiency, (e) transformer utilization factor at $\alpha = 30°$. Assume turn ratio of transformer is 1:1

4.39 A single phase fully controlled bridge converter with R-L load is supplied from 210V, 50Hz ac supply. If the firing angle is $45°$, Calculate average output voltage.

4.40 A single phase fully controlled bridge converter is supplied from 220V, 50Hz ac supply and fed to a load which consists of R=10Ω and large inductance so that the load current is constant. If the firing angle is $45°$, calculate (a) average output voltage, (b) average output current (c) average current of thyristor (d) rms current of thyristor (e) power factor.

PRACTICAL

3. Use CRO to observe the output waveform of half wave controlled rectifier with resistive load and determine the load voltage

Aim:

1. To study the performance and waveforms of half-wave controlled rectifier with resistive load.
2. To plot a graph of variation of output voltage V_o with respect to firing angle α

Apparatus:

SCR TY604/2N 6394/TYN 612; Transformer +12V - 0 -12V, 500mA; Resistance R 1kΩ, 1 Watt; Synchronized UJT triggering circuit; Voltmeter 0-20V DC; Ammeter 0-1A; Digital multi-meter 0-20V DC, 0-1A; Dual Trace 10 MHz CRO

Circuit Diagram:

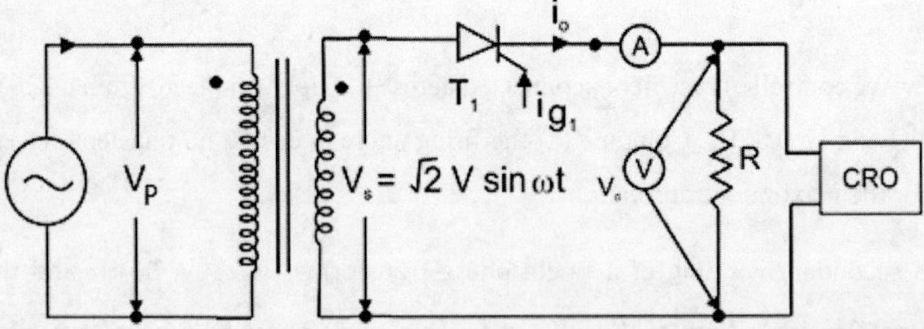

Fig.4.43 Circuit diagram for the test "Use CRO to observe the output waveform of half wave controlled rectifier with resistive load and determine the load voltage"

Theory:

Half wave controlled rectifier is a circuit which converts AC input voltage into controlled DC output voltage using controlled device like SCR by varying firing angle α. In this controlled rectifier, the output voltage can be controlled in only positive half cycle of input ac voltage. The output voltage is unidirectional dc output. The detail operating principle, voltage and current waveforms of half wave controlled rectifier is described in section 4.3.

Single phase half wave controlled rectifier has its main applications in low power DC drives. This controlled rectifier can also be used in portable handheld instruments, flexible speed controlled industrial drives, and battery charger.

Precautions:

6. Initially all switches of the power supplies are at off condition.
7. Do not increase the gate current more than rated value of SCR.
8. The applied voltage and current should not be more than the maximum rating of SCR.
9. Use appropriate heat sink for proper cooling of SCR.
10. Reading of observation should be error free.

Procedure:

7. Make the circuit connection using all required equipment as per the circuit diagram
8. Switch on power supply
9. Measure AC input voltage at secondary of transformer using voltmeter/multi-meter
10. Connect CRO across the input ac supply to observe as well as measure the input voltage
11. Connect CRO across the load to observe as well as measure the output voltage
12. Observe the output voltage waveforms at different firing angles.
13. Draw the input and output voltage waveforms on graph paper at different firing angles observing the waveforms in CRO.
14. Using observation table, plot graph for variation of output voltage with respect to firing angle α

Observation Table:

Table 4.1 Observation Table for Half wave rectifier

Sr.No.	Firing angle α in degree	Measured output voltage (V_o)	Calculated output voltage (V_o)

VIVA Questions Related to Half wave rectifier

1. Write specifications of SCR from manufacturer data sheet
2. State the effect on output voltage when there is no gate pulse
3. State the effect on output voltage when gate pulse is applied at 180°
4. Explain the performance and working principle of half-controlled rectifier with relevant waveforms for resistive load.
5. What will be the output voltage v_o if firing angle α=90°

4. **Draw the output waveform of Full wave-controlled rectifier with R load, RL load, freewheeling diode and determine the load voltage**

Aim:

1. To study the performance and waveforms of full-wave controlled rectifier with resistive load, RL load and freewheeling diode.

2. To plot a graph of variation of output voltage Vo with respect to firing angle α

Apparatus:

SCR TY604/2N6394/TYN 612; Diode 1N4001-T; Transformer +12V - 0 -12V, 500mA; Resistance R 1kΩ, 1 Watt; Inductance L 100mH, 500mA; Synchronized UJT triggering circuit; Voltmeter 0-20V DC; Ammeter 0-1A; Digital multi-meter 0-20V DC, 0-1A; Dual trace 10 MHz CRO

Circuit Diagram:

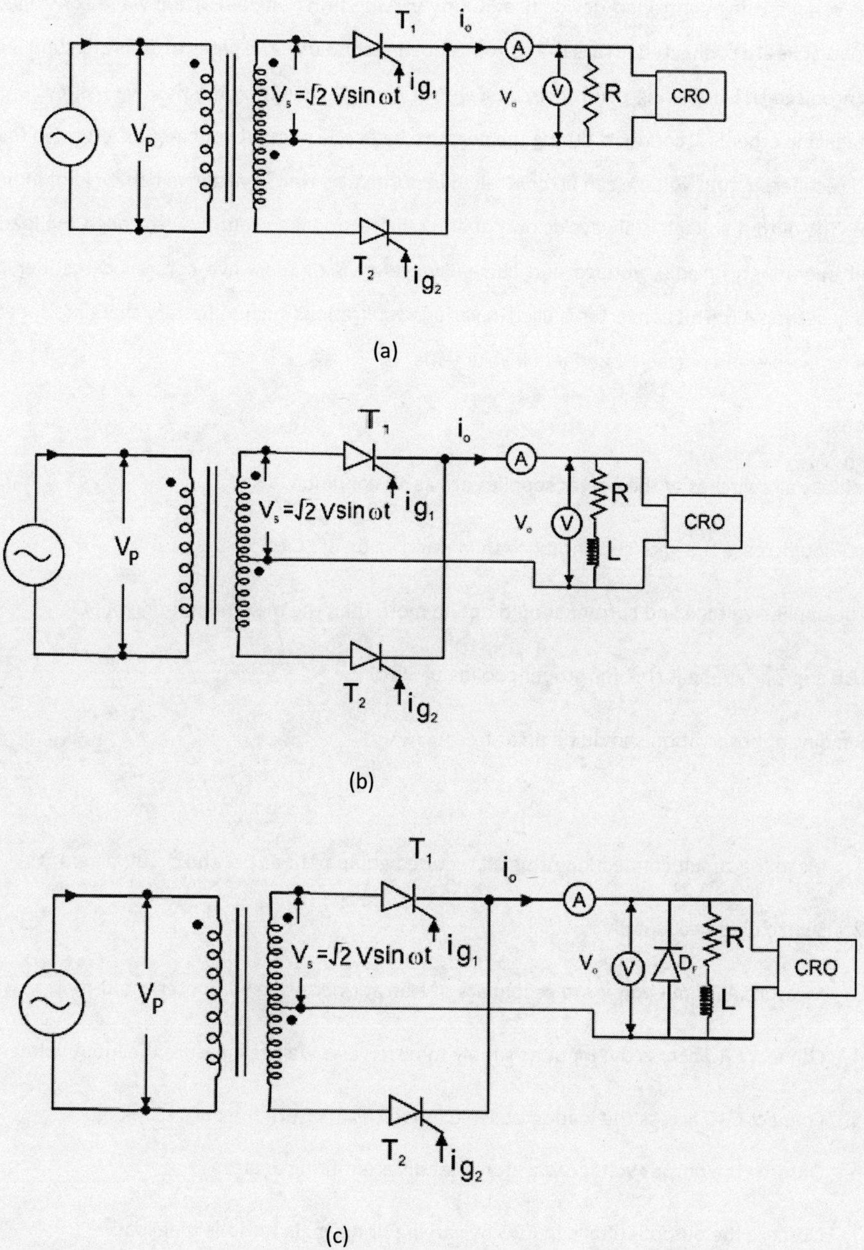

Fig.4.45 Circuit diagram for the experiment "Draw the output waveform of full wave controlled rectifier with (a) R load, (b) R-L load, and (c) R-L load with freewheeling diode and determine the load voltage"

Theory:

Single phase full wave controlled rectifier is a circuit which converts AC input voltage into controlled DC output voltage using controlled device like SCR by varying firing angle. For full-wave controlled rectifier, two SCRs are connected across the center tapped secondary. The gates of SCRs are supplied from synchronized UJT triggering control circuits. One SCR conducts during the positive half cycle of input voltage and other SCR conducts during the negative half cycle of input voltage. Therefore, in this controlled rectifier, output voltage can be controlled in positive as well as negative half cycle of input ac voltage. The output voltage is always dc output and unidirectional current flows through the load. The detail operating principle, voltage and current waveforms of full wave controlled rectifier is described in section 4.6. This converter is used in various applications such as battery charging, speed control of dc motors and portable handheld instruments.

Precautions:

1. Initially all switches of the power supplies are at off condition.
2. Do not increase the gate current more than rated value of SCR.
3. The applied voltage and current should not be more than the maximum rating of SCR.
4. Use appropriate heat sink for proper cooling of SCR.
5. Reading of observation should be error free.

Procedure:

1. Make the circuit connection using all required equipment as per the circuit diagram
2. Switch on power supply
3. Measure AC input voltage at secondary of transformer using voltmeter/multi-meter
4. Connect CRO across the input ac supply to observe as well as measure the input voltage
5. Connect CRO across the load to observe as well as measure the output voltage
6. Observe the output voltage waveforms at different firing angles.
7. Observe the output voltage in CRO by varying firing angle for following load

   a) R Load

   b) R-L Load

c) R-L load with freewheeling diode

8. Draw the input and output voltage waveforms on graph paper at different firing angles observing the waveforms in CRO.

9. Using observation table, plot graph for variation of output voltage with respect to firing angle α

Observation Table:

Table 4.2 Observation Table for Full wave rectifier with R load

Sr.No.	**Firing angle α in degree**	**Measured output voltage (V_o)**	**Calculated output voltage ($V_{o(cal)}$)**

Table 4.3 Observation Table for Full wave rectifier with R-L load

Sr.No.	**Firing angle α in degree**	**Measured output voltage (V_o)**	**Calculated output voltage ($V_{o(cal)}$)**

Table 4.4 Observation Table for Full wave rectifier with R-L load with free-wheeling diode

Sr.No.	Firing angle α in degree	Measured output voltage (V_o)	Calculated output voltage ($V_{o(cal)}$)

VIVA Questions on Full wave-controlled rectifier

1. Write specifications of SCR from manufacturer data sheet
2. State the effect on output voltage when there is no gate pulse
3. State the effect on output voltage when gate pulse is applied at 180°
4. Explain the performance and working principle of full-wave controlled rectifier with relevant waveforms for R-L load with free-wheeling diode.
5. State the effect on output voltage if freewheeling diode is open circuit

KNOW MORE

Thyristor rectifier converts an AC voltage to a DC voltage at the output and the power flow is bidirectional between the AC and the DC side. The circuit operation depends on the state of the AC source and the firing angle α of single phase and three phase controlled rectifier when the source inductance L_s is neglected. When the firing angle of thyristors is zero, the controlled rectifier circuit behaves as diode rectifier with inductive load. When firing angle is more than 0° and less than 90°, converter act as rectifier and when firing angle is more than 90° and less than 180°, converter acts as inverter and power flows from the dc to the ac side. The performance of converter depends on the source inductance L_s. To study the influence of source inductor L_s on converter, please scan the QR code as given below. To study the detailed operation of three phase controlled rectifier, dual converter and commutation overlap, we scan the following QR codes.

For more on the controlled rectifier with source inductance

For more three phase controlled rectifiers

For more on Dual Converter and Communication Overlap

References

1. Sugandhi, Rajendra kumar and Sugandhi, Krishna kumar, Thyristors: Theory and Applications, New Age International (P) ltd. publishers, New Delhi
2. Bhattacharya,S.K., Fundamental of Power Electronics, Vikas Publishing House Pvt. Ltd. Noida
3. Jain & Alok, Power Electronics and its Applications, Penram International Publishing(India) Pvt. Ltd, Mumbai
4. Rashid, Muhammad, Power Electronics Circuits Devices and Applications, Pearson Education India, Noida
5. Sing,M.D. and Khanchandani, K.B., Power Electronics, Tata McGraw Hill Publishing Co. Ltd, New Delhi
6. Zbar, Paul B., Industrial Electronics: A Text-Lab manual, McGraw Hill Publishing Co. Ltd, New Delhi

Dynamic QR Code for Further Reading

-QR codes embedded in the unit.

5 Industrial Control Circuits

UNIT SPECIFIES

Through this unit we have discussed the following aspects:

- Different industrial applications of Power electronics devices
- Burglar's alarm system
- Battery charger using SCR
- Emergency light system
- Temperature controller using SCR
- Illumination control/fan speed control using TRIAC
- Switch mode power supply (SMPS)
- UPS: off-line and online
- SCR based AC and DC circuit breakers

The above practical applications of the topics are discussed for generating curiosity and creativity as well as improving problem solving capacity of learners.

Not only a large number of multiple choice questions are incorporated in this chapter but also questions of short and long answer types marked in two categories following lower and higher order of Bloom's taxonomy. A list of references and suggested readings are illustrated in this unit so that learners can go through them for practice and get detailed knowledge. It is also be noted that some QR codes have been incorporated in different sections of this unit for getting more information on various topics of interest.

After discussion the content related to theory, there is a "Know More" section at end of this unit which is related to laboratory experiments. This section has been carefully designed so that the supplementary information provided in this section becomes beneficial for the learners. Usually, this section highlights the initial activity, examples of some interesting facts, analogy, and history of the development of the power electronics devices focusing the salient observations and finding, timelines starting from the development of the concerned topics up to the recent time, applications of the subject matter for our day-to-day life, and industrial applications of power electronics devices.

RATIONALE

This unit on "Industrial Control Circuits" can be able to provide detail information regarding different industrial applications of power electronics devices. This unit can helps students to get a primary idea about burglar's alarm system, battery charger using SCR, emergency light system, and temperature controller using SCR. The illumination control/fan speed control using TRIAC, switch mode power supply (SMPS), UPS: off-line and online and SCR based AC and DC circuit breakers are also described elaborately to help students to know about the applications of power semiconductor devices.

PRE-REQUISITES

Semiconductor physics

Analog and digital electronics

Electrical circuit theory

Electrical machines and power system

UNIT OUTCOMES

List of outcomes of this unit is as follows:

U5-O1: To explain about industrial control circuits

U5-O2: To describe burglar's alarm system

U5-O3: To learn about construction and working principle of battery charger using SCR and emergency light system

U5-O4: To discuss about construction and working principle of temperature controller using SCR and illumination control/fan speed control using TRIAC

U5-O5: To understand the concept of switch mode power supply (SMPS) and UPS: off-line and online

U5-O6: To explain aspects of SCR based ac and dc circuit breakers

U5-O7: To troubleshoot the burglar's alarm, emergency light system, speed control system, and temperature control system

Unit-5 Outcomes	EXPECTED MAPPING WITH COURSE OUTCOMES (*1-Weak Correlation; 2- Medium Correlation; 3-Strong Correlation*)				
	CO-1	**CO-2**	**CO-3**	**CO-4**	**CO-5**
U5-O1	3	2	2	3	3
U5-O2	3	2	2	3	3
U5-O3	3	2	2	3	3
U5-O4	3	2	2	3	3
U5-O5	3	2	2	3	3
U5-O6	3	2	2	3	3
U5-O7	3	2	2	3	3

5.1 Introduction

Presently power electronics devices have been intensively used in industrial control circuits such as power supplies, converters, inverters, battery chargers, temperature control, variable speed motor drives, renewable energy sources, transmission of electric power in dc and ac, electric vehicles, robot arm drives, washing machines and it is also contributing to the development and evaluation of new structures for processing of electrical energy. With the evaluation of microprocessors and advanced control theories, power electronics are playing an important role in development of our society. Hence the design, development and optimization of power electronics and controller devices are required to achieve the optimal use of electrical energy in consumer electronics circuits. In this unit, circuits based on power electronics devices such as burglar's alarm system, battery charger using SCR, emergency light system, temperature controller using SCR, illumination control and fan speed control using TRIAC, SMPS, UPS: Off-line and Online, and SCR based AC and DC circuit breakers are described elaborately.

5.2 Burglar's Alarm System

Burglar's alarm is an utility system which is used in order to safe-guard property against theft. It uses a very simple sensor like fine wires. Fig. 5.1 shows the circuit diagram of a burglar alarm system. Initially switch S_1 is kept in ON position and the loop circuit A-B consists of a length of fine wire which is fixed on all windows and ventilators of the house at such a height so that the fine wire will be broken whenever anyone want to enter into the premises. The variable resistance R_1 is used for sensitivity control. Whenever the loop circuit A-B is closed, the relay is de-energised. When the loop circuit A-B is closed, transistor T_1 turns ON. Consequently, most of the voltage is dropped across R_5 and base emitter junction of transistor T_2 gets reverse biased. Subsequently transistor T_2 operates in an OFF state and the relay is not energised. Whenever the loop wire A-B is broken, the base of T_1 becomes open and T_1 turns OFF and the voltage across resistance R_5 becomes zero. Consequently the base emitter junction of transistor T_2 becomes forward biased which turns ON the transistor T_2. Subsequently relay is energised and gets closed and bell starts ringing. This circuit can be used to protect the vehicles when it is kept in open space of the premises.

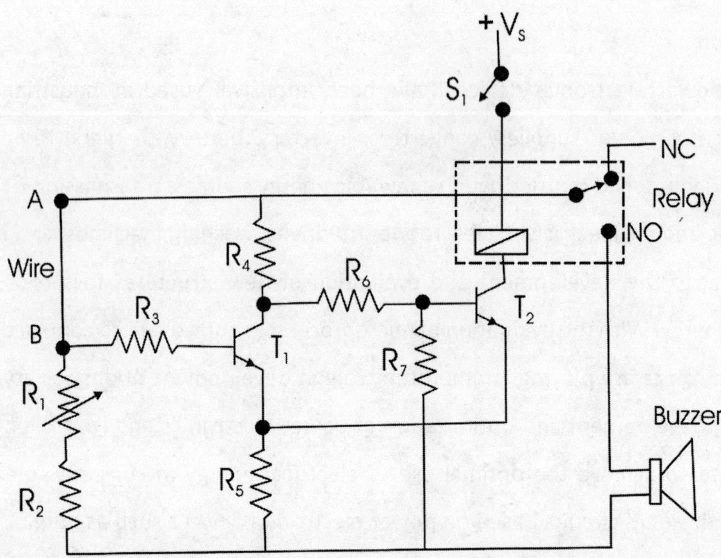

Fig.5.1 Circuit diagram of a burglar alarm

5.2.1 Equipment Required

Sl. No.	Item with specification	Quantity
1	Power Transistor T_1 (BC 147)	1
2	Power Transistor T_2 (SL100)	1

Sl. No.	Item with specification	Quantity
3	9V DC Power supply	1
4	Variable resistance/Potentiometer R_1- 4.7 kΩ, 1/2 Watt	1
5	R_2- 1.2 kΩ, 1/2 Watt	1
6	R_3- 4.7 kΩ, 1/2 Watt	1
7	R_4- 4.7 kΩ, 1/2 Watt	1
8	R_5- 10Ω, 1/2 Watt	1
9	R_6- 1.2 kΩ, 1/2 Watt	1
10	R_7- 560Ω, 1/2 Watt	1
11	Switch S_1	1
12	Relay: 9V, 1A	1
13	Speaker /Electric buzzer, 6V	1
14	Cabinet for alarm	1
15	Bakelite sheet, wires, solder	-

5.2.2 Applications

Burglar alarm has the following applications:

(i) Burglar alarm is used to protect property against theft.

(ii) It can be used as alarm against theft in houses, cars and other premises

(iii) It can also be used as standard equipment in stores and business offices for security purpose.

5.3 Battery Charger using SCR

Usually the battery chargers are used to control the charging current by continuously comparing the terminal voltage of battery with the constant reference voltage across zener diode. When the battery is fully charged, SCR (T_1) will be turned off and also disconnect the battery from the charging circuit. Therefore, the battery will not be over charged and it becomes healthy. Fig.5.2 shows the circuit diagram of battery charger using SCR. This circuit consists of following elements:

(i) Single phase ac supply

(ii) Step down transformer

(iii) Rectifier circuit

(iv) SCR

(v) Zener diode for voltage regulation and

(vi) Battery to be charged

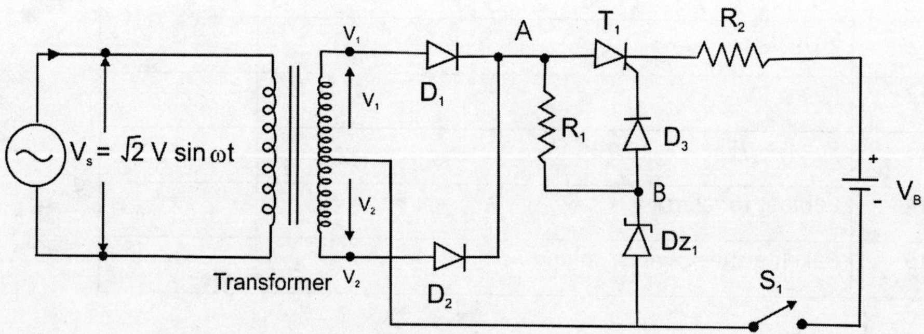

Fig.5.2 Circuit diagram of battery charger using SCR

Here input voltage is 230V, 50Hz single phase ac supply and 230V/15V-0-15V, 2A step down transformer is used to provide ac supply to full wave rectifier comprising diodes D_1 and D_2. Initially 230V, 50Hz ac supply is provided to the step-down transformer which converts the high voltage into low voltage and 15V ac can be obtained from the secondary winding of transformer. During the positive half cycle of supply voltage, D_1 is forward biased and operates in conduction state and D_2 is reverse biased and operates in non-conducting state. In the negative half cycle of supply voltage, D_2 is forward biased and operates in conduction state and D_1 is reverse biased and operates in non-conducting state. Hence the full wave rectified dc output voltage is obtained from point A.

A zener diode (Dz_1) with breakdown voltage of 12V used as a voltage regulator to regulate the voltage level of the circuit. Therefore the terminal voltage at point B will be 12V due to the presence of zener diode. Usually the rectified output voltage at A is more than at terminal B, and the SCR (T_1) is forward biased. When switch S_1 is ON and a gate current flows through gate cathode of SCR, SCR will starts to conduct and current flows through load. As current flows through the battery, battery will be charged and energy can be stored in the battery. Whenever the rectified output voltage is less than the terminal voltage at B, SCR (T_1) will be reverse biased and it will be turned-off and current will not flow through battery. The rating of breakdown voltage of the zener diode and the transformer depends on the charging potential of the battery.

5.3.1 Drawbacks of Battery Charger Using SCR

The drawbacks of battery charger using SCR are given below:

(i) This charging method takes quite a long time during charging of battery.

(ii) As filter circuit is not present, ripples are exist in output of rectifier

(iii) The process of charging and discharging is slow

(iv) This circuit is suitable for charging batteries up to medium ampere-hour rating.

5.3.2 Equipment Required:

Sl. No.	Item with specification	Quantity
1	SCR T_1 TYN 604 , 4A	1
2	Transformer 230V/15-0-15V, 2A, 50 Hz	1
3	12V Battery	1
4	Diodes D_1, D_2 1N 5008	2
5	Zener diode Dz 1A 15 V	1
6	Diode D_3 1N 4001	1
7	R_1- 470Ω, ½ Watt	1
8	Current limiting resistance R_2- 4.7Ω, 10 Watt	1
9	Bakelite sheet, wires, solder, plug On-OFF switch	-
10	DC voltmeter	1
11	DC Ammeter	1

5.3.3 Applications

Battery charger using SCR has the following applications:

`(i) It can be used as an automatic battery charger

(ii) It can be used to charge batteries of electronic toys

(iii) It is available as a portable circuit and can be carried anywhere.

For more on Battery charger using SCR

5.4 Emergency Light System using SCR

Emergency light system is an illumination device which can be designed using SCR. In presence of 230V, 50Hz, single phase ac supply, the lamp will not glow as the SCR operates in turned-off state and simultaneously battery will be charged. Whenever ac supply is disturbed, the lamp will glow immediately. The storage battery will provide the energy to lamp for a certain period depending on the amp-hr rating of battery.

Fig.5.3 shows the circuit diagram of emergency light system using SCR. In this circuit, input voltage is 230V, 50Hz single phase ac supply and 230V/6-0-6V, 1A step down transformer is used to provide ac supply to full wave rectifier, which consists of diodes D_1 and D_2. At first 230V, 50Hz ac supply is provided to the step-down transformer which converts the high voltage into low voltage and 6V ac can be obtained from the secondary winding of transformer. In the positive half cycle of supply voltage, D_1 is forward biased and operates in conduction state and D_2 is reverse biased and operates in non-conducting state. During the negative half cycle of supply voltage, D_2 is forward biased and operates in conduction state and D_1 is reverse biased and operates in non-conducting state. Thus the full wave rectified dc output voltage is obtained from point X.

When the ac supply voltage is available, the 250µF, 25V capacitor get charged through resistance R_1 and its upper plate is positive charged and the lower plate is negatively charged. As the negative polarity is applied at the gate, SCR operates in turned-off sate and lamp will not glow but dc battery should be charge through the diode D_3 and resistance R_3. When the ac supply is disturbed, the 250µF capacitor starts discharging through diode D_3, resistance R_3 and R_2. After certain time, SCR T_1 will be forward biased and gate terminal will be positive polarity, then SCR T_1 will be turned-on and starts conducting. Consequently the lamp is connected to battery and it starts to glow. DC battery also discharges through SCR T_1. At the moment, ac power supply is reinstated, the SCR T_1 will be automatically turned-off and lamp will be switched-off and the battery will start getting charged.

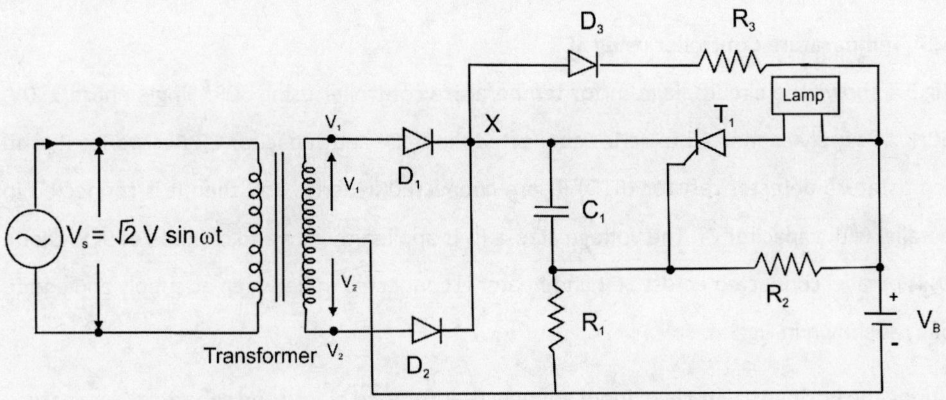

Fig.5.3 Circuit diagram of emergency light system using SCR

5.4.1 Equipment Required

Sl. No.	Item with specification	Quantity
1	SCR T_1 TYN 604 , 4A	1
2	Transformer 230V/6-0-6V, 1A, 50 Hz	1
3	6V Battery	1
4	Diodes D_1, D_2 1N 5008	2
5	Diode D_3 1N 4001	1
6	R_1- 22Ω, 2 Watt	1

Sl. No.	Item with specification	Quantity
7	R_2- 440Ω, ½ Watt	1
8	R_3- 1Ω, 2 Watt	1
9	Capacitor C_1 250μF, 25V	1
10	Lamp 6V, 20W	1
11	6V, 4 amp-hr Battery	1

5.4.2 Applications

Emergency light system using SCR has the following applications:

(i) It can be used as an emergency light

(ii) It is available as a portable circuit and can be carried anywhere.

5.5 Temperature Controller using SCR

Fig.5.4 shows the circuit diagram for temperature controller using SCR. Single phase 230V, 50Hz ac supply connected to series connected diode D_1 and capacitor C_1. Resistance R_1 and temperature detector resistor (RTD) R_2 are connected in series and then it is connected in parallel with capacitor C_1. The voltage across R_2 is applied to gate and cathode of SCR. Diode D_2 is parallel connected across SCR and heater is connected in between ac supply and anode of T_1 as shown in Fig.5.4.

During the positive half cycle of input voltage, D_1 is forward biased and capacitor C_1 is charged up to the positive peak of the supply voltage. As C_1 remains charged, current flows through R_1 and R_2. Then voltage V_g is applied between gate and cathode of SCR. Subsequently SCR will be turned-on as it is forward biased and current flows through heater and SCR. During the negative half cycle of supply voltage, current flows through diode D_2 and heater. Therefore, temperature can be controlled by varying the conduction time of SCR T_1 during positive half cycle of ac supply only.

Resistance R_2 is temperature dependent resistance. The value of R_2 decreases as temperature increases and its value increases as temperature decreases. For any change in temperature, R_2 changes and voltage across R_2 i.e. V_g changes accordingly. Then this voltage controls the SCR triggering which control the heater operation to maintain temperature at certain level.

When the temperature decreases at certain level, R_2 increases, then V_g voltage increases and SCR is switched ON and the conduction period of SCR is increased as the triggering angle is reduced. Consequently, the heater operation is controlled by SCR T_1 to maintain the temperature at certain level.

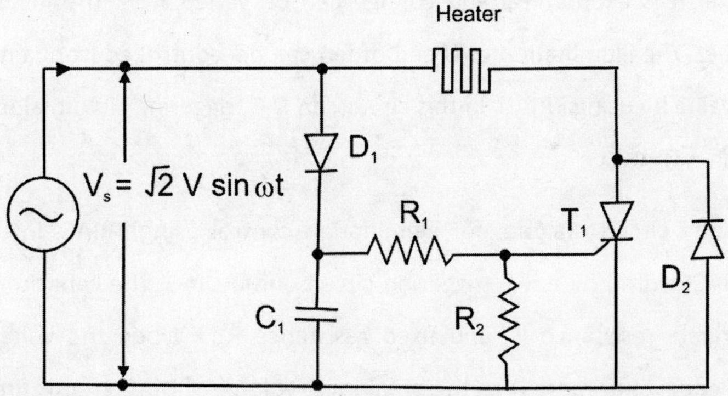

Fig.5.4 Circuit diagram for temperature controller using SCR

5.5.1 Equipment Required

Sl. No.	Item with specification	Quantity
1	SCR T_1 400V, 4A	1
2	Capacitor C_1 0.047μF, 400V	1
3	R_1- 27Ω, ½ Watt	1
4	R_2- 220kΩ, ½ Watt RTD	2
5	Diodes D_1, D_2 1N 5008	1
6	Heater 250V, 100W	1

5.5.2 Applications

Temperature controller using SCR has the following applications:

(i) It can be used as temperature controller

(ii) It can be used for room temperature control

(ii) It is available as a portable circuit and can be carried anywhere.

5.6 Illumination control / fan speed control using TRIAC

AC voltage controller circuits are used in domestic light dimmers, electric fan speed control, small motor controls and control of small ac power domestic appliances. This circuit can replace the conventional dimmerstat. The advantage of ac voltage controller over a dimmerstat that it is economical and voltage can be varied from the minimum value to maximum value. The illumination or speed of fan can be controlled from a minimum value to maximum value by using TRIAC. In this circuit, an R-C triggering circuit along with DIAC is used to turn-on a TRIAC.

Fig.5.5 shows the circuit diagram for illumination control using TRIAC. In this ac voltage controller, TRIAC is fired by a R-C triggering circuit and a DIAC. The capacitor C_2 is charged through a variable resistance R_3 and fixed resistance R_2. When the voltage across the capacitor C_2 is equal to or more than the break-over voltage of DIAC, triac is fired and it starts to conduct. The firing angle of TRIAC can be controlled by changing the value of variable resistance R_3 (potentiometer). In case of fan speed control, lamp load is replaced by electric fan in Fig.5.5.

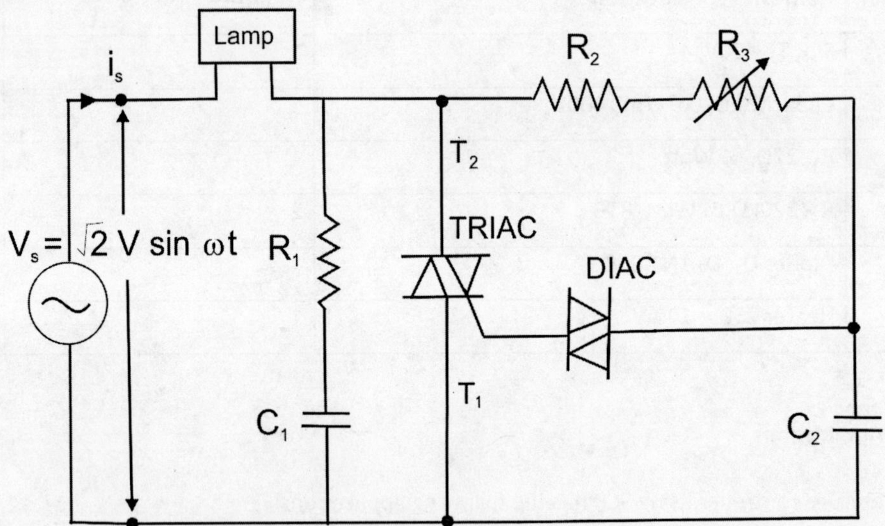

Fig.5.5 Circuit diagram for illumination control using TRIAC

During the positive half cycle of supply voltage, a positive gate signal is applied to TRIAC to turn-on it. When R_3 is significantly large, the voltage across capacitor C_2 (V_{C2}) does not exceed the diac breakdown voltage. Therefore DIAC operates in off-state and TRIAC will also operate in off-state. Subsequently, if R_3 decreases, the voltage across capacitor C_2 increases and V_{C2} exceeds the breakover voltage of DIAC. As V_{C2} exceeds the breakover voltage of DIAC, DIAC operates in on-state and it can trigger the TRIAC. In the same way, the similar operation takes place in the negative half cycle of supply voltage. A R-C snubber circuit (R_1=47Ω, ½ Watt and C_1=0.1μF, 600V) is connected across TRIAC to supress the undesired transients appearing along with ac supply.

When the firing angle of TRIAC is less, the voltage applied across load (lamp/fan) is more and hence intensity of light (illumination) or fan speed will be high. If the firing angle of TRIAC is more, the voltage applied across load (lamp/fan) is less and hence intensity of light (illumination) or fan speed will be low. Hence the illumination of lamp and the fan speed can be controlled using TRIAC. In this circuit DIAC exhibits a negligible resistance in the both directions when it conducts.

5.6.1 Equipment Required

Sl. No.	Item with specification	Quantity
1	TRIAC 400V, 4A BT136	1
2	DIAC 32V, 5mA DB3	1
3	R_1- 47Ω, ½ Watt	1
4	R_2- 1kΩ, ½ Watt	1
5	R_3- 220kΩ, ½ Watt Potentiometer	1
6	Capacitor C_1 0.1μF, 600V	1
7	Capacitor C_2 0.1μF, 400V	1
8	Lamp 250V, 60W	1
9	Electric Fan 250V, 60W	1

5.6.2 Applications

Illumination/fan control using TRIAC has the following applications:

(i) It can be used as illumination controller

(ii) It can be used as fan speed controller

(ii) It is available as a portable circuit and can be carried anywhere.

5.7 SMPS (Switch Mode Power Supply)

In several industrial applications, dc voltage of different voltage levels with respect to ground such as +5V, -5V, +6V, -6V, +9V, -9V, +10V, -10V, +12V, -12V, +15V, and -15V are required. In SMPS, the electrical isolation between the dc source and load is used to provide protection to load. Actually transformer is used for electrical isolation. If normal silicon steel core transformers are used, there will be high hysteresis loss due to high switching frequency of dc to dc converters and the wide hysteresis-loop. Therefore, ferrite-core transformer is used for electrical isolation to reduce hysteresis loss with narrow hysteresis-loop.

Fig.5.6 shows a circuit diagram for an isolated dc to dc converter. This circuit generates two dc voltage V_{o1} and V_{o2} from a same dc supply. The amplitude of output voltage depends upon the turn ratio of isolating transformers and turn-on and turn-off time of switch S. These converters can be act as step up, step-down, buck, boost, buck-boost converters. This type of isolated dc to dc converter is called as switch mode power supply (SMPS).

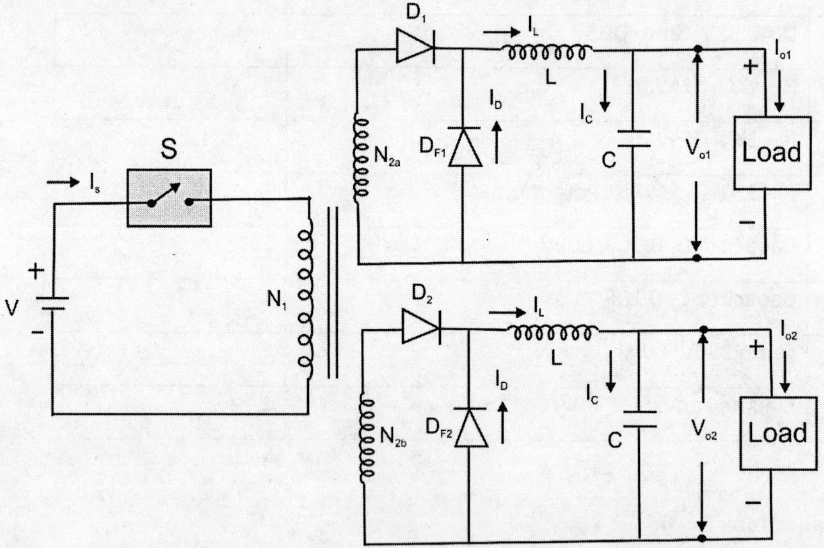

Fig.5.6 Circuit diagram for switch mode power supply (SMPS)

The most commonly used SMPS are

(i) Fly-back converter

(ii) Forward converter

(iii) Push-pull converter

(iv) Half bridge converter

(v) Full bridge converter

5.7.1 Advantages and Disadvantages of SMPS

The advantages and disadvantages of SMPS are given below:

Advantages of SMPS

(i) Due to high frequency operation, SMPS has smaller size, less weight and high efficiency compared to conventional linear dc power supply.

(ii) The output voltage of SMPS is less sensitive with respect to input voltage variation.

Disadvantages of SMPS

(i) SMPS generates both the electromagnetic and radio frequency interference due to high switching frequency.

(ii) To reduce radio frequency noise, filter circuit should be used on both input and output of SMPS

5.8 UPS (Uninterruptible Power Supply) : Off-line and On-line

There are several critical applications where a sudden power failure causes a great deal of public inconvenience and there will be huge economic losses. Some examples of critical load are hospital intensive care units, process control plants, a large computer network systems, safety monitoring system of a plant, and communication systems etc. An uninterruptible power supply (UPS) system is required to provide continuous power without interruption in all critical loads. Usually, the two types of UPS configurations are commonly used such as

(i) Off-line UPS where load normally connected to ac supply and

(ii) On-line UPS where load normally connected to inverter

5.8.1 Off-line UPS

Fig.5.7 shows the schematic block diagram of off-line UPS where load is normally connected to ac supply and rectifier maintains the full charge of battery. Whenever the supply fails, load is switched to the output of inverter that takes over the main supply. In this UPS configuration, the circuit will be break momentarily and the transfer by solid state switches which takes about 4 to 5ms. Inverter operates only during the time when power supply fails.

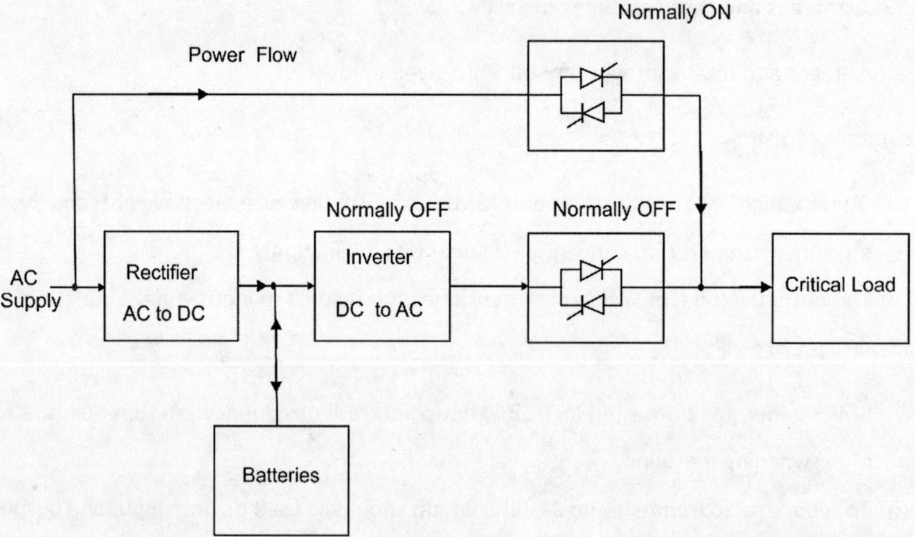

Fig.5.7 Off-line UPS where load is normally connected to ac supply

5.8.2 On-line UPS

Fig.5.8 shows the schematic block diagram of online UPS where inverter operates continuously and it's output is connected to load. Therefore, there is no need for breaking the power supply. The rectifier unit supplies power to inverter and maintains the charge on standby battery. In online UPS configuration, inverter can be used to provide power supply, to protect the load from the transients in the main supply and to maintain the desired frequency at load. Whenever inverter fails to give supply, the load is switched to ac power supply. The advantages of online UPS are

(i) Load will be protected from transients in ac supply

(ii) Inverter is used to provide power supply to load

(ii) Inverter output frequency must be maintained at desired value.

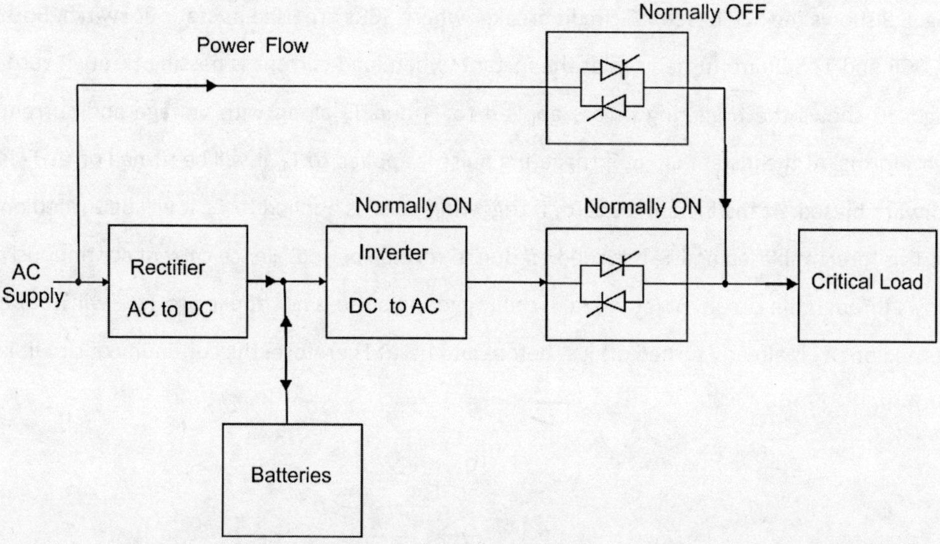

Fig.5.8 Online UPS where load normally connected to inverter

Nickel-cadmium and lead-acid type batteries are used in UPS system. Nickel-cadmium batteries have the following advantages over lead-acid type batteries:

(i) The electrolyte of Nickel-cadmium (NC) batteries does not emit any explosive gas during charging

(ii) The electrolyte of NC batteries is non-corrosive

(iii) NC batteries cannot be damaged by overcharging or discharging and these batteries have longer life.

However the cost of NC batteries is two or three times that of lead-acid batteries. The time period during which a bank of NC batteries can deliver power to load depends on battery size and nature of load.

5.9 SCR Based AC and DC Circuit Breakers

The SCR based static circuit breakers are capable of providing a fast and reliable interruption to a continuous current. Usually static circuit breakers are two types such as (i) SCR based AC Circuit Breakers and (ii) SCR based DC Circuit Breakers. In this section both static circuit breakers are described elaborately.

5.9.1 SCR Based AC Circuit Breakers

Fig. 5.9 shows the SCR based AC circuit breaker where SCRs are used as static ac switch. Both T_1 SCR and T_2 SCR are turned on at the instant when load current is passing through zero. Fig.5.10 shows the triggering pulses applied to T_1 and T_2 along with voltage and current waveforms. At the instant $\omega t=\alpha$, if triggering pulse is applied to T_1, it will be turned on as T_1 is forward biased. At the instant $\omega t=\pi+\alpha$, if triggering pulse is applied to T_2, it will be turned on as T_2 is forward biased and T_1 is turned off due to reverse biased. Hence current continuously flows through the circuit. At $\omega t=4\pi+\alpha$, if the triggering pulse is not applied to T_1, it will not be turned on. T_2 is already turned off just before $\omega t=4\pi+\alpha$. Therefore, the continuity of circuit is broken.

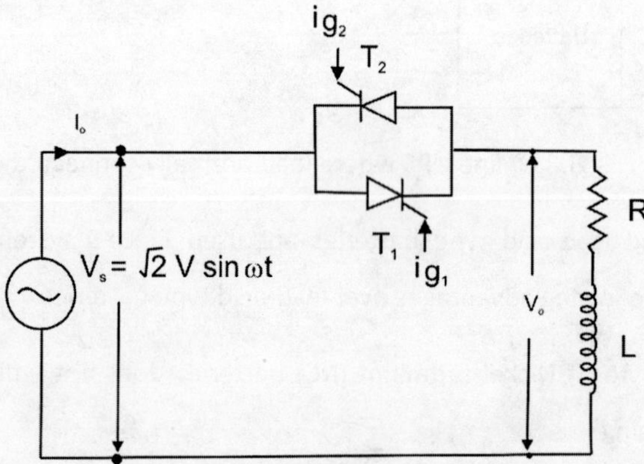

Fig. 5.9 SCR based AC circuit breaker

Whenever turn-off command is received by the control circuit due to fault in the system, the application of triggering pulse will be withdrawn and accordingly circuit will be broken. If the turn-off command is received at $\omega t=5\pi+\alpha$, load current will be broken only at $\omega t=6\pi+\alpha$, and a delay of π radian or a half-cycle is required. When the turn-off command is received at $5\pi+\alpha<\omega t<6\pi+\alpha$, circuit can be broken at $\omega t=6\pi+\alpha$. Consequently, the maximum delay time for breaking the circuit is one half-cycle i.e. π/ω seconds.

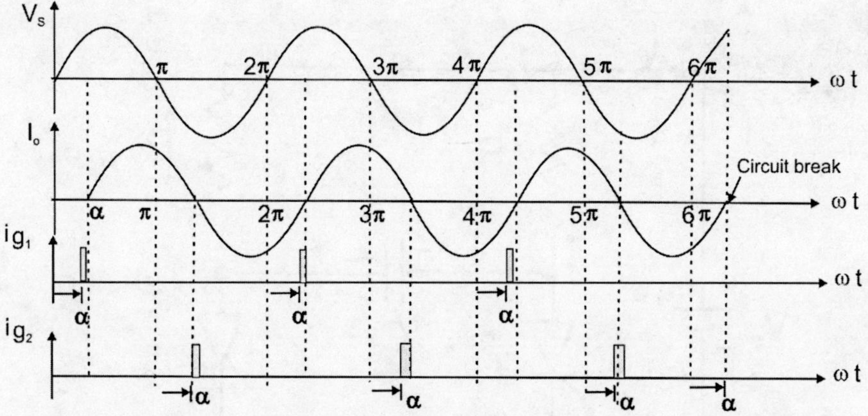

Fig. 5.10 Voltage and current waveforms of SCR based AC circuit breaker

5.9.2 SCR Based DC Circuit Breakers

Fig.5.11 shows the SCR based DC circuit breaker. The operating principle of this circuit is similar to class-C commutation. When the main SCR T_1 is turned-on, load voltage (voltage across load resistance R_1) is equal to supply voltage V_s and capacitor C starts to charge through the path V_s+ – R_2 – C+ – C- – T_1 – V_s-. The capacitor C will be charged with right side plate positive charged. Since the main SCR T_1 is switched ON and operates in conduction state, the load current flows through load resistance R_1. Whenever it is required to break the load current, the auxiliary SCR T_2 should be turned ON. As the auxiliary SCR T_2 is turned ON, SCR T_1 is reverse biased and it should be turned off and the circuit will break.

As T_1 is forced commutated, capacitor C will be charged from $+V_s$ to $-V_s$, through the path V_s+ – R_1 – C^- – C+ – T_2 – V_s-. As soon as C is charged to $-V_s$, the current flow through load becomes zero and simultaneously current flows through R_1 is below the holding current of T_2 and then T_2 will be turned off.

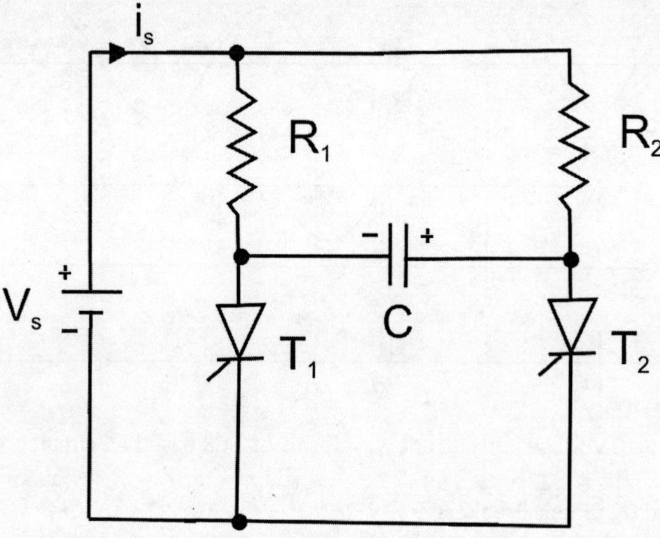

Fig.5.11 SCR based DC circuit breaker

Exercise

Fill in The Blanks

5.1 An utility system which is used in order to safeguard property against theft is known as _______________. (Burglar's alarm)

5.2 __________________are used to control the charging current by continuously comparing the terminal voltage of battery with the constant reference voltage across zener diode.(Battery charger)

5.3 Whenever ac supply is disturbed, the lamp will glow immediately in ______________. (Emergency light system using SCR)

5.4 An utility system which is used for room temperature control is known as____________.(Temperature controller using SCR)

5.5 AC voltage controller circuits are used in___________, and ______________. (domestic light dimmers, electric fan speed control)

5.6 The most commonly used SMPS are__________, ______________ and________________. (Fly-back converter, Forward converter, Push-pull converter)

5.7 _____________UPS where load normally connected to ac supply and _________ UPS where load normally connected to inverter. (Off-line, Online)

5.8 The operating principle of SCR based DC circuit breaker is similar to ___________ commutation. (class-C)

Short Answer Type Questions

5.9 Give a list of industrial applications of power electronics.

5.10 Write the applications of Burglar's alarm

5.11 What are the drawbacks of battery charger using SCR?

5.12 Give a list of applications of emergency light system

5.13 What are the advantages and disadvantages of SMPS? Write applications of SMPS.

5.14 Write the difference between off-line UPS and on-line UPS.

5.15 Why are transformers used in SMPS?

Long Answer Type Questions

5.16 Explain the operating principle of Burglar's alarm system with circuit diagram.

5.17 Draw the circuit diagram of battery charger using SCR and describe operating principle briefly.

5.18 Provide a list of applications of Battery Charger using SCR?

5.19 Describe the operation of emergency light system using SCR with circuit diagram.

5.20 Write the application of temperature controller using SCR.

5.21 Draw the circuit diagram for temperature controller using SCR and describe the circuit operation briefly.

5.22 Write short notes on the following:

i) Fan speed control using TRIAC

ii) Illumination control using TRIAC

iii) SMPS

iv) AC circuit breakers

v) DC circuit breakers

5.23 Explain the operating principle of SMPS with circuit diagram.

5.24 Draw the schematic block diagram of 'off-line UPS' and explain operation of each block briefly.

5.25 Describe the operation of 'on-line UPS' with schematic block diagram.

5.26 What is SCR based AC circuit breakers? Explain it's operating principle and "voltage and current waveforms".

PRACTICAL

1. DIAC and TRIAC Phase Control Circuit Performance

Aim:

1. To study the performance and waveforms of ac voltage controller by using TRIAC and DIAC
2. To plot a graph of variation of output voltage V_o with respect to firing angle α

Apparatus:

TRIAC BT136, DIAC DB3, Single phase 230V, 50Hz AC supply, Resistance R_1=1kΩ, 0.5W; Variable resistance VR_1 470 kΩ, 0.5W; Capacitor C=0.1μF; 60W lamp, Voltmeter 0-400V AC, Ammeter 0-1A, Digital multi-meter 0-400V AC, 0-1A, Dual trace 10 MHz CRO

Circuit Diagram:

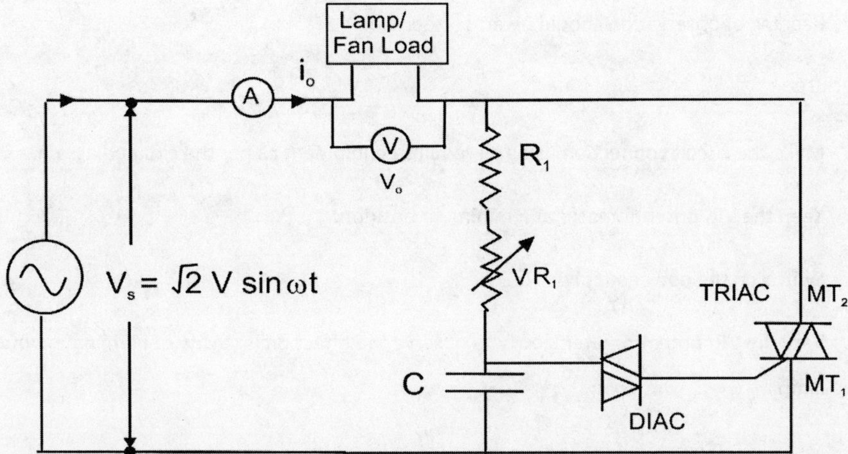

Fig.5.12 Circuit diagram for the test "DIAC and TRIAC Phase Control Circuit Performance"

Theory:

In single phase ac voltage controller, TRIAC is switched ON and OFF in synchronism with the mains supply so that only a part of each half cycle is applied across the load. Fig.5.12 shows the circuit diagram for the test "DIAC and TRIAC Phase Control Circuit Performance'. In this circuit as ac supply voltage V_s increases, voltage across capacitor C increases due to current following through load, resistance R_1, variable resistance VR_1 and capacitor C. Since the voltage drop across C increases, the voltage drop across DIAC also increases until it reaches to it's break over voltage. When DIAC starts to conduct, a large amount of current is injected into the gate of TRIAC and subsequently TRIAC will be turned on and starts conducting. The firing angle of TRIAC can be controlled by varying VR_1. This circuit can be used in domestic light dimmers, electric fan speed control, small motor control and control of small AC powered domestic appliances.

Precautions:

1. Initially all switches of the power supplies are at off condition.

2. Do not increase the gate current more than rated value of TRIAC.

3. The applied voltage and current should not be more than the maximum rating of TRIAC.
4. Use appropriate heat sink for proper cooling of TRIAC.
5. Reading of observation should be error free.

Procedure:

1. Make the circuit connection using all required equipment as per the circuit diagram.
2. Keep the VR_1 potentiometer at a minimum position.
3. Switch on the power supply.
4. Vary the VR_1 potentiometer slowly to observe the effect on intensity of illumination of the lamp.
5. Measure AC input voltage using voltmeter/multi-meter.
6. Connect CRO across the input ac supply to observe as well as measure the input voltage.
7. Connect CRO across the load to observe as well as measure the output voltage.
8. Observe the output voltage waveforms at different firing angles.
9. Draw the input and output voltage waveforms on graph paper at different firing angles observing the waveforms in CRO.
10. Using observation table, plot graph for variation of output voltage with respect to firing angle α.

Observation Table:

Table 5.1 Observation Table for "DIAC and TRIAC Phase Control Circuit Performance"

Sr. No.	Position of Resistance (VR_1)	Firing angle (α)	Measured voltage across DIAC (V_{DIAC})	Measured voltage across TRIAC (V_{TRIAC})	Measured output voltage (V_o or V_{LOAD})

VIVA Questions Related to DIAC and TRIAC Phase Control Circuit Performance

1. Write specifications of TRIAC and DIAC from manufacturer data sheet.
2. State the effect on output voltage when there is no gate pulse.
3. Explain the performance and working principle of DIAC and TRIAC Phase Control Circuit Performance.
4. If capacitor is shorted, state the effect on the TRIAC phase control circuit performance.

2. Simulate the firing angle control for DIAC and TRIAC Phase control circuit in SCILAB software

Aim:

1. To study the performance and waveforms of ac voltage controller by using TRIAC and DIAC

2. To plot a graph of variation of output voltage V_o with respect to firing angle α

Circuit Diagram:

Student should write the code for simulation the firing angle control for DIAC and TRIAC Phase control circuit in SCILAB software as shown in Fig.5.13.

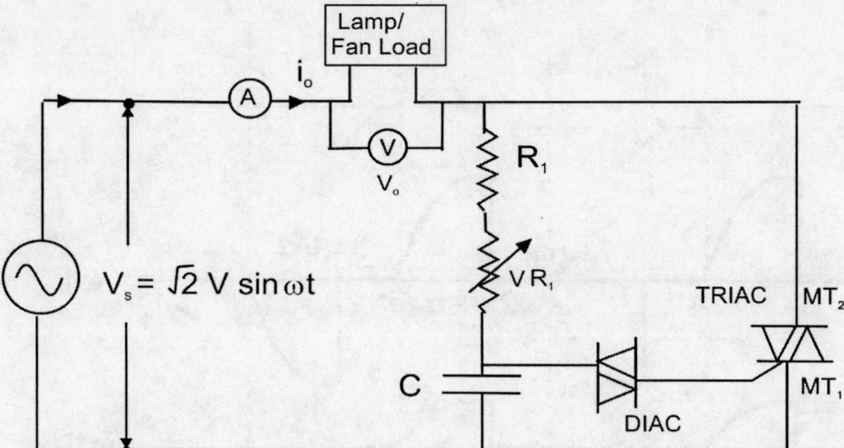

Fig.5.13 Circuit diagram for simulation the firing angle control for DIAC and TRIAC Phase control circuit in SCILAB software

Theory:

Single Phase ac voltage controller circuits are used in domestic light dimmers, electric fan speed control, small ac motor control and control of small ac powered domestic appliances. TRIAC is easy to use and having low cost as compared to two thyristors back to back connection for low power applications. For phase control application, TRIAC is switched on and off in synchronization with ac mains supply so that only a part of each half cycle is applied across the load. The voltage and current waveforms for single phase AC voltage controller with R load are shown in Fig.5.14.

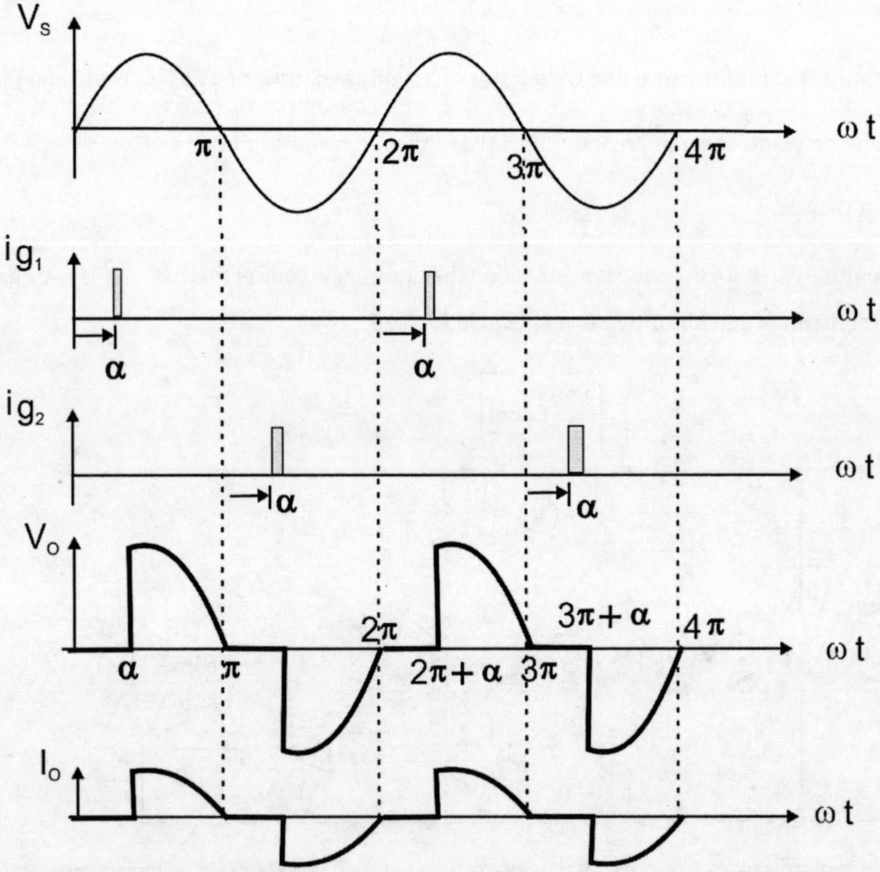

Fig.5.14 Voltage and current waveforms for single phase AC voltage controller with R load

Assume the supply (input) voltage is $V_S = \sqrt{2}V \sin \omega t$

RMS output voltage across load at firing angle α is equal to

$$V_O = \left[\frac{1}{\pi}\int_{\alpha}^{\pi}(\sqrt{2}V\sin\omega t)^2.d(\omega t)\right]^{1/2} = V\left[1-\frac{\alpha}{\pi}+\frac{\sin 2\alpha}{2\pi}\right]^{1/2}$$

The RMS output voltage can be controlled from 0 to V by changing firing angle α from 0° to 180°. The rms value of load current is $I_O = \frac{V_O}{R}$. The code is written in SCILAB to find output voltage and output current at different firing angle.

Resource Required: Open source SCILAB software

SCILAB CODE:

```
// Firing angle control for DIAC and TRIAC Phase control circuit in SCILAB software
// SCILAB 5.4.1
// To plot a graph of variation of output voltage Vo with respect to firing angle α
// clear
// clc
// "--------------Enter data -----------"
V= input ('Enter the rms value of supply voltage')
α = input ('Enter the firing angle of TRIAC')
R= input ('Enter the value of load resistance')
for i= α : 5: 180;
    Vo=V*sqrt((pi-i+(sin2i/2))/pi); // Vo= output voltage
    Io=Vo/R; // Io= Output current
end
plot(Vo, i)
title ('Output voltage –vs – firing angle α of single phase AC voltage controller')
xlabel('Firing angle α ----------→');
ylabel('Output Voltage Vo ----------→');
```

```
xgrid;
plot(Io, i)
title ('Output current –vs – firing angle α of single phase AC voltage controller')
xlabel('Firing angle α  ---------→');
ylabel('Output Current Io ---------→');
xgrid;
```

Procedure:

1. Open SCILAB software
2. Open new project/file
3. Write the SCILAB code
4. Execute the programme
5. Plot the output voltage vs firing angle and the plot output current vs firing angle

Observations and Calculations:

Attach printout of simulation waveforms as results.

Conclusions

Write the effect of firing angle α on output voltage V_o and load current I_o.

VIVA Questions Related to DIAC and TRIAC Phase Control Circuit Performance

1. Analysis the plot output voltage vs firing angle and the plot output current vs firing angle.
2. Write the applications of SCILAB software.
3. State the effect of firing angle α on output voltage V_o and load current I_o
4. What are the difficulties you have faced during simulation in SCILAB software?

3. Test the performance of given SMPS

Aim:

1. To study the performance of given Switch Mode Power Supply (SMPS)

Apparatus:

SMPS: input voltage 220V, current 5A, dc output voltage +5V, +12V, +3.3V, -5V, -12V

Digital Multimeter: 500V dc, 500V ac, current 0-200mA ac, 0-200mA dc

CRO: 20MHz Dual Trace CRO

Circuit Diagram:

Student should draw the schematic block diagram of the given Switch Mode Power Supply (SMPS) and try to draw the detailed diagram of SMPS after disassemble of the given UPS.

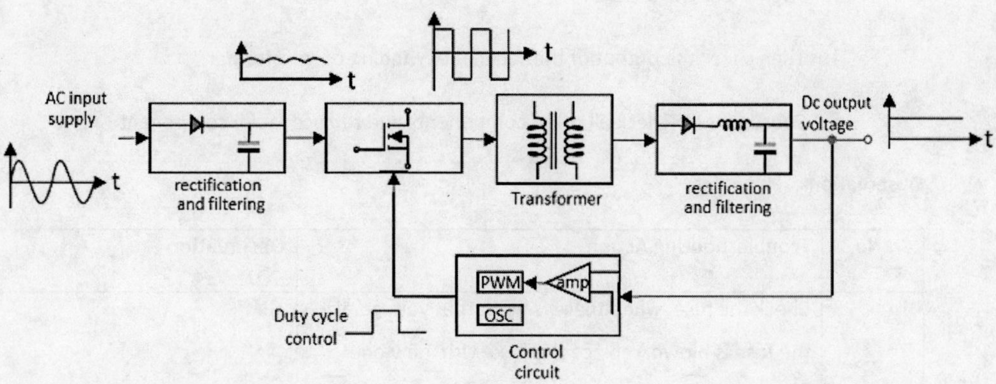

Fig.5.15 Schematic block diagram for the given Switch Mode Power Supply (SMPS)

Theory:

Usually, the power supply of most of the electronic devices are supplied from standard dc power sources. Switch-mode power supplies (SMPS) are frequently used to provide the different voltage levels (+5V, +12V, +3.3V, -5V, -12V) of dc output power which are needed for modern electronic appliances.

Actually SMPS converts a dc input voltage into multiple dc output voltages based on the circuit topology such as step-down (buck), step-up (boost), and step-up/step-down (buck-boost converter).

These converters are designed using MOSFET switch, diode, output capacitor and inductor. MOSFET operation is controlled by a controller which applies a pulse-width modulated (PWM) square wave to the MOSFET's gate. To maintain a constant output voltage, the controller should sense the SMPS output voltage and varies the duty cycle D [$T_{ON}/(T_{ON}+T_{OFF})$] of square-wave. SMPS is used to provide highly efficient and reliable dc to dc power conversion systems.

Precautions:

1. Before touching the SMPS circuit, discharge the capacitor by using suitable resistance.
2. Measure the voltage and it must be zero before starting disassemble the SMPS.

Procedure

When there is no +5V output:

1. Check the fuse
2. Then check the capacitor
3. Then check the MOSFET
4. Then check the output of the secondary winding of transformer.
5. Subsequently check all other components and replace faulty component

Observations

Sr. No.	Troubleshooting Action	Observation
1.	Check the fuse, when there is no output voltage. If the fuse is blown, replace the fuse with new one.	
2.	Then check the MOSFET, when there is no output voltage	
3.	Then check the output of the secondary winding of transformer	

VIVA Questions Related to SMPS

1. What is the effect of change in ac line voltage on output voltage?
2. State the effect of duty cycle on output voltage.
3. Write the specification of SMPS.

4. Test the performance of given UPS

Aim:

To study the performance of given Uninterruptable Power Supply (UPS)

Apparatus:

UPS: Input voltage range 90V-145V, output voltage 230V, output power capacity 500VA

Digital Multimeter: 500V dc, 500V ac, current 0-200mA ac, 0-10A ac, 0-10A dc

Stopwatch: To measure backup time.

Circuit Diagram:

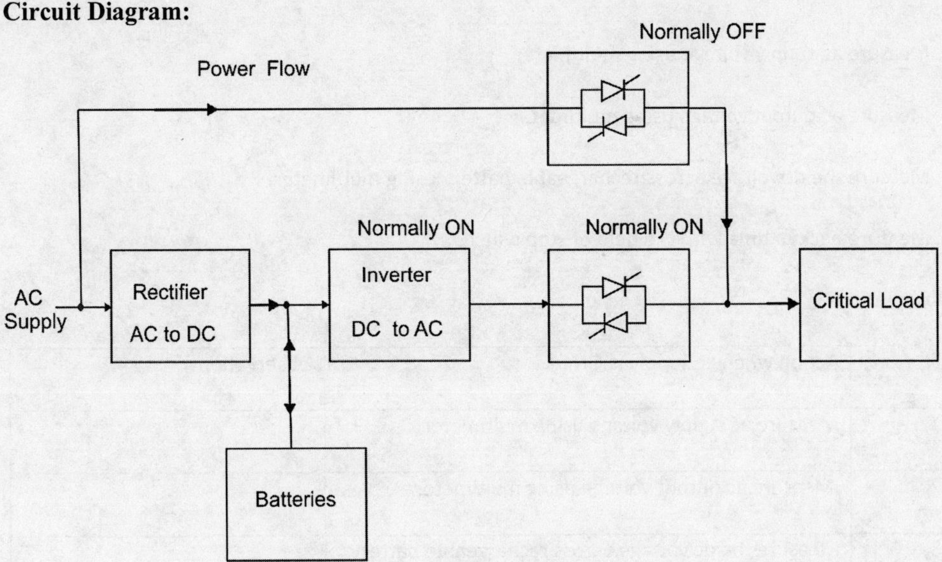

Fig.5.16 Circuit diagram for the given Uninterruptable Power Supply (UPS)

Theory: An Uninterruptable Power Supply (UPS) is an electrical apparatus which is used as emergency power supply to a load. It is most commonly used in computers, data centers, telecommunications equipment and medical equipment or any critical load. Usually, the two types of UPS configurations are commonly used such as (i) Offline UPS where load normally connected to ac supply and (ii) Online UPS where load normally connected to inverter.

In an off-line UPS system, the load is powered directly by the input power and the backup power circuitry is only invoked when the utility power fails. In on-line UPS, firstly ac supply is converted into dc and it passes through rechargeable battery for battery strings then inverting back to 230V AC for powering the protected equipment.

Precautions:

1. UPS must be properly ground.
2. Before opening the UPS, all capacitors of UPS have to be discharged.
3. There is a possibility to have high leakage current. Therefore, the grounding conductor should be connected first.
4. UPS should be installed on a non-inflammable surface.

Procedure

1. Measure ac supply voltage using multimeter
2. Measure ac output voltage using multimeter
3. Measure the dc voltage across rechargeable battery using multimeter.
4. Measure backup time with the help of stop watch.

Observations

Sr. No.	Action when ac supply is ON	Observation
1.	Measure ac supply voltage using multimeter	
2.	Measure ac output voltage using multimeter	
3.	Measure the dc voltage across rechargeable battery using multimeter.	
4.	Measure backup time with the help of stop watch.	

VIVA Questions Related to UPS

1. What is the effect of change in ac line voltage on output voltage?
2. State the effect of duty cycle on output voltage.
3. Write the specification of UPS

6. Troubleshoot the Burglar's alarm, Emergency light system, Speed control system, and Temperature control system

AIM :

1. Diagnose the fault of Burglar's alarm system and rectify the problem
2. Diagnose the fault of Emergency light system and rectify the problem
3. Diagnose the fault of Speed control system and rectify the problem
4. Diagnose the fault of Temperature control system and rectify the problem

Theory:
Burglar alarm is a standard equipment which is used in stores, offices, houses, cars and other premises for the security purpose. Emergency light is an illumination device specially designed for emergency purposes. It operates in case of power failure. The on- time of emergency light system depends on the ampere-rating of the battery. Thyristor/Triac based speed control system is widely used in industrial applications such as control of machine tool and process control system. Temperature control system is used in industrial furnace to maintain specified temperature by controlling the ON and OFF time of heater coil. The schematic circuit diagram and operating principle of Burglar's alarm system, Emergency light system, Speed control system and Temperature control system are described in detail in section 5.2, 5.4, 5.6 and 5.5 respectively.

Problem statement for

a) **Burglar's alarm system**

   1. Student should draw the schematic block diagram and detail circuit diagram of the given Burglar's alarm system.
   2. The Burglar's alarm system is not operating properly, student should diagnose the fault and rectify the problem.

b) **Emergency light system**

   1. Student should draw the schematic block diagram and detail circuit diagram of the given Emergency light system.
   2. The Emergency light system is not working properly, student should diagnose the fault and rectify the problem.

3. The intensity of emergency light starts decreasing, find out the fault and rectify the problem.

c) Speed control system

1. Student should draw the schematic block diagram and detail circuit diagram of the given Speed control system.

2. The Speed control system is not working properly, student should diagnose the fault and rectify the problem.

3. The speed of motor is suddenly drops when load is applied, find out the fault and rectify the problem.

d) Temperature control system

1. Student should draw the schematic block diagram and detail circuit diagram of the given Temperature control system

2. The Temperature control system is not operating properly, student should diagnose the fault and rectify the problem.

3. Diagnose and rectify the problem when temperature sensor is not working properly in Temperature control system

Procedure

a) To Repair Burglar's alarm system

1. Usually five main faults namely battery failure, faulty sensor, loose cable connections, faulty transistor in control panel, fault in relay and faulty alarm occur in burglar alarm system.

2. Check the battery. If it is not working, replace the old battery with new one.

3. Check the cable. If there is any cable damage, replace the damaged cable. If there is any loose connection, rectify it.

4. Check the sensor. Replace the faulty sensor.

5. Check all transistors in control panel. Replace the faulty transistor.

6. Check the relay. If it is faulty, try to repair it or replace by new one.

b) To Repair Emergency light system

1. The most common faults of emergency light system are battery failure, faulty lamp, loose cable connections, faulty SCR and diodes in control panel, faulty capacitor and fault in transformer.

2. Check the battery. If it is not at working condition, replace the old battery with new one

3. Check the cable. If there is any cable damage, replace the damaged cable. If there is any loose connection, rectify it.
4. Check the bulb. If it is damaged, replace the old bulb with new one
5. Check all diodes and SCR in control panel. Replace the faulty diodes and SCR.
6. Check the capacitor. If it is not at good condition, replace it.
7. Check the transformer. If it is faulty, try to repair it otherwise replace by new one.

c) To Repair Speed control system

1. The main faults of speed control system are fault in motor, triggering circuit not generating triggering pulse, faulty SCR, loose cable connections, and fault in control panel.
2. Check the motor. If it is not at normal operating condition, try to repair it or replace by new one.
3. Check the triggering circuit. If the triggering circuit is not generating triggering pulse, try to rectify it or replace by new one.
4. Check the cable. If there is any cable damage, replace the damaged cable. If there is any loose connection, rectify it.
5. Check the SCR. If it is faulty, replace the faulty SCR.
6. Check the control panel. If it is faulty, try to repair it or replace by new one.

d) To Repair Temperature control system

1. The most common faults of temperature control system are faulty heater, faulty temperature sensor, loose cable connections, faulty SCR, and triggering circuit not generating triggering pulse in control panel.
2. Check the heater. If it is not working properly, try to repair it or replace by new one.
3. Check the temperature sensor. If it is faulty, replace the faulty sensor.
4. Check the cable. If there is any cable damage, replace the damaged cable. If there is any loose connection, rectify it.
5. Check the SCR. If it is faulty, replace the faulty SCR.
6. Check the triggering circuit. If the triggering circuit is not generating triggering pulse, try to rectify it or replace by new one.

Practical Related Questions:

1. What are the different faults you have found during routine servicing of Burglar's alarm system?
2. Write specification of the given Burglar's alarm system.

3. When the battery charging is low, explain the effect on intensity of light in Emergency light system.
4. Write specification of the given Emergency light system.
5. Lamp is not glowing in emergency light system after ac supply is interrupted, state the all possible reasons.
6. Write the specification of thyristor which is used in speed control system.
7. While triggering pulse is not generated by triggering circuit to turn-on SCR, how much voltage is applied across motor.
8. If gate terminal of SCR becomes open circuit, sate the effect on the output voltage.
9. What is the effect of load on SCR in Temperature control system?
10. When temperature sensor is not working properly in Temperature control system, diagnose and rectify the problem.
11. Write the specification of SCR and temperature sensor which are used in given Temperature control system.

KNOW MORE

Recently power electronics devices have been intensively used in process control industry for various applications such as power supplies, converters, inverters, battery chargers, temperature control, variable speed motor drives, robot arm drives and electric vehicles etc. The design, development, and optimization of power electronics and controller devices are highly essential for optimum utilization of power, power system stability and control, safety of power system operation and overall improvement of power quality. Now a days Flexible AC Transmission System (FACTS) controllers, static VAR compensator (SVC) and high-voltage direct current (HVDC) converters are used in power system for power quality improvement. To study the applications of power electronics devices in FACTS controllers, SVC, and HVDC converters, we scan the following QR codes.

For more on Static var compensator

For more on FACTS

References

1. Bhattacharya,S.K and Chatterji, S., Industrial Electronics and Control, Tata McGraw Hill Publishing Co. Ltd, New Delhi.
2. Zbar, Paul B., Industrial Electronics: A Text-Lab manual, McGraw Hill Publishing Co. Ltd, New Delhi.
3. Grafham D.R., SCR Manual, General Electric Co.
4. Ramamoorty M., An Introduction to Thyristors and their applications, East-West Press Pvt. Ltd., New Delhi.
5. Sugandhi, Rajendra Kumar and Sugandhi, Krishna Kumar, Thyristors: Theory and Applications, New Age International (P) ltd. Publishers, New Delhi.
6. Jain & Alok, Power Electronics and its Applications, Penram International Publishing (India) Pvt. Ltd, Mumbai.
7. Rashid, Muhammad, Power Electronics Circuits Devices and Applications, Pearson Education India, Noida.

Dynamic QR Code for Further Reading

-QR codes embedded in the unit.

CO AND PO ATTAINMENT TABLE

Course outcomes (COs) for this course can be mapped with the programme outcomes (POs) after the completion of the course and a correlation can be made for the attainment of POs to analyze the gap. After proper analysis of the gap in the attainment of POs necessary measures can be taken to overcome the gaps.

Table for CO and PO attainment

Course Outcomes	**Attainment of Programme Outcomes** *(1- Weak Correlation; 2- Medium correlation; 3- Strong Correlation)*						
	PO-1	**PO-2**	**PO-3**	**PO-4**	**PO-5**	**PO-6**	**PO-7**
CO-1							
CO-2							
CO-3							
CO-4							
CO-5							

The data filled in the above table can be used for gap analysis.